Set Theory and its Philosophy

Set Theory and its Philosophy

A Critical Introduction

Michael Potter

OXFORD
UNIVERSITY PRESS

Great Clarendon Street, Oxford OX2 6DP

Oxford University Press is a department of the University of Oxford.
It furthers the University's objective of excellence in research, scholarship,
and education by publishing worldwide in

Oxford New York

Auckland Bangkok Buenos Aires Cape Town Chennai
Dar es Salaam Delhi Hong Kong Istanbul Karachi Kolkata
Kuala Lumpur Madrid Melbourne Mexico City Mumbai Nairobi
São Paulo Shanghai Taipei Tokyo Toronto

Oxford is a registered trade mark of Oxford University Press
in the UK and certain other countries

Published in the United States
by Oxford University Press Inc., New York

First published 2004

British Library Cataloguing in Publication Data
Data available

Library of Congress Cataloging in Publication Data
Data available

ISBN 0–19–926973–4 (hbk.)
ISBN 0–19–927041–4 (pbk.)

1 3 5 7 9 10 8 6 4 2

Typeset by the author
Printed in Great Britain
on acid-free paper by
T J International Ltd.,
Padstow, Cornwall

Preface

This book was written for two groups of people, philosophically informed mathematicians with a serious interest in the philosophy of their subject, and philosophers with a serious interest in mathematics. Perhaps, therefore, I should begin by paying tribute to my publisher for expressing no nervousness concerning the size of the likely readership. Frege (1893–1903, I, p. xii) predicted that the form of his *Grundgesetze* would cause him to

> relinquish as readers all those mathematicians who, if they bump into logical expressions such as 'concept', 'relation', 'judgment', think: *metaphysica sunt, non leguntur*, and likewise those philosophers who at the sight of a formula cry: *mathematica sunt, non leguntur*.

And, as he rightly observed, 'the number of such persons is surely not small'.

As then, so now. Any book which tries to form a bridge between mathematics and philosophy risks vanishing into the gap. It is inevitable, however hard the writer strives for clarity, that the requirements of the subject matter place demands on the reader, sometimes mathematical, sometimes philosophical. This is something which anyone who wants to make a serious study of the philosophy of mathematics must simply accept. To anyone who doubts it there are two bodies of work which stand as an awful warning: the philosophical literature contains far too many articles marred by elementary technical misunderstandings,[1] while mathematicians have often been tempted, especially in later life, to commit to print philosophical reflections which are either wholly vacuous or hopelessly incoherent.

Both mathematicians and philosophers, then, need to accept that studying the philosophy of mathematics confronts them with challenges for which their previous training may not have prepared them. It does not follow automatically, of course, that one should try, as I have done here, to cater for their differing needs in a single textbook. However, my main reason for writing the book was that I wanted to explore the constant *interplay* that set theory seems to exemplify between technical results and philosophical reflection, and this convinced me that there was more than expositional economy to be said for trying to address both readerships simultaneously.

[1] I should know: I have written one myself.

In this respect the book differs from its predecessor, *Sets: An Introduction* (1990), which was directed much more at philosophically ignorant mathematicians than at philosophers. The similarities between the new book and the old are certainly substantial, especially in the more technical parts, but even there the changes go well beyond the cosmetic. Technically informed readers may find a quick summary helpful.

At the formal level the most significant change is that I have abandoned the supposition I made in *Sets* that the universe of collections contains a sub-universe which has all the sets as members but is nevertheless capable of belonging to other collections (not, of course, sets) itself. The truth is that I only included it for two reasons: to make it easy to embed category theory; and to sidestep some irritating pieces of pedantry such as the difficulty we face when we try, without going metalinguistic, to say in standard set theory that the class of all ordinals is well-ordered. Neither reason now strikes me as sufficient to compensate for the complications positing a sub-universe creates: the category theorists never thanked me for accommodating them, and if they want a sub-universe, I now think that they can posit it for themselves; while the pedantry is going to have to be faced at some point, so postponing it does no one any favours.

The style of set-theoretic formalism in which the sets form only a sub-universe (sometimes in the literature named after Grothendieck) has never found much favour. (Perhaps category theory is not popular enough.) So in this respect the new book is the more orthodox. In two other respects, though, I have persisted in eccentricity: I still allow there to be individuals; and I still do not include the axiom scheme of replacement in the default theory.

The reasons for these choices are discussed quite fully in the text, so here I will be brief. The first eccentricity — allowing individuals — seems to me to be something close to a philosophical necessity: as it complicates the treatment only very slightly, I can only recommend that mathematicians who do not see the need should regard it as a foible and humour me in it. The second — doing without replacement — was in fact a large part of my motivation to write *Sets* to begin with. I had whiled away my student days under the delusion that replacement is needed for the formalization of considerable amounts of mathematics, and when I discovered that this was false, I wanted to spread the word (spread it, that is to say, beyond the set theorists who knew it already). On this point my proselytizing zeal has hardly waned in the intervening decade. One of the themes which I have tried to develop in the new book is the idea that set theory is a measure (not the only one, no doubt) of the degree of abstractness of mathematics, and it is at the very least a striking fact about mathematical practice, which many set theory textbooks contrive to obscure, that even before we try to reduce levels by clever use of coding, the overwhelming majority of mathematics sits comfortably inside the first couple of dozen levels of the hierarchy above the natural numbers.

The differences from standard treatments that I have described so far affect our conception of the extent of the set-theoretic hierarchy. One final difference affecting not our conception of it but only the axiomatization is that the axioms I have used focus first on the notion of a *level*, and deduce the properties of *sets* (as subcollections of levels) only derivatively. The pioneer of this style of axiomatization was Scott (1974), but my earlier book used unpublished work of John Derrick to simplify his system significantly, and here I have taken the opportunity to make still more simplifications.

Lecturers on the look-out for a course text may feel nervous about this. No executive, it is said, has ever been sacked for ordering IBM computers, even if they were not the best buy; by parity of reasoning, I suppose, no lecturer is ever sacked for teaching ZF. So it is worth stressing that ZU, the system which acts as a default throughout this book, is interpretable in ZF in the obvious manner: the theorems stated in this book are, word for word, theorems of ZF. So teaching from this book is not like teaching from Quine's *Mathematical Logic*: you will find no self-membered sets here.

Most of the exercises have very largely been taken unchanged from my earlier book. I recommend browsing them, at least: it is a worthwhile aid to understanding the text, even for those students who do not seriously attempt to do them all.

It is a pleasure, in conclusion, to record my thanks: to Philip Meguire and Pierre Matet for correspondence about the weaknesses of my earlier book; to the members of the Cambridge seminar which discussed chapters of the new book in draft and exposed inadequacies of exposition; and especially to Timothy Smiley, Eric James, Peter Clark and Richard Zach, all of whom spent more time than I had any right to expect reading the text and giving me perceptive, judicious comments on it.

M. D. P.

Contents

Part I

Sets

Introduction to Part I

This book, as its title declares, is about sets; and sets, as we shall use the term here, are a sort of aggregate. But, as a cursory glance at the literature makes clear, aggregation is far from being a univocal notion. Just what sort of aggregate sets are is a somewhat technical matter, and indeed it will be a large part of the purpose of the book merely to get clear about this question. But in order to begin the task, it will help to have an idea of the reasons one might have for studying them.

It is uncontroversial, first of all, that set-theoretic language can be used as a vehicle for communication. We shall concentrate here on the mathematical case, not because this is the only context in which set talk is useful — that is far from being so — but rather because ordinary language cases do not seem to need a theory of anything like the same complexity to underpin them. But in this role set theory is being used *merely* as a language. What will interest us here much more are those uses of set theory for which a substantial theory is required. Three strands can be distinguished.

The first of these is the use of set theory as a tool in understanding the infinite. This strand will lead us to develop in part III of the book the theory of two distinct types of infinite number, known as cardinals and ordinals. These two theories are due in very large part to one man, Georg Cantor, who worked them out in the last quarter of the 19th century. This material is hardly controversial nowadays: it may still be a matter of controversy in some quarters whether infinite sets exist, but hardly anyone now tries to argue that their existence is, as a matter of pure logic, contradictory. This has not always been so, however, and the fact that it is so now is a consequence of the widespread acceptance of Cantor's theories, which exhibited the contradictions others had claimed to derive from the supposition of infinite sets as confusions resulting from the failure to mark the necessary distinctions with sufficient clarity.

The revolution in attitudes to the infinitely large which Cantor's work engendered was thus as profound as the roughly simultaneous revolution caused by the rigorous development of the infinitesimal calculus. In the 20th century, once these two revolutions had been assimilated, the paradoxes of the infinitely small (such as Zeno's arrow) and of the infinitely large (such as the correspondence between a set and its proper subset) came to be regarded not as serious philosophical problems but only as historical curiosities.

The second role for set theory, which will occupy us throughout the book but especially in part II, is the foundational one of supplying the subject matter of mathematics. Modern textbooks on set theory are littered with variants of this claim: one of them states baldly that 'set theory is the foundation of mathematics' (Kunen 1980, p. xi), and similar claims are to be found not just (as perhaps one might expect) in books written by set theorists but also in many mainstream mathematics books. Indeed this role for set theory has become so familiar that hardly anybody who gets as far as reading this book can be wholly unaware of it. Yet it is worth pausing briefly to consider how surprising it is. Pre-theoretically we surely feel no temptation whatever to conjecture that numbers might 'really' be sets — far less sets built up from the empty set alone — and yet throughout the 20th century many mathematicians did not merely conjecture this but said that it was so.

One of the themes that will emerge as this enquiry progresses, however, is that what mathematicians say is no more reliable as a guide to the interpretation of their work than what artists say about *their* work, or musicians. So we certainly should not automatically take mathematicians at their set-theoretically reductionist word. And there have in any case been notable recusants throughout the period, such as Mac Lane (1986) and Mayberry (1994). Nevertheless, we shall need to bear this foundational use for set theory in mind throughout, both because it has been enormously influential in determining the manner in which the theory has been developed and because it will be one of our aims to reach a position from which its cogency can be assessed.

A third role for set theory, closely related to the second but nonetheless distinguishable from it (cf. Carnap 1931), is to supply for diverse areas of mathematics not a common subject matter but common modes of reasoning. The best known illustration of this is the axiom of choice, a set-theoretic principle which we shall study in part IV.

Once again, the historical importance of this role for set theory is unquestionable: the axiom of choice was the subject of controversy among mathematicians throughout the first half of the 20th century. But once again it is at least debatable whether set theory can indeed have the role that is ascribed to it: it is far from clear that the axiom of choice is correctly regarded as a set-theoretic principle at all, and similar doubts may be raised about other purported applications of set-theoretic principles in mathematics.

These three roles for set theory — as a means of taming the infinite, as a supplier of the subject matter of mathematics, and as a source of its modes of reasoning — have all been important historically and have shaped the way the subject has developed. Most of the book will be taken up with presenting the technical material which underpins these roles and discussing their significance.

In this first part of the book, however, we shall confine ourselves to the seemingly modest goal of setting up an elementary theory of sets within which to frame the later discussion. The history of such theories is now a century old — the theory we shall present here can trace its origins to Zermelo (1908*b*) — and yet there is, even now, no consensus in the literature about the form they should take. Very many of the theories that have been advanced seek to formalize what has come to be known as the *iterative* conception of sets, and what we shall be presenting here is one such theory. However, it is by no means a trivial task to tease out what the iterative conception amounts to. So the goal of this part of the book will turn out not to be quite as modest as it seems.

Chapter 1

Logic

This book will consist in very large part in the exposition of a mathematical theory — the theory (or at any rate *a* theory) of sets. This exposition will have at its core a sequence of *proofs* designed to establish *theorems*. We shall distinguish among the theorems some which we shall call *lemmas*, *propositions* or *corollaries*. Traditionally, a lemma is a result of no intrinsic interest proved as a step towards the proof of a theorem; a proposition is a result of less independent importance than a theorem; and a corollary is an easy consequence of a theorem. The distinctions are of no formal significance, however, and we make use of them only as a way of providing signposts to the reader as to the relative importance of the results stated.

One central element in the exposition will be explicit *definitions* to explain our use of various words and symbols. It is a requirement of such a definition that it should be formally eliminable, so that every occurrence of the word defined could in principle be replaced by the phrase that defines it without affecting the correctness of the proof. But this process of elimination must stop eventually: at the beginning of our exposition there must be mathematical words or symbols which we do not define in terms of others but merely take as given: they are called *primitives*. And proof must start somewhere, just as definition must. If we are to avoid an infinite regress, there must be some propositions that are not proved but can be used in the proofs of the theorems. Such propositions are called *axioms*.

1.1 The axiomatic method

The method for expounding a mathematical theory which we have just described goes back at least to Euclid, who wrote a textbook of geometry and arithmetic in axiomatic form around 300 B.C. (It is difficult to be certain quite how common the axiomatic method was before Euclid because his textbook supplanted previous expositions so definitively that very little of them survives to be examined today.)

The axiomatic method is certainly not universal among mathematicians even now, and its effectiveness has been overstated in some quarters, thereby providing an easy target for polemical attack by empiricists such as Lakatos (1976). It is nevertheless true that pure mathematicians at any rate regard its use as routine. How then should we account for it?

Responses to this question fall into two camps which mathematicians have for some time been wont to call realist and formalist. This was not an altogether happy choice of terminology since philosophers had already used both words for more specific positions in the philosophy of mathematics, but I shall follow the mathematicians' usage here.

At the core of attitudes to the axiomatic method that may be called *realist* is the view that 'undefined' does not entail 'meaningless' and so it may be possible to provide a meaning for the primitive terms of our theory in advance of laying down the axioms: perhaps they are previously understood terms of ordinary language; or, if not, we may be able to establish the intended meanings by means of what Frege calls elucidations — informal explanations which suffice to indicate the intended meanings of terms. But elucidation, Frege says, is inessential. It merely

> serves the purpose of mutual understanding among investigators, as well as of the communication of science to others. We may relegate it to a propaedeutic. It has no place in the system of a science; no conclusions are based on it. Someone who pursued research only by himself would not need it. (1906, p. 302)

If the primitive terms of our theory are words, such as 'point' or 'line', which can be given meanings in this manner, then by asserting the axioms of the theory we commit ourselves to their *truth*. Realism is thus committed to the notion that the words mathematicians use already have a meaning independent of the system of axioms in which the words occur. It is for this reason that such views are described as realist. If the axioms make existential claims (which typically they do), then by taking them to be true we commit ourselves to the existence of the requisite objects.

Nevertheless, realism remains a broad church, since it says nothing yet about the nature of the objects thus appealed to. Two sorts of realist can be distinguished: a platonist takes the objects to exist independently of us and of our activities, and hence (since they are certainly not physical) to be in some sense abstract; a constructivist, on the other hand, takes the objects to exist only if they can be constructed, and hence to be in some sense mental. But 'in some sense' is usually a hedging phrase, and so it is here. To say that a number owes its existence to my construction of it does not of itself make the number mental any more than my bookcase is mental because I built it: what is distinctive of constructivism in the philosophy of mathematics (and hence distinguishes numbers, as it conceives of them, from bookcases) is the idea that numbers are *constituted* by our constructions of them. I said earlier that philosophers of mathematics use the word 'realist' differently, and this is the point where the difference emerges, since constructivism would not be counted realist on their usage: it counts here as a species of realism because it interprets mathematical existence theorems as truths about objects which do not owe their existence to the signs used to express them.

During the 19th century, however, there emerged another cluster of ways of regarding axioms, which we shall refer to as *formalist*. What they had in common was a rejection of the idea just mentioned that the axioms can be regarded simply as true statements about a subject matter external to them. One part of the motivation for the emergence of formalism lay in the different axiom systems for geometry — Euclidean, hyperbolic, projective, spherical — which mathematicians began to study. The words 'point' and 'line' occur in all, but the claims made using these words conflict. So they cannot all be true, at any rate not unconditionally. One view, then, is that axioms should be thought of as assumptions which we suppose in order to demonstrate the properties of those structures that exemplify them. The expositor of an axiomatic theory is thus as much concerned with truth on this view as on the realist one, but the truths asserted are conditional: *if* any structure satisfies the axioms, *then* it satisfies the theorem. This view has gone under various names in the literature — implicationism, deductivism, if-thenism, eliminative structuralism. Here we shall call it *implicationism*. It seems to be plainly the right thing to say about the role axioms play in the general theories — of groups, rings, fields, topological spaces, differential manifolds, or whatever — which are the mainstay of modern mathematics. It is rather less happy, though, when it is applied to axiomatizations of the classical theories — of natural, real or complex numbers, of Euclidean geometry — which were the sole concern of mathematics until the 19th century. For by conditionalizing all our theorems we omit to mention the existence of the structure in question, and therefore have work to do if we are to explain the applicability of the theory: the domain of any interpretation in which the axioms of arithmetic are true is infinite, and yet we confidently apply arithmetical theorems within the finite domain of our immediate experience without troubling to embed it in such an infinite domain as implicationism would require us to do. Implicationism seems capable, therefore, of being at best only *part* of the explanation of these classical cases.

Nonetheless, the axiomatic method was by the 1920s becoming such a mathematical commonplace, and implicationism such a common attitude towards it, that it was inevitable it would be applied to the recently founded theory of sets. Thus mathematicians such as von Neumann (1925) and Zermelo (1930) discussed from a metatheoretic perspective the properties of structures satisfying the set-theoretic axioms they were considering. One of the evident attractions of the implicationist view of set theory is that it obviates the tedious requirement imposed on the realist to justify the axioms as *true* and replaces it with at most the (presumably weaker) requirement to persuade the reader to be interested in their logical consequences. Even in the extreme case where our axiom system turned out to be inconsistent, this would at worst make its consequences uninteresting, but we could then convict the implicationist only of wasting our time, not of committing a *mistake*.

There is an evident uneasiness about this way of discussing set theory, however. One way of thinking of a structure is as a certain sort of set. So when we discuss the properties of *structures* satisfying the axioms of set theory, we seem already to be presupposing the notion of set. This is a version of an objection that is sometimes called Poincaré's *petitio* because Poincaré (1906) advanced it against an attempt that had been made to use mathematical induction in the course of a justification of the axioms of arithmetic.

In its crudest form this objection is easily evaded if we are sufficiently clear about what we are doing. There is no direct circularity if we presuppose sets in our study of sets (or induction in our study of induction) since the first occurrence of the word is in the metalanguage, the second in the object language. Nevertheless, even if this is all that needs to be said to answer Poincaré's objection in the general case, matters are not so straightforward in the case of a theory that claims to be foundational. If we embed mathematics in set theory and treat set theory implicationally, then mathematics — all mathematics — asserts only conditional truths about structures of a certain sort. But our metalinguistic study of set-theoretic structures is plainly recognizable as a species of mathematics. So we have no reason not to suppose that here too the correct interpretation of our results is only conditional. At no point, then, will mathematics assert anything unconditionally, and any application of any part whatever of mathematics that depends on the unconditional existence of mathematical objects will be vitiated.

Thoroughgoing implicationism — the view that mathematics has no subject matter whatever and consists solely of the logical derivation of consequences from axioms — is thus a very harsh discipline: many mathematicians profess to believe it, but few hold unswervingly to what it entails. The implicationist is never entitled, for instance, to assert unconditionally that no proof of a certain proposition exists, since that is a generalization about proofs and must therefore be interpreted as a conditional depending on the axioms of proof theory. And conversely, the claim that a proposition *is* provable is to be interpreted only as saying that according to proof theory it is: a further inference is required if we are to deduce from this that there is indeed a proof.

One response to this difficulty with taking an implicationist view of set theory is to observe that it arises only on the premise that set theory is intended as a foundation for mathematics. Deny the premise and the objection evaporates. Recently some mathematicians have been tempted by the idea that other theories — topos theory or category theory, for example — might be better suited to play this foundational role.

Maybe so, but of course this move is only a postponement of the problem, not a solution. Those inclined to make it will have to address just the same difficulties in relation to the axioms of whatever foundational theory they favour instead (cf. Shapiro 1991). Perhaps it is for this reason that some mathematicians (e.g. Mayberry 1994) have tried simply to deny that mathematics *has* a foundation. But plainly more needs to be said if this is to be anything more substantial than an indefinite refusal to address the question.

Another response to these difficulties, more popular among mathematicians than among philosophers, has been to espouse a stricter formalism, a version, that is to say, of the view that the primitive terms of an axiomatic theory refer to nothing outside of the theory itself. The crudest version of this doctrine, pure formalism, asserts that mathematics is no more than a game played with symbols. Frege's demolition of this view (1893–1903, II, §§86–137) is treated by most philosophers as definitive. Indeed it has become popular to doubt whether any of the mathematicians Frege quotes actually held a view so stupid. However, there are undoubtedly some mathematicians who claim, when pressed, to believe it, and many others whose stated views entail it.

Less extreme is *postulationism* — which I have elsewhere (Potter 2000) called axiomatic formalism. This does not regard the sentences of an axiomatic theory as meaningless positions in a game but treats the primitive terms as deriving their meaning from the role they play in the axioms, which may now be thought of as an implicit definition of them, to be contrasted with the *explicit* definitions of the non-primitive terms. 'The objects of the theory are defined *ipso facto* by the system of axioms, which in some way generate the material to which the true propositions will be applicable.' (Cartan 1943, p. 9) This view is plainly not as daft as pure formalism, but if we are to espouse it, we presumably need some criterion to determine whether a system of axioms *does* confer meaning on its constituent terms. Those who advance this view agree that no such meaning can be conferred by an inconsistent system, and many from Hilbert on have thought that bare consistency is sufficient to confer meaning, but few have provided any argument for this, and without such an argument the position remains suspect. Moreover, there is a converse problem for the postulationist if the axiom system in question is not complete: if the language of arithmetic has its meaning conferred on it by some formal theory T, for instance, what explanation can the postulationist give of our conviction that the Gödel sentence of T, which is expressed in this language, is true?

Nevertheless, postulationism, or something very like it, has been popular among mathematicians, at least in relation to those parts of mathematics for which the problem of perpetual conditionalizing noted above makes implicationism inappropriate. For the great advantage of postulationism over implicationism is that if we are indeed entitled to postulate objects with the requisite properties, anything we deduce concerning these objects will be true unconditionally. It may be this that has encouraged some authors (Balaguer 1998; Field 1998) to treat a position very similar to postulationism as if it were a kind of *realism* — what they called *full-blooded* (Balaguer) or *plenitudinous* (Field) platonism — but that seems to me to be a mistake. Field, admittedly, is ready enough to concede that their position is in one sense 'the antithesis of platonism' (1998, p. 291), but Balaguer is determined to classify the view as realist, not formalist, because it

> simply doesn't say that 'existence and truth amount to nothing more than consistency'. Rather it says that all the mathematical objects that logically possibly could exist actually do exist, and then it *follows* from this that all consistent purely mathematical theories truly describe some collection of actually existing mathematical objects. (1998, p. 191)

In order to make full-blooded platonism plausible, however, Balaguer has to concede that mathematical theories have a subject matter only in what he calls a 'metaphysically thin' sense, a sense which makes it wholly unproblematic how we could 'have beliefs about mathematical objects, or how [we] could dream up stories about such objects' (p. 49). It is this that makes me classify the view as formalist despite Balaguer's protestations: for a view to count as realist according to the taxonomy I have adopted here, it must hold the truth of the sentences in question to be metaphysically constrained by their subject matter more substantially than Balaguer can allow. A realist conception of a domain is something we win through to when we have gained an understanding of the nature of the objects the domain contains and the relations that hold between them. For the view that bare consistency entails existence to count as realist, therefore, it would be necessary for us to have a quite general conception of the whole of logical space as a domain populated by objects. But it seems quite clear to me that we simply have no such conception.

1.2 The background logic

Whichever we adopt of the views of the axiomatic method just sketched, we shall have to make use of various canons of logical reasoning in deducing the consequences of our axioms. Calling this calculus 'first-order' marks that the variables we use as placeholders in quantified sentences have *objects* as their intended range. Very often the variables that occur in mathematical texts are intended to range over only a restricted class of objects, and in order to aid readability mathematicians commonly press into service all sorts of letters to mark these restrictions: m, n, k for natural numbers, z, w for complex numbers, $\mathfrak{a}$, $\mathfrak{b}$ for cardinal numbers, G, H for groups, etc. In the first two chapters of this book, however, the first-order variables are intended to be completely unrestricted, ranging over any objects whatever, and to signal this we shall use only the lower-case letters x, y, z, t and their decorated variants x', x'', x_1, x_2, etc.

First-order predicate calculus is thus contrasted with the second-order version which permits in addition the use of quantified variables ranging over *properties* of objects. It is a common practice, which we shall follow at least until the end of the next chapter, to use the upper case letters X, Y, Z, etc. for such variables.

When we need to comment on the features that certain formulae have in common, or to introduce uniform abbreviations for formulae with a certain pattern, we shall find it helpful to use upper-case Greek letters such as Φ, Ψ, etc. to represent arbitrary formulae; if the formula Φ depends on the variables $x_1, \ldots, x_n$, we can write it $\Phi(x_1, \ldots, x_n)$ to highlight that fact. These Greek letters are to be thought of as part of the specification language (metalanguage) which we use to describe the formal text, not as part of the formal language itself.

At this point one possible source of confusion needs to be highlighted. Bradman was the greatest batsman of his time; 'Bradman' is a name with seven letters. Confusion between names and the objects which they denote (or between formulae and whatever it is that they denote) can be avoided by such careful use of quotation marks. Another strategy, which any reader attuned to the distinction between use and mention will already have observed in the last few paragraphs, is to rely on common sense to achieve the same effect. We shall continue to employ this strategy whenever the demands of readability dictate.

In this book the canons of reasoning we use will be those of the first-order predicate calculus with identity. It is common at this point in textbooks of set theory for the author to set down fully formal formation and inference rules for such a calculus. However, we shall not do this here: from the start we shall use ordinary English to express logical notions such as negation ('not'), disjunction ('or') and conjunction ('and'), as well as the symbols '$\Rightarrow$' for the conditional, '$\Leftrightarrow$' for the biconditional, and '$\forall$' and '$\exists$' for the universal and existential quantifiers. We shall use '$=$' for equality, and later we shall introduce other binary relation symbols: if R is any such symbol, we shall write $x\,R\,y$ to express that the relation holds between x and y, and $x \not R\, y$ to express that it does not.

There are several reasons for omitting the formal rules of logic here: they can be found in any of a very large number of elementary logic textbooks; they are not what this book is about; and their presence would tend to obscure from view the fact that they are being treated here only as a codification of the canons of reasoning we regard as correct, not as themselves constituting a formal theory to be studied from without by some *other* logical means.

The last point, in particular, deserves some emphasis. I have already mentioned the popularity of the formalist standpoint among mathematicians, and logicians are not exempt from the temptation. If one formalizes the rules of inference, it is important nonetheless not to lose sight of the fact that they remain rules of *inference* — rules for reasoning from meaningful premises to meaningful conclusions.

It is undoubtedly significant, however, that a formalization of first-order logic is available at all. This marks a striking contrast between the levels of logic, since in the second-order case only the formation rules are completely formalizable, not the inference rules: it is a consequence of Gödel's first incompleteness theorem that for each system of formal rules we might propose there is a second-order logical inference we can recognize as valid which is not justified by that system of rules.

Notice, though, that even if there is some reason to regard formalizability as a requirement our logic should satisfy (a question to which we shall return shortly), this does not suffice to pick out first-order logic uniquely, since there are other, larger systems with the same property. An elegant theorem due to Lindström (1969) shows that we must indeed restrict ourselves to reasoning in first-order logic if we require our logic to satisfy in addition the Löwenheim/Skolem property that any set of sentences which has a model has a countable model. But, as Tharp (1975) has argued, it is hard to see why we should wish to impose this condition straight off. Tharp attempts instead to derive it from conditions on the quantifiers of our logic, but fails in turn (it seems to me) to motivate these further conditions satisfactorily.

1.3 Schemes

By eschewing the use of second-order variables in the presentation of our theory we undoubtedly follow current mathematical fashion, but we thereby limit severely the strength of the theories we are able to postulate. On the classical conception going back to Euclid, the axioms of a system must be finite in number, for how else could they be written down and communicated? But it is — to take only one example — a simple fact of model theory that no finite list of first-order axioms has as its models all and only the infinite sets. So if we wish to axiomatize the notion of infinity in a first-order language, we require an infinite list of axioms.

But how do we specify an infinite list with any precision? At first it might seem as though the procedure fell victim to a variant of Poincaré's *petitio* since we presuppose the notion of infinity in our attempt to characterize it. But once again Poincaré's objection can be met by distinguishing carefully between object language and metalanguage. We cannot in the language itself assert an infinite list of object language sentences, but we can in the metalanguage make a commitment to assert any member of such a list by means of a finite description of its syntactic form. This is known as an *axiom scheme* and will typically take the following form.

If Φ is any formula of the language of the theory, then this is an axiom:

$$\ldots \Phi \ldots$$

(Here '$\ldots \Phi \ldots$' is supposed to stand for some expression such that if every 'Φ' in it is replaced by a formula, the result is a sentence of the object language.)

The presence of an axiom scheme of this form in a system does nothing to interfere with the system's formal character: it remains the case that the theorems of a first-order theory of this kind will be recursively enumerable, in contrast to the second-order case, because it will still be a mechanical matter to check whether any given finite string of symbols is an instance of the scheme or not.[1]

Recursive enumerability comes at a price, however. Any beginner in model theory learns a litany of results — the Löwenheim/Skolem theorems, the existence of non-standard models of arithmetic — that testify to the unavoidable weakness of first-order theories. Kreisel (1967*a*, p. 145) has suggested — on what evidence it is unclear — that this weakness 'came as a surprise' when it was discovered by logicians in the 1910s and 1920s, but there is a clear sense in which it should not have been in the least surprising, for we shall prove later in this book that if there are infinitely many objects, then (at least on the standard understanding of the second-order quantifier) there are uncountably many properties those objects may have. A first-order scheme, on the other hand, can only have countably many instances (assuming, as we normally do, that the language of the theory is countable). So it is to be expected that the first-order theory will assert much less than the second-order one does.

So much, then, for the purely formal questions. And for the formalist those are presumably the only questions there are. But for a realist there will be a further question as to how we finite beings can ever succeed in forming a commitment to the truth of all the infinitely many instances of the scheme. One view, common among platonists, is that we do not in fact form a commitment to the scheme at all, but only to the single second-order axiom

$$(\forall X) \ldots X \ldots$$

If we state the much weaker first-order scheme, that is only because it is the nearest approximation to the second-order axiom which it is possible to express in the first-order language.

Notice, though, that even if we abandoned the constraint of first-order expression and did state the second-order axiom, that would not magically enable us to prove lots of new theorems inaccessible to the first-order reasoner: in

[1] Although schemes are one way of generating infinite sets of axioms without destroying a system's formal character, they are not the only way: a theory whose set of finite models is not recursive will not be axiomatizable by schemes, even though it may be axiomatizable (see Craig and Vaught 1958). However, a result of Vaught (1967) shows that this distinction is irrelevant in formalizing set theory, where examples of this sort cannot arise.

order to make use of the second-order axiom, we need a *comprehension scheme* to the effect that if Φ is any first-order formula in which the variables $x_1, \ldots, x_n$ occur free, then

$$(\exists X)(\forall x_1, \ldots, x_n)(X(x_1, \ldots, x_n) \Leftrightarrow \Phi).$$

The difference is that *this* scheme is categorized as belonging to the background logic (where schematic rules are the norm rather than the exception) and, because it is logical and hence topic-neutral, it will presumably be taken to hold for any Φ, of any language. If we at some point enlarge our language, no fresh decision will be needed to include the new formulae thus generated in the intended range of possible substitutions for Φ in the comprehension scheme, whereas no corresponding assumption is implicit in our commitment to a scheme within a first-order theory. On the view now under consideration, therefore, we are all really second-order reasoners in disguise. Our commitment to a first-order scheme is merely the best approximation possible in the particular first-order language in question to the second-order axiom that expresses what we genuinely believe.

But it is important to recognize that this view is by no means forced on us simply by the presence of a scheme. It may, for instance, be axiomatic that if a dog Φs a man, then a man is Φed by a dog; but there is surely no temptation to see this scheme as derived from a single second-order axiom — if only because it is hard to see what the second-order axiom would be.

The issue is well illustrated by the arithmetical case. Here the second-order theorist states the principle of mathematical induction as the single axiom

$$(\forall X)((X(0) \text{ and } (\forall x)(Xx \Rightarrow X(\mathrm{s}x)) \Rightarrow (\forall x)Xx),$$

whereas the first-order reasoner is committed only to the instances

$$(\Phi(0) \text{ and } (\forall x)(\Phi(x) \Rightarrow \Phi(\mathrm{s}x)) \Rightarrow (\forall x)\Phi(x)$$

for all replacements of Φ by a formula in the first-order language of arithmetic. It is by no means obvious that the only route to belief in all these instances is via a belief in the second-order axiom. Isaacson (1987), for example, has argued that there is a stable notion of arithmetical truth which grounds only the first-order axioms, and that all the familiar examples of arithmetical facts not provable on their basis require higher-order reflection of some kind in order to grasp their truth. If this is right, it opens up the possibility that someone might accept all the first-order axioms formulable in the language of arithmetic but regard higher-order reflection as in some way problematic and hence resist some instance of the second-order axiom.

1.4 The choice of logic

The distinction between first- and second-order logic that we have been discussing was originally made by Peirce, and was certainly familiar to Frege, but neither of them treated the distinction as especially significant: as van Heijenoort (1977, p. 185) has remarked, 'When Frege passes from first-order logic to a higher-order logic, there is hardly a ripple.' The distinction is given greater primacy by Hilbert and Ackermann (1928), who treat first- and higher-order logic in separate chapters, but the idea that first-order logic is in any way privileged as having a radically different status seems not to have emerged until it became clear in the 1930s that first-order logic has a complete formalization but second-order logic does not. The result of this was that by the 1960s it had become standard to state mathematical theories in first-order form using axiom schemes. Since then second-order logic has been very little studied by mathematicians (although recently there seems to have been renewed interest in it, at least among logicians).

So there must have been a powerful reason driving mathematicians to first-order formulations. What was it? We have already noted the apparent failure of attempts to supply plausible constraints on inference which characterize first-order logic uniquely, but if first- and second-order logic are the only choices under consideration, then the question evidently becomes somewhat simpler, since all we need do is to find a single constraint which the one satisfies but not the other. Yet even when the question is simplified in this manner, it is surprisingly hard to say for sure what motivated mathematicians to choose first-order over second-order logic, as the texts one might expect to give reasons for the choice say almost nothing about it.

Very influential on this subject among philosophers (at least in America) was Quine: he argued that the practice of substituting second-order variables for predicates is incoherent, because quantified variables ought to substitute, as in the first-order case, for names; and he queried whether there is a well-understood domain of entities (properties, attributes or whatever) that they can be taken to refer to. Now Quine's criticisms are not very good — for a persuasive demolition see Boolos 1975 — but in any case they are evidently not of the sort that would have influenced mathematicians even if they had read them.

A more likely influence is Bourbaki (1954), who adopted a version of first-order logic. It is unquestionable that Bourbaki's works were widely read by mathematicians, in contrast to Quine's: Birkhoff (1975), for instance, recalls that their 'systematic organization and lucid style mesmerized a whole generation of American graduate students'. But the mere fact that Bourbaki adopted a first-order formulation certainly cannot be the whole of an explanation: many other features of Bourbaki's logical system (his use of Hilbert's ε-operator, for instance, or his failure to adopt the axiom of foundation) sank without trace. What may have influenced many mathematicians was the philosophy of mathematics which led Bourbaki to adopt a first-order formulation of his system. This philosophy has at its core a conception of rigour that is essentially formalist in character. In an unformalized mathematical text, he says,

> one is exposed to the danger of faulty reasoning arising from, for example, incorrect use of intuition or argument by analogy. In practice, the mathematician who wishes to satisfy himself of the perfect correctness or 'rigour' of a proof or a theory hardly ever has recourse to one or another of the complete formalizations available nowadays ... In general, he is content to bring the exposition to a point where his experience and mathematical flair tell him that translation into formal language would be no more than an exercise of patience (though doubtless a very tedious one). If, as happens again and again, doubts arise as to the correctness of the text under consideration, they concern ultimately the possibility of translating it unambiguously into such a formalized language: either because the same word has been used in different senses according to the context, or because the rules of syntax have been violated by the unconscious use of modes of argument which they do not specifically authorize. Apart from this last possibility, the process of rectification, sooner or later, invariably consists in the construction of texts which come closer and closer to a formalized text until, in the general opinion of mathematicians, it would be superfluous to go any further in this direction. In other words, the correctness of a mathematical text is verified by comparing it, more or less explicitly, with the rules of a formalized language. (Bourbaki 1954, Introduction)

This conception of the formalism as an ultimate arbiter of rigour has certainly been influential among mathematicians.

> I think there is clear evidence that the way in which doubts (about a piece of mathematics) are resolved is that the doubtful notions or inferences are refined and clarified to the point where they can be taken as proofs and definitions from existing notions, within some first order theory (which may be intuitionistic, non-classical, or category-theoretical, but in mainstream mathematics is nowadays usually some part of set theory, at least in the final analysis). (Drake 1989, p. 11)

The attraction of this view is that in principle it reduces the question of the correctness of a purported proof to a purely mechanical test. Adopting a fully formalized theory thus has the effect of corralling mathematics in such a way that nothing within its boundary is open to philosophical dispute. This seems to be the content of the observation, which crops up repeatedly in the mathematical literature, that mathematicians are platonists on weekdays and formalists on Sundays: if a mathematical problem is represented as amounting to the question whether a particular sentence is a theorem of a certain formal system, then it is certainly well-posed, so the mathematician can get on with the business of solving the problem and leave it to philosophers to say what its significance is.

> On foundations we believe in the reality of mathematics, but of course when philosophers attack us with their paradoxes we rush behind formalism and say: 'Mathem-

atics is just a combination of meaningless symbols.' ... Finally we are left in peace to go back to our mathematics and do it as we have always done, with the feeling each mathematician has that he is working with something real. (Dieudonné 1970, p. 145)

But this view gives rise to a perplexity as to the role of the formalism in grounding our practice. For Gödel's incompleteness theorem shows that no formalism encompasses all the reasoning we would be disposed to regard as correct; and even if we restrict ourselves to a fixed first-order formal theory, Gödel's completeness theorem shows only that what is formally provable coincides *extensionally* with what follows from the axioms (see Kreisel 1980, pp. 161–2).

The role the formal rules play in actual reasoning is in fact somewhat opaque, and indeed the authors of Bourbaki were uncomfortably aware of this even as they formulated the view just quoted: the minutes of their meetings report that Chevalley, one of the members involved in writing the textbooks, 'was assigned to mask this as unhypocritically as possible in the general introduction' (Corry 1996, pp. 319–20). At the beginning of the text proper (Bourbaki 1954) they state a large number of precise rules for the syntactic manipulation of strings of symbols, but then, having stated them, immediately revert to informal reasoning. So 'the evidence of the proofs in the main text depends on an *understood* notion of logical inference' (Kreisel 1967*b*, p. 210), not on the precise notion defined by the formal specification of syntax. At the time they were writing the book, this was a matter of straightforward necessity: they realized that to formalize even simple mathematical arguments using the formalism they had chosen would take too long to be feasible. They seriously under-estimated just *how* long, though: they claimed that the number of characters in the unabbreviated term for the cardinal number 1 in their formal system was 'several tens of thousands' (Bourbaki 1956, p. 55), but the actual number is about 10^{12} (Mathias 2002). Only much more recently has it become possible to contemplate using computers to check humanly constructed mathematical arguments against formal norms of correctness; but this is still no more than an ongoing research project, and even if it is carried out successfully, it will remain unclear why the fact that a proof can be formalized should be regarded as a *criterion* of its correctness.

1.5 Definite descriptions

If $\Phi(x)$ is a formula, let us abbreviate $(\forall y)(\Phi(y) \Leftrightarrow x = y)$ as $\Phi!(x)$. The formula $(\exists x)\Phi!(x)$ is then written $(\exists! x)\Phi(x)$ and read 'There exists a unique x such that $\Phi(x)$'. Strictly speaking, though, this definition of $\Phi!(x)$ is unsatisfactory as it stands. If $\Phi(x)$ were the formula '$x = y$', for example, we would find that $\Phi!(x)$ is an abbreviation for $(\forall y)(y = y \Leftrightarrow x = y)$, which is true iff x is the only object in existence, whereas we intended it to mean 'x is the unique object equal to y', which is true iff $x = y$. What has gone wrong is that the variable y occurs in the formula $\Phi(x)$ and has therefore become accidentally bound. This sort of collision of variables is an irritating feature of quantified logic which we shall ignore from now on. We shall assume whenever we use a variable that it is chosen so that it does not collide with any of the other variables we are using: this is always possible, since the number of variables occurring in any given formula is finite whereas the number of variables we can create is unlimited. (We can add primes to x indefinitely to obtain x', x'', etc.).

If $\Phi(x)$ is a formula, then we shall use the expression $\imath!x\Phi(x)$, which is read 'the x such that $\Phi(x)$', to refer to the unique object which is Φ if there is one, and to nothing otherwise. Expressions of this form are called *definite descriptions*. More generally, expressions of the sort that denote objects are called *terms*. If $\Phi(x, x_1, \ldots, x_n)$ is a formula depending on the variables $x, x_1, \ldots, x_n$, then $\imath!x\Phi(x, x_1, \ldots, x_n)$ is a term depending on $x_1, \ldots, x_n$. *Proper names* are also terms, but they do not depend on any variables. We shall use lower-case Greek letters such as σ, τ, etc. to stand for arbitrary terms; if the term σ depends on the variables $x_1, \ldots, x_n$, then we can write it $\sigma(x_1, \ldots, x_n)$ to highlight that fact. These schematic lower-case Greek letters are, like the upper-case Greek letters we brought into service earlier to stand for formulae, part of the metalanguage, not of the object language: they stand in a schematic sentence in the places where particular terms stand in an actual sentence.

If a term 'σ' denotes something, then we shall say that σ *exists*. It is a convention of language that proper names always denote something. The same is not true of definite descriptions: consider for example the description $\imath!x(x \neq x)$. In general, $\imath!x\Phi(x)$ exists iff $(\exists!x)\Phi(x)$. We adopt the convention that if σ and τ are terms, the equation $\sigma = \tau$ is to be read as meaning 'If one of σ and τ exists, then they both do and they are equal'.

Because definitions are formally just ways of introducing abbreviations, the question of their correctness is simply one of whether they enable us mechanically and unambiguously to eliminate the expression being defined from every formula in which it occurs; the correctness in this sense of the definitions in this book will always (I hope) be trivially apparent. (The question of their psychological potency is of course quite a different matter.)

The things for which definitions introduce abbreviations will be either formulae or terms. If they are terms, then apart from their formal correctness (i.e. unambiguous eliminability) there is the question of the existence of objects for them to refer to. It is not wrong to use terms which do not denote anything; but it may be misleading, since the rules of logic are not the same for them as they are for ones which do. (For example, the move from $\Phi(\sigma)$ to $(\exists x)\Phi(x)$ is fallacious if σ does not exist.) So the introduction of a new symbol to abbreviate a term will sometimes be accompanied by a justification to show that this term denotes something; if no such justification is provided, the reason may well be that the justification in question is completely trivial.

Notes

The logical spine of this book will be the informal exposition of a particular axiomatic theory. No knowledge of metalogic is needed in order to follow that exposition, but only what Bourbaki famously called 'a certain capacity for abstract thought'. The commentary that surrounds this spine, which aims to flesh out the exposition by means of various more or less philosophical reflections on its intended content, does occasionally allude to a few metalogical results, however. Most readers will no doubt be familiar with these already, but those who are not will find enough for current purposes in the enjoyably opinionated sketch by Hodges (1983): they should pay special attention to the limitative results such as the Löwenheim/Skolem theorems and the existence of non-standard models of first-order theories.

I have cautioned against regarding formalizability as a *criterion* of the correctness of mathematical reasoning. Nonetheless it is of considerable importance to note that the theory which forms the spine of this book is capable of formalization as a first-order theory, since it is this that ensures the applicability to it of the metalogical results just alluded to. Implicit throughout the discussion, therefore, will be the distinction between object language and metalanguage, and the related distinction between use and mention. Although I have promised to ignore these distinctions whenever it aids readability to do so, it is important to be aware of them: Quine (1940, §§ 4–6) offers a bracing lecture on this subject.

Goldfarb (1979) illuminates the context for the rise to dominance of first-order logic without really attempting an explanation for it. Some further clues are offered by G.H. Moore (1980). However, there is much about this matter that remains obscure.

Chapter 2

Collections

2.1 Collections and fusions

The language of aggregation is everywhere: a library consists of books, a university of scholars, a parliament of crooks. Several words are commonly used, depending on context, to describe these formations of one thing from many: my library is a *collection* of books on a *manifold* of different subjects; it includes a *set* of Husserl texts, which belong, alas, to the *class* of books I have never quite got round to reading; the *extension* of my library is not that of all the books that I own, since I keep many others at home, but in *sum*, I was surprised to discover recently, they weigh well over a ton.

It will turn out that there are several different concepts here sheltering under one umbrella, and we shall need quite shortly to press several of these words — set, class, extension, collection — into technical service to express them. In the meantime, therefore, we shall reserve the word 'aggregate' as our umbrella term for all such notions.

But what *is* an aggregate? What, that is to say, is the subject matter of the theory we wish to set up? We might start by saying that an aggregate is, at least in the standard cases of which ordinary language usually treats, a single entity which is in some manner composed of, or formed from, some other entities. But the standard cases have a tendency to obscure the distinction between two quite different ways in which it has been taken that things can be aggregated — collection and fusion. Both are formed by bundling objects together, but a fusion is no more than the sum of its parts, whereas a collection is something more. *What* more it is is disconcertingly hard to say, and this has inclined some philosophers, especially those with nominalist sympathies, to prefer fusions: a fusion, they say, is no more than an alternative way of referring, in the singular, to the objects that make it up, which we might otherwise refer to in the plural.

To be sure, if we accept mereology [the science of this sort of aggregation], we are committed to the existence of all manner of mereological fusions. But given a prior commitment to cats, say, a commitment to cat-fusions is not a *further* commitment. The fusion is nothing over and above the cats that compose it. It just *is* them. They just *are* it. Take them together or take them separately, the cats are the same portion of Reality either way. Commit yourself to their existence all together or one at a time, it's the same commitment either way. If you draw up an inventory of Reality according to your scheme of things, it would be double counting to list the cats and then also list their fusion. In general, if you are already committed to some things, you incur no further commitment when you affirm the existence of their fusion. The new commitment is redundant, given the old one. (Lewis 1991, pp. 81–2)

A collection, by contrast, does not merely lump several objects together into one: it keeps the things distinct and is a further entity over and above them. Various metaphors have been used to explain this — a collection is a sack containing its members, a lasso around them, an encoding of them — but none is altogether happy.[1] We need to be aware straightaway, therefore, that collections are metaphysically problematic entities if they are entities at all, and need to be handled with care.

The contrast between collections and fusions becomes explicit when we consider the notion of membership. This is fundamental to our conception of a collection as consisting of its members, but it gets no grip at all on the notion of a fusion. The fusion of the cards in a pack is made up out of just those cards, but they cannot be said to be its members, since it is also made up out of the four suits. A collection has a determinate number of members, whereas a fusion may be carved up into parts in various equally valid (although perhaps not equally interesting) ways.

The distinction between collections and fusions is at its starkest when we consider the trivial case of a single object such as my goldfish Bubble. The collection whose only member is Bubble is usually called a *singleton* and written {Bubble}. It is not the same object as Bubble, since it has exactly one member (Bubble), whereas Bubble itself, being a goldfish, does not have any members at all. The fusion of Bubble, by contrast, is just Bubble itself, no more and no less.

And what if we try to make something out of nothing? A container with nothing in it is still a container, and the empty collection is correspondingly a collection with no members. But a fusion of nothing is an impossibility: if we try to form a fusion when there is nothing to fuse, we obtain not a trivial object but no object at all.

The distinction between collections and fusions, and the corresponding one between membership and inclusion, were not clearly drawn until the end of the 19th century. In textbooks the distinction between membership and inclusion is sometimes attributed to Peano, who introduced different notations for the two concepts in 1889, §4. But it is perhaps a little generous to give Peano *all* the credit: only a page or two later we find him asserting that if k is contained in s, then k is also an element of s just in case k has exactly

[1] See Lewis 1991 for an excellent discussion of the difficulty of making good metaphysical sense of such metaphors.

one member, which is to make precisely the blunder he had apparently just avoided. A year later (1890, p. 192) he did indeed introduce a notation to distinguish between a set b and the singleton $\{b\}$ (which he denoted ιb), but his motivation for this was somewhat quaint: 'Let us decompose the sign = into its two parts *is* and *equal to*; the word *is* is already denoted by ϵ; let us also denote the expression *equal to* by a sign, and let ι (the first letter of $\iota\sigma o\varsigma$) be that sign; thus instead of $a = b$ one can write $a \epsilon \iota b$.' Evidently, then, Peano's motivation was overwhelmingly notational. The language of classes was for him, at this stage at least, just that — a language — and there is little evidence that he conceived of classes as entities in their own right.

It seems in fact to be Frege who deserves credit for having first laid out the properties of fusions clearly. A fusion, he said, 'consists of objects; it is an aggregate, a collective unity, of them; if so it must vanish when these objects vanish. If we burn down all the trees of a wood, we thereby burn down the wood. Thus there can be no empty fusion.' (1895, pp. 436–7, modified) But the work Frege was reviewing when he made this remark (Schröder 1890–5) was very influential for a time: it gave rise to a tradition in logic which can be traced through to the 1920s. And it was plainly fusions, not collections, that Dedekind had in mind in *Was sind und was sollen die Zahlen?* when he avoided the empty set and used the same symbol for membership and inclusion (1888, nos. 2–3) — two tell-tale signs of a mereological conception. He drafted an emendation adopting the collection-theoretic conception only much later (see Sinaceur 1973).

Given the early popularity of fusions, then, it is striking how complete and how quick the mathematical community's conversion to collections was. In practical terms it was no doubt of great significance that Zermelo chose collections, not fusions, as the subject of his axiomatization in 1908*b*. And the distinctions which mereology elides, such as that between the cards in a pack and the suits, are ones which mathematicians frequently wish to make: so talk of collections has something to recommend it over talk of fusions, where mathematics is concerned. But this cannot be the whole explanation: as in the parallel case of first- and second-order logic that we noted earlier, more remains to be said.

2.2 Membership

Collections, unlike fusions, can always be characterized determinately by their membership. In our formalism we shall therefore treat the relation of membership as primitive. In other words, the language of the theory of collections has in it as a non-logical primitive a binary relation symbol '$\in$'. The formula '$x \in y$' is read 'x belongs to y'.

But of course we must not lose sight of the fact that the theory of collections is of no use in itself: its point is to let us talk about *other* things. We shall not make any presuppositions here about what those other things are. We shall simply assume that we start with a theory T about them. We shall call the objects in the domain of interpretation of T *individuals*. Some other authors call them 'atoms', and many, in tribute to the dominance of German writers in the development of the subject, call them 'Urelemente' (literally 'original elements'). Before we start talking about collections, we need to ensure that we do not unthinkingly treat the claims made by T about the individuals as if they applied to collections. (If T were a formalization of Newtonian mechanics, for instance, we would not want to find ourselves claiming without argument that collections are subject to just the same physical laws as their members.) To ring-fence the individuals, then, we introduce a predicate $\mathrm{U}(x)$ to mean that x is an individual, and we relativize all the axioms of T to U. That is to say, we replace every universal quantifier '$(\forall x)\ldots$' in an axiom of T with '$(\forall x)(\mathrm{U}(x) \Rightarrow \ldots)$' and every existential quantifier '$(\exists x)\ldots$' with '$(\exists x)(\mathrm{U}(x)$ and $\ldots)$'; and for every constant 'a' in the language of T we add $\mathrm{U}(a)$ as a new axiom.

Having relativized T to the individuals in this manner, we are now in a position to introduce the central idea of the *collection* of objects satisfying a property Φ.

Definition. *If $\Phi(x)$ is a formula, the term $\imath! y($not $\mathrm{U}(y)$ and $(\forall x)(x \in y \Leftrightarrow \Phi(x)))$ is abbreviated $\{x : \Phi(x)\}$ and read 'the collection of all x such that $\Phi(x)$'.*

In words: $\{x : \Phi(x)\}$, if it exists, is the unique non-individual whose elements are precisely the objects satisfying Φ. We shall also use variants of this notation adapted to different circumstances: for instance, we often write $\{x \in a : \Phi(x)\}$ instead of $\{x : x \in a$ and $\Phi(x)\}$; we write $\{y\}$ for $\{x : x = y\}$, $\{y, z\}$ for $\{x : x = y$ or $x = z\}$, etc.; and we write $\{\sigma(x) : \Phi(x)\}$ for $\{y : (\exists x)(y = \sigma(x)$ and $\Phi(x))\}$. The objects satisfying Φ are said to be *elements* or *members* of $\{x : \Phi(x)\}$; they may, but need not, be individuals. The collection $\{x : \Phi(x)\}$, on the other hand, is constrained by definition not to be an individual.

(2.2.1) **Lemma.** *If $\Phi(x)$ is a formula such that $a = \{x : \Phi(x)\}$ exists, then $(\forall x)(x \in a \Leftrightarrow \Phi(x))$.*

Proof. This follows at once from the definition. □

It would, I suppose, have been more accurate to call this a 'lemma scheme', since it cannot be formalized as a single first-order proposition but has to be thought of as describing a whole class of such propositions. We shall not be this pedantic, however, and will continue to describe such schemes as lemmas (or propositions, or corollaries, or theorems, as the case may be).

(2.2.2) **Lemma.** *If* $\Phi(x)$ *and* $\Psi(x)$ *are formulae, then*

$$(\forall x)(\Phi(x) \Leftrightarrow \Psi(x)) \Rightarrow \{x : \Phi(x)\} = \{x : \Psi(x)\}.$$

Proof. If $(\forall x)(\Phi(x) \Leftrightarrow \Psi(x))$, then $(\forall x)(x \in y \Leftrightarrow \Phi(x)) \Leftrightarrow (\forall x)(x \in y \Leftrightarrow \Psi(x))$, and so $\{x : \Phi(x)\} = \{x : \Psi(x)\}$. □

Note that this lemma has to be interpreted according to the convention we introduced in §1.5: the collections derived from logically equivalent formulae are equal *if they exist*.

2.3 Russell's paradox

Let us call a property *collectivizing* if there is a collection whose members are just the objects which have it. One of the matters that will interest us is to try to settle which properties are collectivizing: not all of them can be, for that supposition rapidly leads to a contradiction.

(2.3.1) **Russell's paradox (absolute version).** $\{x : x \notin x\}$ *does not exist.*

Proof. Suppose that $a = \{x : x \notin x\}$ exists. Then $(\forall x)(x \in a \Leftrightarrow x \notin x)$ [lemma 2.2.1]. Therefore in particular $a \in a \Leftrightarrow a \notin a$. Contradiction. □

The first thing to notice about this result is that we have proved it before stating *any* axioms for our theory. This serves to emphasize that Russell's paradox is not a challenge to, or refutation of, any one theory of collections, but a feature that has to be taken account of in *any* such theory. It is worth noting, too, how elementary is the logic that is used to derive the paradox. That is not to say, of course, that the paradox is derivable in any logical system whatever. By the simple device of writing out the proof in full we could no doubt identify a restricted logic which blocks its derivation. This heroic course has indeed been recommended by some authors, but it is an *extremely* desperate strategy, since the restrictions these authors have to impose (e.g. denying the transitivity of implication) risk crippling logic irreparably.

Non-self-membership was not the first instance of a non-collectivizing property to be discovered: Cantor told Hilbert in 1897 that 'the set of all alephs ... cannot be interpreted as a definite, well-defined finished set'. Nor was the sad appendix which Frege added to the second volume of the *Grundgesetze* the first reference in print to a paradox of this kind: Hilbert referred in his (1900) lecture on the problems of mathematics to 'the system of all cardinal numbers or even of *all* Cantor's alephs, for which, as may be shown, a consistent system of axioms cannot be set up'. What is particularly striking about Russell's paradox, however, is that it is in essence more purely logical than the others: for any binary relation R it is a first-order logical truth that

$$\text{not}\,(\exists y)(\forall x)(x\ R\ y \Leftrightarrow \text{not}\ x\ R\ x).$$

By contrast, the other set-theoretic paradoxes involve cardinal or ordinal numbers in some way.[2] Curiously, Hilbert described these other paradoxes (in a letter to Frege) as 'even more convincing': it is not clear why.

2.4 Is it a paradox?

A paradox is a fact which is contrary to expectation (from the Greek $\pi\alpha\rho\alpha$ + $\delta o\xi\alpha$, 'beyond expectation'). Whether Russell's result is a paradox presumably depends on how one understands the notion of a collection. It was certainly a surprise to Frege to be told that one of the axioms of his formal system led to contradiction, but what he was trying to capture was probably not the notion of a collection but a notion more closely connected to logic which we shall here call a *class*. So although there was undoubtedly a serious flaw in Frege's understanding (a flaw which we shall discuss in appendix B), that does not yet give us any reason to think that the flaw infects the notion of a collection.

Indeed it is quite common nowadays to find books presenting the notion of a collection in such a way that what Russell discovered appears not to be a paradox at all, but rather something we should have expected all along. However, this Panglossian view should be treated with a little scepticism, since even if all parties now agree that there is a coherent notion of collection, they do not yet seem to agree on what it is.

In any case, the idea that the paradoxes are not really so paradoxical if we only think about them in the right way is hard to find in print before Gödel (1944) and did not become widespread until much later. In 1940 Quine thought the paradoxes 'were implicit in the inferential methods of uncritical common sense' (1940, p. 166) and as late as 1951 he could still assert that 'common sense is bankrupt for it wound up in contradiction. Deprived of his tradition, the logician has had to resort to myth-making.' (1951, p. 153) Even now there are still many who would agree with him (e.g. Weir 1998*a*).

It is also worth being cautious about the significance of the set-theoretic paradoxes for the foundations of mathematics. The 30 years following their discovery are often referred to as the period of the crisis of foundations, but it is not obvious that they deserve this title. Certainly there was a crisis in the foundations of *set theory*, but even here many mathematicians continued to work informally in ways that did not depend on one resolution or other of

[2] We shall look at the technicalities of these other paradoxes in later sections when we have defined the notions of cardinal and ordinal numbers on which they depend.

the paradoxes: Hausdorff's (1914) textbook on set theory, for example, barely mentions them. And there is no reason to regard even this as a crisis in the foundations of mathematics more generally unless one already accepts the set-theoretic reduction of mathematics; but in advance of a satisfying resolution of the paradoxes the more natural reaction would be simply to treat them as a refutation of the possibility of such a reduction.

There is nonetheless a more general threat which the set-theoretic paradoxes have sometimes been interpreted as presenting, namely that the canons of reasoning which we 'naively' (i.e. pre-theoretically) find compelling are in fact contradictory. Those who have argued this have not, however, agreed on what follows from it. One moral, favoured by the more logically inclined, is that as common sense can lead us astray, we should not trust it but rely instead on the outputs of a formal system; but others (e.g. Priest 1995; Restall 1992) have thought instead that we should learn to accept that common sense (which is in the end all we have) is inevitably contradictory — learn, in short, to tolerate contradictions.

Now this last issue is to some extent tangential to the concerns of this book, since it does not arise only in relation to the set-theoretic paradoxes: the moral that common sense is contradictory could just as well have been drawn from a consideration of the liar paradox or the paradox of the unexpected hanging. So it is hard to see why this is any more pressing an issue for set theorists than for anyone engaged in rational argument. I shall therefore confine myself here to recording my faith in tutored common sense as a tool for deductive reasoning, and continue to make free use of this tool in what follows.

But even if the set-theoretic paradoxes are not *uniquely* troubling, they are troubling nevertheless: simply saying that we ought never to have expected any property whatever to be collectivizing, even if true, leaves us well short of an account which will settle which properties are collectivizing and which are not. What we must do if we are to achieve this last objective is evidently to refine our conception of what a collection is, and it will be a large part of the task of this book to explore the prospects for a satisfactory resolution of the paradoxes by this means.

2.5 Indefinite extensibility

One idea that has cropped up repeatedly (e.g. Lear 1977; Dummett 1993) is that it is central to the correct resolution of the paradoxes to limit ourselves to intuitionistic logic, i.e. to forswear the law of the excluded middle as a general principle applicable to any proposition that concerns collections. It needs to be emphasized straightaway that this proposal does not belong to the class we mentioned earlier of attempts to resolve the paradoxes simply by weakening the logic while clinging onto the naive view that every property is collectivizing. The argument we gave in §2.3 to show that the property of non-self-membership is not collectivizing makes no use of excluded middle and is therefore intuitionistically valid. So the intuitionistic reasoner is barred from the naive view just as firmly as the classicist.[3]

A rejection of intuitionistic logic when reasoning about collections can therefore only be one part of a resolution of the paradoxes and not the whole. It is unsurprising, therefore, that most of the arguments that may be given for this are intimately linked to particular conceptions of what a collection is. For this reason we shall at present confine ourselves to exploring one argument, due to Dummett, which does not on its face depend on any particular conception. His argument is that the correct logic to use when reasoning about what he calls an indefinitely extensible concept is intuitionistic. He takes the paradoxes to show that the concept of a collection is indefinitely extensible, but this is not, according to him, the only example. So his argument would, if correct, lead to intuitionism in a variety of spheres of discourse, of which mathematics is only the most prominent.

But not in all. Although indefinite extensibility is endemic in mathematics and logic, Dummett is not claiming that it infects our ordinary sublunary discourse. According to Heck (1993, p. 233), this is

> a *new* argument for intuitionism, quite different in character from the meaning-theoretic arguments for which Dummett is well known. It is a local argument for anti-realism about mathematics, one which depends upon considerations peculiarly mathematical in character; it therefore has not the propensity to generalize which the meaning-theoretic arguments have.

In one respect Heck is not quite right: the argument was not in fact new when Dummett presented it in 1991 but can already be found in his earlier writings, for instance in 1973, pp. 529–30 and 568–9. What *is* new in his publications of the 1990s is rather his conception of how this local argument for intuitionism in mathematics is related to the global considerations to which Heck alludes. What has changed is that Dummett no longer imagines that a wholly general argument from considerations of meaning *could* deliver a global anti-realism — a reason, that is to say, to abandon classical logic *tout court*. Instead he now sees these considerations as doing no more than provide a scheme whose instances are the outlines of arguments for anti-realism local to particular spheres of discourse. How such an outline is to be fleshed out will then depend on parochial features of the sphere of discourse in question. What this framework does, therefore, is permit us to see Dummett's 'new'

[3]The reason this needs some emphasis is that one commonly finds the argument establishing Russell's paradox expressed as a dilemma: the collection a either belongs to itself or not, but either supposition leads to contradiction. Put like this, the argument does indeed depend on the law of the excluded middle; but as we have seen, this is not the only way to put it.

argument for intuitionism as indicating a way in which the outline argument can be fleshed out in the case where the sphere of discourse is mathematics.

What, then, is Dummett's argument? As I have suggested, Dummett usually argues via an intermediate stage. First he presents an analysis of the failure of a concept to be collectivizing as owing to its *indefinite extensibility*. Then he argues that the correct logic to use in any reasoning that involves quantification over the range of an indefinitely extensible concept is intuitionistic. Let us consider these two stages of Dummett's argument in turn.

The first stage, the analysis of collectivization failure as due to indefinite extensibility, was originated not by Dummett but by Russell, who as early as 1906*b* called such properties *self-reproductive*. A property is self-reproductive or indefinitely extensible if, 'given any class of terms all having [the] property, we can always define a new term also having the property' (Russell 1973*a*, p. 144). More precisely, a property F is indefinitely extensible if there is a process which, when applied to some Fs, gives rise to another object which is not among them but is nevertheless an F.

As an analysis of the known paradoxes, this is quite persuasive. Not only Russell's paradox but others we shall come across later, such as Burali-Forti's (§11.2), can without too much distortion be put into a form whereby what they exhibit is that a certain property — the one whose instances we are attempting to collect — is indefinitely extensible in Dummett's sense. It is less clear, though, what reason there is to think that *any* non-collectivizing property will be indefinitely extensible. We might well think that vagueness is a second, distinct reason why some collections — the collection of balding men, or of short words — fail to exist. If we exclude vague properties by *fiat*, we have, of course, an inductive argument for the claim, since, as just observed, all the proofs we have found so far of failure to collectivize go via indefinite extensibility; but we can hardly regard that as compelling, especially since some of the paradoxes did not arise independently but were discovered by analysing the forms of argument employed in those already known: if there is a set-theoretic paradox that is not analysable in terms of indefinite extensibility, then by the nature of the case it involves a wholly new idea, and we plainly have little idea in advance how likely this is to occur.

Notice, too, that even if we agree that the failure to collectivize is always a result of indefinite extensibility, the route from there via Dummett's second claim to his conclusion — that indefinite extensibility leads to the abandonment of classical logic in all mathematical reasoning — depends on some sort of set-theoretic reductionism. For that reason, perhaps, Dummett has also attempted to argue directly (i.e. independent of any such reduction) that indefinite extensibility is a feature not just of the concept *set* but of *natural number* and *real number*. These direct arguments are problematic, but this is not the place to discuss them. Instead let us move on to Dummett's second main claim, namely that the correct logic to use when reasoning about the range of an indefinitely extensible concept is intuitionistic and not classical.

It is no exaggeration to say that commentators have found Dummett's argument for this claim obscure. Boolos (1993), Clark (1993*a*; 1998), Oliver (1998) and Wright (1999) are among those who have struggled to understand it. At times the argument has seemed to centre on the idea that indefinite extensibility is a species of vagueness, and this has understandably puzzled commentators, since it is hard to see how the paradoxes lend encouragement to the thought that it is vague which objects are sets.

What seems most likely is that Dummett's failure to articulate clearly why indefinite extensibility should lead inexorably to intuitionism is explained by the level of generality at which the argument is couched. Precisely because this move cannot resolve the paradoxes on its own, the reason for making it cannot be independent of the conception driving the other parts of the resolution. We should therefore return to this issue once we have fleshed out in more detail some strategies for resolving the paradoxes.

2.6 Collections

We have defined 'collection of . . . ', but not 'collection' on its own. The naive idea is that collections are precisely objects of the form $\{x : \Phi(x)\}$ for some formula $\Phi(x)$; but this does not work under the formal restrictions we imposed on ourselves in chapter 1 (since the phrase 'for some formula' is not first-order). It turns out, though, that we can get what we want from the following definition, which *is* first-order.

Definition. *We say that* b *is a* collection *if* $b = \{x : x \in b\}$.

(2.6.1) **Lemma.** *No collection is an individual.*

Proof. This follows at once from the definitions. □

So nothing is both a collection and an individual. We have not yet said, and will not make any assumption that commits us to saying, that everything is one or the other. Something else which we have not assumed, but which we might have added for the sake of tidiness, is that individuals do not have members.

(2.6.2) **Lemma.** *Suppose that* $\Phi(x)$ *is a formula. If* $\{x : \Phi(x)\}$ *exists, then it is a collection.*

Proof. If $b = \{x : \Phi(x)\}$ exists, then

$$(\forall x)(x \in b \Leftrightarrow \Phi(x)) \text{ [lemma 2.2.1]},$$

so that $\{x : x \in b\}$ exists and

$$b = \{x : \Phi(x)\} = \{x : x \in b\} \text{ [lemma 2.2.2].}$$ □

In the remainder of this chapter and in the following one, a, b, c, a', b', c', etc. will always denote collections.

In particular the quantifiers '$(\forall a)$' and '$(\exists a)$' should be read 'For every collection a' and 'For some collection a' respectively. We continue to use x, y, z, etc. as variables ranging over everything — collections, individuals, whatever.

We write $(\forall x \in a)\Phi$ instead of $(\forall x)(x \in a \Rightarrow \Phi)$, and $(\exists x \in a)\Phi$ instead of $(\exists x)(x \in a \text{ and } \Phi)$. We also write $\Phi^{(a)}$ for the result of replacing every quantifier '$(\forall x)$' or '$(\exists x)$' in Φ with the corresponding relativized quantifier '$(\forall x \in a)$' or '$(\exists x \in a)$' respectively.

(2.6.3) **Lemma.** *Suppose that $\Phi(x)$ is a formula.*

$$(\exists a)(\forall x)(x \in a \Leftrightarrow \Phi(x)) \Leftrightarrow \{x : \Phi(x)\} \textit{ exists.}$$

Proof. If a is a collection such that $(\forall x)(x \in a \Leftrightarrow \Phi(x))$, then

$$a = \{x : x \in a\} = \{x : \Phi(x)\} \text{ [lemma 2.2.2]}$$

and so $\{x : \Phi(x)\}$ exists. Conversely, if $a = \{x : \Phi(x)\}$ exists, then it is a collection [lemma 2.6.2] and $(\forall x)(x \in a \Leftrightarrow \Phi(x))$ [lemma 2.2.1]. □

(2.6.4) **Extensionality principle.**

$$(\forall x)(x \in a \Leftrightarrow x \in b) \Rightarrow a = b.$$

Proof. Suppose that a and b are collections. Then $a = \{x : x \in a\}$ and $b = \{x : x \in b\}$. But if $(\forall x)(x \in a \Leftrightarrow x \in b)$, then

$$\{x : x \in a\} = \{x : x \in b\} \text{ [lemma 2.2.2]},$$

and so $a = b$. □

In words: a collection is determined by its elements.

The extensionality principle is taken as an axiom by Zermelo (1908*b*) — he calls it the axiom of definiteness (*Axiom der Bestimmtheit*) — and by most treatments since. The presentation we have given here, which makes it a theorem rather than an axiom, emphasizes its purely definitional character: a collection is just the sort of thing that is determined by its elements.

The formula $(\forall x)(x \in a \Rightarrow x \in b)$ is abbreviated $a \subseteq b$ and read 'a is contained in b' or 'b contains a' or 'a is a subcollection of b'; the formula '$a \subseteq b$ and $a \neq b$' is abbreviated $a \subset b$ and read 'a is strictly contained in b' or 'a is a proper subcollection of b'.

We say that a collection a is *empty* if $(\forall x)(x \notin a)$.

Definition. $\varnothing = \{x : x \neq x\}$.

If it exists, $\emptyset$ is evidently empty; but as we have not stated any axioms, we cannot yet hope to be able prove formally that any collections exist, and so in particular we cannot prove that an empty collection exists. What we *can* do already is to prove that if such a collection exists, it is unique. For if a and a' are both empty, i.e. if $(\forall x)(x \notin a)$ and $(\forall x)(x \notin a')$, then

$$(\forall x)(x \in a \Leftrightarrow x \in a'),$$

from which it follows by the extensionality principle that $a = a'$.

Definition. $a \setminus b = \{x : x \in a \text{ and } x \notin b\}$ *('relative complement of b in a')*.

Definition. $\mathfrak{P}(a) = \{b : b \subseteq a\}$ *('power of a')*.

Definition. $\bigcap a = \{x : (\forall b \in a)(x \in b)\}$ *('intersection of a')*.

Definition. $\bigcup a = \{x : (\exists b \in a)(x \in b)\}$ *('union of a')*.

These last two notations are often varied in particular cases: we write $a \cup b$ instead of $\bigcup\{a, b\}$, $\bigcup_{\Phi} \sigma$ instead of $\bigcup\{\sigma : \Phi\}$; and correspondingly for intersections. Thus, for instance, $a \cup b = \{x : x \in a \text{ or } x \in b\}$ and $\bigcup_{x \in a} \tau(x) = \{y : (\exists x \in a) y = \tau(x)\}$.

Two collections a and b are said to be *disjoint* if they have no members in common. A collection of collections is said to be *pairwise disjoint* if every pair of them is disjoint, i.e. if no object belongs to more than one of them.

One feature of these definitions needs to be stressed, though: no claim is being made yet that the terms we are introducing denote anything. For this reason they should be treated with caution. For example, we cannot even prove yet that two collections a and b are disjoint if and only if $a \cap b = \emptyset$ since the latter statement is trivially true if it happens that neither $a \cap b$ nor $\emptyset$ exists, whether or not a and b are disjoint.

Notes

The theory of fusions has been almost totally neglected by mathematicians over the last century. Interest has been somewhat greater among philosophers and philosophical logicians: Lesniewski's studies (see Fraenkel, Bar-Hillel and Levy 1958, pp. 200ff.) led to work by Lejewski (1964), Henry (1991) and others, as well as to a calculus of individuals (Leonard and Goodman 1940) of particular interest to nominalist metaphysicians. But the target of this work has been different. Very little investigation has been done into the adequacy of the theory of fusions as a foundation for mathematics. Indeed the collection-theoretic way of thinking is so entrenched among mathematicians that it is easy for them to forget how natural it is to think of a line, say, as the sum of its points rather than as the collection of them.

Some of the resolutions of the paradoxes that have been attempted place restrictions on the principle of extensionality. The reason we shall not give such resolutions house room here is that according to the approach we have adopted, which takes extensionality as a definition, they are guilty of simply changing the subject: whatever else it is, a theory which denies extensionality is not a theory of *collections*. This view is widely shared: Boolos (1971, p. 230), for example, hesitates to call extensionality analytic of the concept set only because of Quinean doubts as to whether *anything* is analytic.

The other kind of solution we shall devote no space to is that which retains naive comprehension. The difficulty this strategy faces is encapsulated in the fact noted earlier that Russell's paradox arises directly from the sentence

$$\text{not}(\exists y)(\forall x)(x \ R \ y \Leftrightarrow \text{not } x \ R \ x),$$

which is both a classical and an intuitionistic logical truth. A solution to the paradoxes which clings onto naive comprehension will therefore have to be one which makes this not a logical truth. A valiant attempt to motivate this desperate strategy is made by Weir (1998*b*; 1999).

However, there is a respectable tradition in the subject that denies the existence of a hierarchical notion of collection and recognizes only fusions. Frege, for instance, drew the distinction between the two notions precisely in order to deny the coherence of the first. Russell similarly found the empty collection and singletons problematic. More recently, Lewis (1991) has investigated at length the idea of splitting the notion of collectionhood into two parts: the mereological notion of fusion and the distinctively collection-theoretic operation of singleton formation. Lewis thus claims only to isolate the metaphysical problem, not to solve it, since he purports to be mystified as to what this last operation could be. I have questioned elsewhere (1993) whether he has located the difficulty correctly, and there remains a doubt in any case whether the relation of inclusion between collections can be assimilated to the mereological relation of part to whole (see Oliver 1994).

Chapter 3

The hierarchy

In the last chapter we encountered the fact that some properties, such as non-self-membership, are not collectivizing (i.e. do not give rise to collections). I discouraged the notion that this is particularly surprising in itself, but I did not provide a diagnosis. Now that we have the task of selecting axioms that characterize some properties as collectivizing, we need to consider the matter in more detail.

3.1 Two strategies

We can distinguish two broad strategies that have guided realists in formulating axiomatizations not just of the concept of a collection but of many other notions in mathematics: I shall call these strategies the *regressive* and the *intuitive*.

The regressive strategy takes the purpose of the theory to be the formation of a theoretical foundation for mathematics and regards an axiomatic base as successful if it is strong enough to generate as theorems those results we already believe on other grounds to be true, but not so strong as to prove those we believe to be false. 'The attitude is frankly pragmatic; one cures the visible symptoms [of the paradoxes] but neither diagnoses nor attacks the underlying disease.' (Weyl 1949, p. 231)

According to this view, then, the object of a good axiomatization is to retain as many as possible of the naive set-theoretic arguments which we remember with nostalgia from our days in Cantor's paradise, but to stop just short of permitting those arguments which lead to paradox. 'There is at this point nothing left for us to do but to proceed in the opposite direction and, starting from set theory as it is historically given, to seek out the principles required for establishing the foundations of this mathematical discipline.' (Zermelo 1908*b*, p. 261) Note that we can make use of the regressive method only if we have a prior commitment to some species of realism, since it enjoins us to assess the plausibility of a putative axiom on the basis of whether it has consequences we *already* believe to be true. This means of justification is not open to the postulationists, since they hold that the terms occurring in a coherent theory get their meaning from the axioms; so if we formulate a new theory, we cannot test its consequences for truth, since we cannot have any prior understanding of what they mean, far less whether they are true.

The literature on the foundations of mathematics is littered with more or less explicit examples of the application of the regressive method in the selection of axioms, but we need to exercise care in identifying such cases, since formalism is so widespread among mathematicians that what appears at first sight to be a regressive justification for an axiom may in fact be a formalist one. Thus, for example, when Bourbaki (1949*a*, p. 3) says that absence of contradiction is to be regarded 'as an empirical fact rather than as a metaphysical principle', it is not immediately obvious whether this is a regressive or a formalist remark.

But although the adoption of the regressive method is quite distinct from formalism, it shares many of its disadvantages. For one thing, the security of a theory that is justified by either method seems to depend on no more than its failure so far to lead to contradiction.

> During the 40 years since we have formulated with sufficient precision the axioms of [set theory] and drawn their consequences in the most varied branches of mathematics, we have never come across a contradiction, and we are entitled to hope that one will never be produced. (Bourbaki 1954, p. 8)

Of course, it is impossible to deny that the century which has elapsed without a contradiction being found is psychologically influential in engendering confidence in the system. But this can scarcely be regarded as amounting to very much. The claim that the system is formally consistent is in principle refutable simply by exhibiting a proof of a contradiction in it. But mathematicians routinely use only a tiny fragment of the generality permitted by the theory, and it would presumably only be by pushing the theory to its limits that a contradiction could be obtained. So the lapse of time can contribute significantly to confidence only if attempts are being made to produce such a refutation: I know of no such attempts.

It may be significant that the position is different for certain other theories, notably Quine's NF (1937) and ML (1940): the emphasis on syntactic analysis in the studies which have been undertaken of Quine's systems is no doubt to some extent a consequence of the syntactic paradox-barring which motivated them, but the question of their consistency is nonetheless regarded by many mathematicians as having a genuinely open character that the corresponding question for ZF does not share.[1]

A second, and even more serious, difficulty which the regressive method inherits from formalism is that it seems powerless to justify a theory that aspires to be epistemologically foundational, since it depends on our having *another*

[1] One might wonder whether it is coincidental that ML was contradictory as initially formulated (Rosser 1942) and needed emendation to reach its current form.

method already available to us for assessing the truth or falsity of our axiomatic theory's consequences. If our knowledge that $2+2=4$ lends support to a theory that has it as a theorem, then it is not clear how the theory can play a significant part in our account of that knowledge.

If the regressive strategy has its difficulties, what is the alternative? The intuitive method invites us instead to clarify our understanding of the concepts involved to such an extent as to determine (some of) the axioms they satisfy. The aim should be to reach sufficient clarity that we become confident in the *truth* of these axioms and hence, but only derivatively, in their consistency.

If the intuitive method is successful, then, it holds out the prospect of giving us greater confidence in the truth of our theorems than the regressive method. We shall therefore do what we can in pursuit of the intuitive method here. The aim, then, will be to supply a motivation for a conception of collections that gives us reason to believe that the axioms we shall be stating are true independent of their consequences.

3.2 Construction

The first sort of motivation for the theory of collections to become popular among mathematicians was the one now known as the limitation of size conception. We shall consider it in §13.5. In the meantime we shall focus on another possible motivation based on the idea that there is a fundamental relation of presupposition, priority or, as we shall usually say, *dependence* between collections. The conception of the theory it gives rise to is now known as the *iterative* conception. In contrast with the limitation of size conception, it took a long time to emerge. It is not mentioned directly in Russell's (1906*a*) discussion of possible solutions to the paradoxes, and the writings of the 1920s supply no more than glimmerings of it. However, in an attempt to make the history of the subject read more like an inevitable convergence upon the one true religion, some authors have tried to find evidence of the iterative conception quite far back in the history of the subject. Wang, for instance, says rather implausibly that it is 'close to Cantor's original idea' (1974, p. 187) and seems convinced (p. 193) that it is implicit in Zermelo's 1908*b*, an article which makes no mention of it at all. Traces of the idea are to be found in Bernays (1935, p. 55), which refers to 'iterating the use of the quasi-combinatorial concept of a function and adding methods of collection', although it is perhaps a little generous of Wang to claim that Bernays here 'develops and emphasizes' (1974, p. 187) the iterative conception. It is popular for modern writers to locate the iterative conception in Zermelo (1930), but for reasons that I will explain below (§3.9) I am sceptical about this. It is not really until Gödel (1947) that we find a clear description of one version of the iterative conception in print (although Gödel mentioned it in lectures several times in the 1930s). Even then it was slow to catch on: there is no mention of the iterative conception in Bourbaki (1954), for example, and as late as (1971, p. 218n.) Boolos could remark that although it is well known among logicians, 'authors of set theory texts either omit it or relegate it to back pages; philosophers, in the main, seem unaware of it'.

It is only quite recently, then, that the idea has emerged of deriving our conception of collections from a relation of dependence between them. When we come to consider the properties of this relation, we have to distinguish carefully the rather different understanding of it provided by the constructivist and the platonist.

For the constructivist a collection depends on the objects from which it is formed. This gives us, at least in outline, a criterion for the existence of a collection: it is possible to construct a collection if and only if the objects which it presupposes are available. Presumably, then, the relation of dependence must be transitive and irreflexive: no object can be used in its own construction. Moreover, the construction of a collection is supposed by the constructivist to take place in thought, and we might take this to imply that the relation of dependence is well-founded — that, in other words, dependence must terminate eventually. The argument for this is presumably something to do with our conception of what it amounts to (at any rate for a finite being) to understand something.

But if the structural properties the constructivist's relation of dependence would need to have are tolerably clear, it is far less clear what the relation should be. The objects a collection depends on are, we might say, those that we require in order to construct it. But which are they?

In the case of a finite collection, all we need, at least in principle, are its members. The same might also be said of a countable collection if we permit supertasks, i.e. tasks which can be performed an infinite number of times in a finite period by the device of speeding up progressively (so that successive performances might take 1 second, 1/2 second, 1/4 second, etc., and the supertask would be complete in 2 seconds). But there seems to be little hope of extending this idea to the uncountable case (see §11.1).

What we need in general is a method for comprehending infinite collections by means of something that is finite and hence capable of being grasped by a finite mind, namely the property which the members of the collection satisfy. Moreover, if this property involves other objects, they too might be presupposed by collections comprehended in this way. But what does 'involve' mean here? One natural answer would be that a property expressed by a formula $\Phi(x)$ involves all the objects in the range of the quantifiers occurring in Φ. The suggestion would then be that the relation of dependence between collections, which is our primary target, is in some way parasitic on a relation of involvement between intensionally individuated properties.

There is, however, a severe difficulty with this conception, and it arises because of the difference in individuation conditions between collections and properties. Any one collection *could*, it seems, be comprehended in any number of ways involving any number of other objects. For instance, if $a = \{x : \Phi(x)\}$, then also $a = \{x : \Phi(x) \text{ and } y = y\}$, no matter what y is, but we presumably do not wish to say that a presupposes y. The only obvious alternative is to say that a collection presupposes only those objects involved in *all* ways of comprehending it, but now the difficulty is that it is very hard to see why this should be true. There is no obvious reason why there should not be a collection such that we require for its comprehension *some* object other than its members but it does not matter which one: if so, then although we are quite clear what it is that the collection presupposes — namely some object not a member of it — we cannot express that presupposition relationally in the manner that our account requires.

3.3 Metaphysical dependence

This is not a difficulty faced by the platonist, for whom the existence of a collection is in no way dependent on our ability to comprehend it: a collection therefore presupposes not the objects needed to think about it but only those needed to constitute it, namely its members. These in turn will presuppose their members, and so on down. The platonist's relation of presupposition or dependence is thus what we shall later call the ancestral of the membership relation. This relation is then supposed to act as a metaphysical constraint on existence: no collection exists if its doing so would come into conflict with this constraint.

The difficulty the platonist now faces is to say what the relation of priority amounts to. One possible strategy which seems to be implicit in the thinking of many mathematicians is to regard platonism as a sort of *limiting case* of constructivism: it is, roughly, what constructivism would become if we removed all the constraints on the creating subject. An account of the iterative conception on these lines has been given by Wang (1974, pp. 81–90), for example. The most urgent doubt concerning this proposal, however, is whether it even makes sense. The constructivist conception of the creating subject is of a finite, reflective, thinking being in time. Which parts of this conception are to be regarded as 'constraints' to be thrown off? Presumably not the 'thinking' or the 'reflective' parts. But do we understand what it would be for a non-temporal being to think? And what about finiteness? Later parts of this book will be very largely concerned with tracing the rapidly growing confidence with which mathematicians were able to handle the infinite, but it remains to this day a real philosophical perplexity to say how we finite beings achieve this feat of comprehending the infinite (or, indeed, whether we fully do so). Even if the idea of an infinite *set* is unproblematic, it certainly does not follow that the idea of an infinite thinking being is. The point is not that we should be concerned because the account has taken a sudden turn towards the theistic and left the atheist mathematician in the lurch, but rather that the appeal to an infinite being is of quite the wrong shape for the task in hand.

Suppose, then, that we reject the idea of treating platonism as a limiting case of constructivism, and try instead to give an account of dependency as a relation holding between collections independent of our (or God's) thinking of them. What should this account be?

The etymology of 'prior' is of course temporal, and one popular method of explaining the dependence of a collection on its members has been in terms of time: my 1990 was perhaps a more or less typical example of the genre. But this is to make illicit appeal to the constructivist conception: according to the platonist a collection does not exist in time and hence cannot be subject to temporal relations. If they do not want their position to reduce to the limiting case of constructivism we have just rejected, platonists are therefore forced to admit hastily that the appeal to time is a 'mere metaphor'. No doubt we should not dismiss it out of hand for that reason alone: without metaphor, philosophy would be a hard subject to do, and a much harder one to communicate about. Nevertheless, it is difficult to see how this particular metaphor helps: to be told that collections are subject to a time-like structure that is not time is not to be told very much (cf. Lear 1977).

Another route we could try would be to see the modality involved as some kind of necessity. We might say, perhaps, that one object presupposes another if the one would not have existed without the other. This notion of dependence dates back to Plato, to whom Aristotle in the *Metaphysics* (1971, 1019a1-4) attributes the view that a thing is 'prior in respect of its nature and substance when it is possible for it to *be* without other things, but not them without it'. What this amounts to if the thing is a collection is that it would not have existed if its members had not. But once again we run into difficulties very soon, since in the case of pure collections — the empty collection and other collections depending on it alone — the platonist presumably believes that the members exist necessarily and so the antecedent of the conditional cannot be realized. We might hope to deal with applied collections by the proposed route and then treat pure collections by analogy as some sort of special case, but even this does not work, for although it is no doubt true that the singleton of my goldfish would not have existed if Bubble had not existed, the platonist is equally committed to the converse: Bubble would not have existed if its singleton had not. So the platonist cannot, even in the applied case, appeal to counterfactual reasoning to explain the relation of priority between collections (cf. Fine 1995).

There is therefore little choice but to conclude that priority is a modality distinct from that of time or necessity, a modality arising in some way out of the manner in which a collection is constituted from its members. But *if* it is distinct, we cannot rely on our understanding of these other modalities in determining its structural properties. The relation of dependence is transitive by definition, of course, and it is presumably irreflexive, but ought we now to suppose that it is *well-founded*, i.e. that every chain of dependencies terminates in a finite number of steps? The argument that it is well-founded has a long history, since it is in effect a version of one of the classical arguments for substance: any chain of ontological presupposition must terminate, it is claimed, in entities which do not presuppose anything else for their existence, and in traditional metaphysics these entities are termed 'substance'.

The most renowned proponent of substance among 20th century philosophers was Wittgenstein, who based his argument for the existence of substance on the requirement that sense be determinate. In the context in which he deployed the argument this assumption is legitimate: the world whose substance he wished to demonstrate was intrinsically a *represented* world, a world standing in a certain relation to the thoughts by means of which its state is represented. And there is a perspective, which I shall call *internal platonism*, from which the argument is available in the current context too.

No corresponding argument is available to the uncritical platonists with whose position internal platonism is here being contrasted: if there were a bar to our *grasping* any collection not obtainable from individuals in a finite number of steps, that would still not be, for them, an argument against the *existence* of such collections. But the internal platonist regards mathematics as part of our attempt to represent the world and thinks that this imposes constraints on the form it can take. One of these constraints is that no set can lie at the head of an infinite descending $\in$-chain. This is not an epistemological issue, even in an idealized sense: the point is not that we could not know about a set at the top of an infinite descending $\in$-chain. Nor is it a point about what we can construct or imagine, even in an idealized sense. The point is rather that any conceptual scheme which genuinely represents a world cannot contain infinite backwards chains of meaning, and so collections which mirror such chains could only be idle wheels in such a scheme.

3.4 Levels and histories

If we are to examine these issues further, it will be helpful to develop a way of classifying collections according to their presuppositions. Once this classification is in place, we will be able to distinguish two quite distinct aspects to the problem of collection existence that confronts us.

The classification goes like this. The initial level has as its members the objects which already exist independently, i.e. the individuals, and in each subsequent level are the collections which presuppose only those collections which occur on a lower level. So in general the elements of each level will be precisely the individuals plus the elements and subcollections of all lower levels. The collection of all levels prior to a given level is called a 'history', and the whole structure is called the *cumulative iterative hierarchy* ('iterative' because the account proceeds by describing the levels successively, starting with the individuals; 'cumulative' because a collection contained in one level is also contained in all succeeding levels).

Notice, though, that what we have just said by way of introduction, however accurately it describes the intended hierarchy, cannot constitute a definition of it *ab initio* because it involves the word 'level' which we have not defined explicitly. It would therefore be natural at this point to suppose that we need to treat 'level' as an extra primitive in our axiomatized theory, but in fact there is an elegant trick (due to Scott 1974) which allows us to *define* what a level is explicitly in terms of the membership relation. The trick is to define what we mean by a history first: it turns out that we can do this independently without using the concept of a level, and then we can define levels in terms of histories.

Definition. $\mathrm{acc}(a) = \{x : x \text{ is an individual or } (\exists b \in a)(x \in b \text{ or } x \subseteq b)\}$ *('the* accumulation *of a').*

In words, the accumulation of a collection a has as members all the individuals together with all the members and subcollections of all the members of a. (The same caution is required in working with this definition as with those at the end of §2.6, since we cannot yet prove that $\mathrm{acc}(a)$ always exists.)

Definition. *$\mathcal{V}$ is called a* history *if* $(\forall V \in \mathcal{V})(V = \mathrm{acc}(\mathcal{V} \cap V))$.

Definition. *The accumulation of a history is called a* level. *More precisely, if $\mathcal{V}$ is a history,* $\mathrm{acc}(\mathcal{V})$ *(if it exists) is called the* level *with history $\mathcal{V}$.*

If $\varnothing$ exists, it is trivially a history since it has no elements: if its accumulation exists, it is the collection of all the individuals.

Definition. *A collection is said to be* grounded, *or to be a* set, *if it is a subcollection of some level.*

(3.4.1) **Proposition.** *If V is a level with history $\mathcal{V}$, then any member V' of $\mathcal{V}$ is a level belonging to V with history $\mathcal{V} \cap V'$.*

Proof. Suppose that V is a level with history $\mathcal{V}$ and $V' \in \mathcal{V}$. Certainly $V' \subseteq V' \in \mathcal{V}$ and so $V' \in \mathrm{acc}(\mathcal{V}) = V$. Also $V' = \mathrm{acc}(\mathcal{V} \cap V')$ since $\mathcal{V}$ is a history. So V' will be a level provided that $\mathcal{V} \cap V'$ is a history. But if $V'' \in \mathcal{V} \cap V'$, then $V'' \subseteq \mathrm{acc}(\mathcal{V} \cap V') = V'$, so that $\mathcal{V} \cap V'' = (\mathcal{V} \cap V') \cap V''$ and hence

$$\mathrm{acc}((\mathcal{V} \cap V') \cap V'') = \mathrm{acc}(\mathcal{V} \cap V'') = V''$$

since $\mathcal{V}$ is a history. Hence $\mathcal{V} \cap V'$ is indeed a history as required. □

In the remainder of this chapter and the next V, V', V_1, etc. will always be levels. In particular, the quantifiers $(\forall V)$ and $(\exists V)$ should be read 'For every level V' and 'For some level V' respectively.

3.5 The axiom scheme of separation

A level is the accumulation of its history and thus contains, in addition to the individuals, all the members and subcollections of all the lower levels. We may think of it as being justified by the account of dependency between collections which we gave earlier, since the collections we include in a level are just those which depend only on collections which occur at lower levels in the hierarchy. We shall occasionally refer to this as the *first principle of plenitude*.[2]

A platonist might well feel, however, that this principle is deficient when taken on its own: we have so far said nothing about *what* subcollections there are. What prevents us from doing so, according to at least one variety of platonist, is merely our self-denying insistence on formalizing everything in a first-order language. If we abandoned that constraint, we could say that b *strongly accumulates* a if:

(1) $(\forall x)(x \in b \Leftrightarrow x$ is an individual or $(\exists c \in a)(x \in c$ or $x \subseteq c))$; and

(2) $(\forall X)(\forall c \in a)(\{x \in c : Xx\} \in b)$.

We could then use strong accumulations in our definition of 'level' in place of accumulations, and it would be trivial to prove the following second-order principle:

Separation principle.[3] $(\forall X)(\forall V)(\{x \in V : Xx\}$ *exists*$)$.

In the first-order system, however, none of this is open to us. So instead we introduce as axioms all the instances of the second-order separation principle that are formulable in our first-order system.

Axiom scheme of separation. *If* $\Phi(x)$ *is a formula, then the following is an axiom:*

$$(\forall V)(\{x \in V : \Phi(x)\} \textit{ exists}).$$

It is no accident, incidentally, that we have found ourselves introducing an axiom scheme at some stage in our axiomatization: the theory we are aiming towards cannot be finitely axiomatized in the first-order language we are working in, and so the presence in our system of at least one scheme is inevitable. It will turn out, in fact, that separation is our *only* scheme, and so part of our interest in it will be as the focus for the concerns we raised in §1.3 about what is involved in asserting a scheme at all.

The difference between the first-order scheme and second-order separation is important, since the scheme falls well short of giving effect to what the platonist might be thought to intend. If we regard the issue metatheoretically,

[2]The second principle of plenitude, which we shall state in §4.1, addresses the question of how many levels there are in the hierarchy.

[3]Whenever we set down an axiom, as here, for discussion without intending to add it to our default theory, we signal this by not emboldening its name.

the reason for this inadequacy is clear enough: in the case where the level V is infinite, second-order separation encompasses uncountably many instances (cf. Cantor's theorem 9.2.6), whereas the first-order scheme has only countably many instances since the language of set theory is countable. So the first-order scheme is only at best an inadequate surrogate for second-order separation.

Notice, though, that all of this applies only on the platonist understanding of the dependence relation. On the stricter constructivist understanding, we are entitled at each stage to construct only those collections which we can specify in terms of collections already constructed. To capture the first-order content of this idea, it would be necessary to restrict the quantifiers in our formula to the level V in question. We would then be left with the following much weaker separation principle.

Predicative separation. *If $\Phi(x)$ is a formula and $x_1, \ldots, x_n$ are the variables other than x on which it depends, then*

$$(\forall V)(\forall x_1, \ldots, x_n \in V)(\{x \in V : \Phi^{(V)}(x)\} \textit{ exists})$$

is an axiom.

Such a predicative form of the axiom scheme of separation would restrict very substantially what we could prove in our theory.

3.6 The theory of levels

In §2.3 we proved the absolute version of Russell's paradox: this showed that we cannot consistently assume what is sometimes called the *naive comprehension principle*, namely that every property is collectivizing. Now that we have the axiom scheme of separation, we can relativize this argument so as to show that no set has all its own subsets as members.

(3.6.1) **Proposition.** *There is no set b such that $(\forall a)(a \subseteq b \Rightarrow a \in b)$.*

Proof. Suppose that b is a set. So there is a level V such that $b \subseteq V$. Let $a = \{x \in b : x \notin x\}$. Then $a = \{x \in V : x \in a \text{ and } x \notin x\}$, which exists by the axiom scheme of separation. If $a \in b$ then $a \in a \Leftrightarrow a \notin a$, which is absurd. So a is a subcollection of b but not a member of it. □

(3.6.2) **Russell's paradox (relative version).** *There is no set of all sets.*

Proof. Every subcollection of a set is a set. So the set of all sets would, if it existed, have all its own subcollections as members, contrary to what we have just proved. □

Russell actually discovered the relative version of the paradox first, as an easy consequence of a theorem proved by Cantor (theorem 9.2.6). It was by analysing the proof of this result that he arrived at the absolute version some time in 1901 (see Schilpp 1944, p. 13). Zermelo had derived the same contradiction independently in 1900 or 1901 (Rang and Thomas 1981).

We turn now to the task of proving theorem 3.6.4 below, which asserts (in the set-theoretic jargon) that the membership relation is well-founded on each history. What is remarkable is that Russell's paradox actually provides the key to the proof. With this goal in mind let us (temporarily) write $a \prec b$ just in case every subcollection of a belongs to b. Proposition 3.6.1 then becomes the statement that there is no set b such that $b \prec b$.

(3.6.3) **Lemma.** *If $\mathcal{V}$ is a history and $V, V' \in \mathcal{V}$, then $V \in V' \Leftrightarrow V \prec V'$.*

Proof. Suppose that $V, V' \in \mathcal{V}$. If $V \prec V'$, then trivially $V \in V'$. So suppose conversely that $V \in V'$. Then $V \in \mathcal{V} \cup V'$. So if $a \subseteq V$, then $a \in \text{acc}(\mathcal{V} \cap V') = V'$. It follows that $V \prec V'$. □

(3.6.4) **Theorem.** *If $\mathcal{V}$ is a history and a is a non-empty subcollection of $\mathcal{V}$, then there is a member of a that is disjoint from it.*

Proof. Suppose on the contrary that a has no $\in$-minimal member. Let $b = \bigcap a$, which exists by the axiom scheme of separation. Suppose that $V \in a$. Then by hypothesis there exists $V' \in a$ such that $V' \in V$. Hence $V' \prec V$ [lemma 3.6.3]. Now $b \subseteq V'$. So every subset of b is a subset of V' and hence belongs to V. Since V was arbitrary, it follows that every subset of b belongs to b, i.e. $b \prec b$. But this contradicts proposition 3.6.1. □

Definition. *A collection a is* transitive *if* $(\forall b \in a)(\forall x \in b)(x \in a)$.

(3.6.5) **Proposition.** *Every level is transitive.*

Proof. Let $\mathcal{V}$ be a history of V and suppose that $x \in b \in V$. If

$$a = \{V' \in \mathcal{V} : b \subseteq V' \text{ or } b \in V'\},$$

then a is non-empty by definition, and so there exists $V' \in a$ such that V' is disjoint from a [theorem 3.6.4]. So either $b \subseteq V'$ or $b \in V'$. But if $b \in V'$, then because b is not an individual, there exists $V'' \in \mathcal{V} \cap V'$ such that $b \in V''$ or $b \subseteq V''$ [proposition 3.4.1], and so $V'' \in V' \cap a$, contradicting the choice of V'. Hence $b \subseteq V'$, whence $x \in V' \in \mathcal{V}$ and so $x \in V$. □

(3.6.6) **Proposition.** $a \in V \Rightarrow a \subseteq V$.

Proof. Trivial from proposition 3.6.5. □

(3.6.7) **Proposition.** $a \subseteq b \in V \Rightarrow a \in V$.

Proof. Suppose that $a \subseteq b \in V$. As in the proof of proposition 3.6.5, we can obtain $V' \in \mathcal{V}$ such that $b \subseteq V'$. But then $a \subseteq V'$ and so $a \in V$. □

(3.6.8) **Proposition.** $V = \operatorname{acc}\{V' : V' \in V\}$.

Proof. Suppose that $\mathcal{V}$ is a history of V. Then

$$
\begin{aligned}
x \in V &\Leftrightarrow x \in \operatorname{acc}(\mathcal{V}) \\
&\Leftrightarrow x \text{ is an individual or } (\exists V' \in \mathcal{V})(x \in V' \text{ or } x \subseteq V') \\
&\Rightarrow x \text{ is an individual or } (\exists V' \in V)(x \in V' \text{ or } x \subseteq V') \\
&\qquad\qquad\qquad\qquad\qquad\qquad\qquad\qquad\text{[proposition 3.4.1]} \\
&\Leftrightarrow x \text{ is an individual or } (\exists V' \in V)(x \subseteq V') \text{ [proposition 3.6.6]} \\
&\Rightarrow x \in V \text{ [proposition 3.6.7].}
\end{aligned}
$$

□

If $V_1 \in V_2$, we shall sometimes say that V_1 is *lower than* V_2. With this terminology a level may be said to be the accumulation of all lower levels.

The hierarchy of levels is cumulative: if an object belongs to a particular level, then it belongs to all subsequent levels.

(3.6.9) **Lemma.** *If* V *is a level, then* $\{V' : V' \in V\}$ *is a history whose level is* V.

Proof. Let $\mathcal{V} = \{V' : V' \in V\}$ and suppose that $V' \in \mathcal{V}$. Then $V'' \in V' \Rightarrow V'' \in V$ [proposition 3.6.5] and so $\mathcal{V} \cap V' = \{V'' : V'' \in V'\}$. Now V' is a level and so

$$
\begin{aligned}
V' &= \operatorname{acc}\{V'' : V'' \in V'\} \text{ [proposition 3.6.8]} \\
&= \operatorname{acc}(\mathcal{V} \cap V').
\end{aligned}
$$

This shows that $\mathcal{V}$ is a history. Moreover, $\operatorname{acc}(\mathcal{V}) = V$ [proposition 3.6.8], i.e. $\mathcal{V}$ is a history of V. □

(3.6.10) **Proposition.** *If* Φ *is a formula, this is a theorem:*

$$(\exists V)\Phi(V) \Rightarrow (\exists V_0)(\Phi(V_0) \textit{ and not } (\exists V' \in V_0)\Phi(V')).$$

Proof. Suppose that $\Phi(V)$ and let $a = \{V' \in V : \Phi(V')\}$. If a is empty, then we can simply let $V_0 = V$. If not, then note that a is a subset of the history $\{V' : V' \in V\}$ [lemma 3.6.9] and so there exists $V_0 \in a$ such that V_0 is disjoint from a [theorem 3.6.4]; therefore $\Phi(V_0)$ and

$$V' \in V_0 \Rightarrow V' \in V \Rightarrow \text{not } \Phi(V').$$

□

(3.6.11) **Proposition.** $V_1 \in V_2$ *or* $V_1 = V_2$ *or* $V_2 \in V_1$.

Proof. Suppose not. So we can find levels V_1 and V_2 such that $V_1 \notin V_2$, $V_1 \neq V_2$ and $V_2 \notin V_1$: hence more particularly we can choose V_1 in such a way that

$$(\forall V \in V_1)(\forall V')(V \in V' \text{ or } V = V' \text{ or } V' \in V) \tag{1}$$

[proposition 3.6.10] and, having chosen V_1, we can then choose V_2 in such a way that

$$(\forall V \in V_2)(V \in V_1 \text{ or } V = V_1 \text{ or } V_1 \in V) \tag{2}$$

[proposition 3.6.10 again]. We shall now prove that with these choices of V_1 and V_2 we have

$$(\forall V)(V \in V_1 \Leftrightarrow V \in V_2). \tag{3}$$

Suppose first that $V \in V_1$. Then $V \neq V_2$ since $V_2 \notin V_1$. Also $V_2 \notin V$ since otherwise $V_2 \in V_1$ [proposition 3.6.5], contrary to hypothesis. Since $V \in V_1$, we must therefore have $V \in V_2$ by (1). If $V \in V_2$, on the other hand, then the same arguments as before show that $V_1 \neq V$ and $V_1 \notin V$. Hence $V \in V_1$ by (2). This proves (3). But then

$$\begin{aligned} x \in V_1 &\Leftrightarrow x \text{ is an individual or } (\exists V \in V_1)(x \subseteq V) \text{ [proposition 3.6.8]} \\ &\Leftrightarrow x \text{ is an individual or } (\exists V \in V_2)(x \subseteq V) \text{ by (3)} \\ &\Leftrightarrow x \in V_2 \text{ [proposition 3.6.8]}, \end{aligned}$$

and so $V_1 = V_2$ as required. □

These two propositions provide us with what turns out to be a useful method of definition: if $(\exists V)\Phi(V)$ then there is exactly one level V such that $\Phi(V)$ but not $(\exists V' \in V)\Phi(V)$. This unique level is called the *lowest* V such that $\Phi(V)$.

(3.6.12) **Proposition.** $V \notin V$.

Proof. If there exists a level V such that $V \in V$, then there exists a lowest such V, but this leads immediately to a contradiction. □

(3.6.13) **Proposition.** *If $\mathcal{V}$ is a history of the level V, then $\mathcal{V} = \{V' : V' \in V\}$.*

Proof. Suppose $\mathcal{V}$ is a history of V. Certainly $V' \in \mathcal{V} \Rightarrow V' \in V$. So suppose now that $V' \notin \mathcal{V}$. Then for every $V'' \in \mathcal{V}$ we have $V'' \neq V'$ and $V' \notin V''$ (since if $V' \in V'' \in \mathcal{V}$ then $V' \in \mathcal{V}$), and so $V'' \in V'$ [proposition 3.6.11]. So $\mathcal{V} \subseteq \{V'' : V'' \in V'\}$, whence $V = \text{acc}(\mathcal{V}) \subseteq \{V'' : V'' \in V'\} = V'$ and therefore $V' \notin V$ (since otherwise $V' \in V'$). □

(3.6.14) **Proposition.** $V \subseteq V' \Leftrightarrow (V \in V'$ *or* $V = V')$.

Proof. If $V = V'$, then trivially $V \subseteq V'$; if $V \in V'$, then again $V \subseteq V'$ [proposition 3.6.6]. If, on the other hand, neither $V \in V'$ nor $V = V'$, then $V' \in V$ [proposition 3.6.11] and $V' \notin V'$ [proposition 3.6.12], so that $V \nsubseteq V'$. □

(3.6.15) **Proposition.** *$V \subseteq V'$ or $V' \subseteq V$.*

Proof. If $V \nsubseteq V'$, then $V \notin V'$ and $V \neq V'$ [proposition 3.6.14], whence $V' \in V$ [proposition 3.6.11] and so $V' \subseteq V$ [proposition 3.6.14]. □

(3.6.16) **Proposition.** *$V \subset V' \Leftrightarrow V \in V'$.*

Proof. $V \subset V' \Leftrightarrow (V \subseteq V'$ and $V \neq V') \Leftrightarrow V \in V'$ [propositions 3.6.12 and 3.6.14]. □

Exercise

Show that these two assertions are equivalent:

(i) $(\forall x)(\exists V)(x \in V)$;

(ii) $(\forall a)(\exists V)(a \subseteq V)$ and $(\forall V)(\exists V')(V \in V')$.

3.7 Sets

A set, according to our earlier definition, is a collection which occurs somewhere in the iterative hierarchy. One of the great advantages of this conception of set is that it generates a simple criterion to determine which formulae define sets. A set has to be located in the hierarchy at a level which is after the levels of all its members. It follows that things form a set only if there *is* a level to which they all belong: for if there is not, there will be nowhere in the hierarchy for a set of them to be located. We can give this thought precise expression within our theory as follows.

(3.7.1) **Proposition.** *If $\Phi(x)$ is a formula, then $\{x : \Phi(x)\}$ is a set iff there is a level V such that $(\forall x)(\Phi(x) \Rightarrow x \in V)$.*

Necessity. If $a = \{x : \Phi(x)\}$ is a set, then there is a level V such that $a \subseteq V$ and so for all x

$$\Phi(x) \Rightarrow x \in a \Rightarrow x \in V.$$

Sufficiency. If there is a level V such that $(\forall x)(\Phi(x) \Rightarrow x \in V)$, then

$$\{x : \Phi(x)\} = \{x \in V : \Phi(x)\},$$

which exists [axiom scheme of separation] and is evidently a set. □

(3.7.2) **Proposition.** *The members of a set are all either sets or individuals.*

Proof. Immediate [proposition 3.6.8]. □

Definition. *If a is a set, the lowest level V such that $a \subseteq V$ is called the* birthday *of a and denoted* $\mathrm{V}(a)$.

(3.7.3) **Proposition.** *If a is a set, $a \notin a$.*

Proof. If $a \in a$, then $a \in \mathrm{V}(a)$ (since $a \subseteq \mathrm{V}(a)$). Hence $(\exists V \in \mathrm{V}(a))(a \subseteq V)$ [proposition 3.6.8], contradicting the definition of $\mathrm{V}(a)$. □

(3.7.4) **Proposition.** *If Φ is a formula and a is a set, then $\{x \in a : \Phi(x)\}$ is a set.*

Proof. Evidently $\{x \in a : \Phi(x)\} = \{x \in \mathrm{V}(a) : \Phi(x)\}$, which exists by the axiom scheme of separation, and is a set since it is contained in the level $\mathrm{V}(a)$. □

(3.7.5) **Foundation principle.** *If a is a non-empty set, then it has a member which is either an individual or a set b such that a and b are disjoint.*

Proof. Suppose that the non-empty set a has no individuals as members. So all its members are sets [proposition 3.7.2], and we can choose a set $b \in a$ of lowest possible birthday: if $c \in b \cap a$, then $c \in \mathrm{V}(b)$ and so, since c is a set, there exists $V' \in \mathrm{V}(b)$ such that $c \subseteq V'$ [proposition 3.6.8], i.e. $\mathrm{V}(c)$ is lower than $\mathrm{V}(b)$, which contradicts the fact that $c \in a$; therefore $b \cap a = \emptyset$ as required. □

(3.7.6) **Proposition.** *If a is a non-empty set of sets, then $\bigcap a$ is a set.*

Proof. If c is a set belonging to a, then

$$a' = \{x \in c : (\forall b \in a)(x \in b)\}$$

is a set [proposition 3.7.4]. But then

$$x \in a' \Leftrightarrow (x \in c \text{ and } (\forall b \in a)(x \in b)) \Leftrightarrow (\forall b \in a)(x \in b)$$

since $c \in a$. So $a' = \{x : (\forall b \in a)(x \in b)\} = \bigcap a$. □

(3.7.7) **Proposition.** *If a and b are sets, then $a \cap b$ is a set.*

Proof. $a \cap b = \{x \in a : x \in b\}$, which is a set [proposition 3.7.4]. □

(3.7.8) **Proposition.** *If a and b are sets, then $a \setminus b$ is a set.*

Proof. $a \setminus b = \{x \in a : x \notin b\}$, which is a set [proposition 3.7.4]. □

(3.7.9) **Proposition.** *If a is a set, then $\bigcup a$ is a set.*

Proof. If a is a set,

$$x \in b \in a \Rightarrow x \in b \in \mathrm{V}(a) \Rightarrow x \in \mathrm{V}(a)$$

[proposition 3.6.5]; hence

$$\bigcup a = \{x \in \mathrm{V}(a) : (\exists b \in a)(x \in b)\}$$

exists [axiom scheme of separation] and is a set. □

(3.7.10) **Proposition.** *If a and b are sets, then $a \cup b$ is a set.*

Proof. Either $\mathrm{V}(a) \subseteq \mathrm{V}(b)$ or $\mathrm{V}(b) \subseteq \mathrm{V}(a)$ [proposition 3.6.15]: suppose the latter for the sake of argument. Then $a, b \in \mathrm{V}(a)$ and so

$$(x \in a \text{ or } x \in b) \Rightarrow x \in \mathrm{V}(a).$$

Hence

$$a \cup b = \{x \in \mathrm{V}(a) : x \in a \text{ or } x \in b\}$$

exists [axiom scheme of separation] and is a set. □

(3.7.11) **Proposition.** *If Φ is a formula,*

$$(\exists a)\Phi(a) \Rightarrow (\exists a)(\Phi(a) \textit{ and not } (\exists b \in a)\Phi(b)).$$

Proof. Consider the lowest V such that $(\exists a \in V)\Phi(a)$. □

Definition. *The* transitive closure *of a is*

$$\mathrm{tc}(a) = \{x : x \in b \textit{ for every transitive } b \supseteq a\}.$$

(3.7.12) **Proposition.** *If a is a set, then $\mathrm{tc}(a)$ is a set.*

Proof. The birthday $\mathrm{V}(a)$ of a is a transitive set containing a [proposition 3.6.5]. So $\mathrm{tc}(a)$ is a set by separation. □

Transitive closures are much used by set theorists as a tool for studying how properties may be transmitted up the hierarchy. As an instance of this let us mention a common usage among set theorists according to which for any property F a set is said to be *hereditarily* F if both it and all the members of its transitive closure have F. In this book, though, we shall not be studying the hierarchy from the set theorist's perspective, and a symptom of this is that transitive closures will play hardly any part in what follows.

Exercises

1. (a) Show that $a \subseteq b \Rightarrow \mathrm{V}(a) \subseteq \mathrm{V}(b)$.
   (b) Show that $a \in b \Rightarrow \mathrm{V}(a) \in \mathrm{V}(b)$.
2. Show that there do not exist sets a and b such that $a \in b$ and $b \in a$.

3.8 Purity

The main purpose of defining the transitive closure here is to facilitate the formal definition of the notion of a pure set.

Definition. *A set is said to be* pure *if no individuals belong to its transitive closure.*

An alternative way to define this notion would have been to develop a theory of *pure levels*: these are defined just as levels were in §3.4 except that the accumulations are replaced throughout by *pure accumulations*, defined as

$$\text{acc}^{\text{P}}(a) = \{c : (\exists b \in a)(c \in b \text{ or } c \subseteq b)\}.$$

By mimicking the proofs in §3.6 it is easy to show that the pure levels form a hierarchy in the same way as the hierarchy of levels: the difference is only that in the formation of pure levels individuals are omitted, so that the earliest pure level is $\emptyset$, the pure level after U is $\mathfrak{P}(U)$, and a limit pure level is the union of all the lower pure levels. Thus a set is pure iff it is a subset of some pure level.

Consider now the following axiom candidate.

Axiom of purity. *Every set is pure.*

This axiom will certainly be true if there are no individuals. In fact, if a theory in which this is assumed had been our target all along, we could have simplified our presentation right at the beginning: there would have been no need to bring in the predicate $\text{U}(x)$ since nothing in the theory would satisfy it. We could then have simplified our definition of collection terms in §2.2 by deleting the condition that a collection should not be an individual, so that the definition would now read simply

$$\{x : \Phi(x)\} = \imath! y(\forall x)(x \in y \Leftrightarrow \Phi(x)),$$

and we could have simplified our account of the theory of levels by omitting the references to individuals.

The axiom of purity, or something equivalent to it, is assumed in almost every modern treatment of set theory in the literature. The main reason for this is that, as was discovered fairly early, it is not necessary to assume the existence of individuals in order that set theory should act as a foundation for mathematics, while if we rule them out from the outset, we can simplify the theory, getting rid of one primitive and tidying up the development considerably. If the only objective is to give mathematicians a theory which *can* act as a foundation, it is inevitable that they will choose the one that seems simplest to them.

Indeed individuals would probably have made an exit from set theory earlier than they did if it had not been for an accident in the progress of work in the metatheory. In 1922*a* Fraenkel discovered a method for showing the independence of the axiom of choice from theories such as **ZU** that allow individuals. This method was refined (Lindenbaum and Mostowski 1938) and then exploited by others (e.g. Mendelson 1956; Mostowski 1945) to prove the independence from such theories of various other set-theoretic claims. It was not until 1963 that Cohen showed how to convert this into a method that works for theories like **Z** which ban individuals. In the intervening period there was therefore a reason for set theorists (who tend, after all, to be the people who write set theory books) to regard permitting individuals as worth the extra effort. After 1963, however, not even set theorists had any use for individuals. Worse, there are proofs in set theory that do not work if we have to allow for them. So it is unsurprising that in the last 40 years individuals have largely disappeared from view.

However, we shall not follow this trend here. The reason is that to do so would cut our theory off from at least one of its intended applications. It is by no means obvious what justifies the applicability of mathematics in general to what lies outside it, and it may well be that the reduction of mathematics to set theory does not supply such a justification. But even if set theory's role as a foundation for mathematics turned out to be wholly illusory, it would earn its keep through the calculus it provides for counting infinite sets. The most natural, if not literally the only, way to ensure that that calculus is available to be applied to counting non-mathematical things — chairs, electrons, thoughts, angels — is to allow such things into the theory as individuals.

3.9 Well-foundedness

The particular treatment of the hierarchy of levels adopted here did not become known until the 1970s. But the grounded collections were first singled out for study much earlier. Mirimanoff's (1917) remarkable treatment called them 'ordinary' collections: they are often called 'well-founded' in the literature, but we avoid that usage here in order to reserve that word for a closely related property of the membership relation, namely that expressed in proposition 3.7.11.

The well-foundedness of the membership relation gives the grounded collections an especially simple structure which makes it inevitable that they would be singled out for special study eventually. But Mirimanoff did not make the further move of suggesting that *every* collection is a set. He did not, that is to say, propose to add the following axiom to the theory of sets.

Axiom of foundation. *Every collection is grounded.*

One way of putting this informally would be to use the language of classes: let **M** be the class of all collections and **V** the *grounded part* of **M**, i.e. the class of all sets; the axiom of foundation then asserts that **V** = **M**. As matters stand, however, this way of putting the matter *is* informal, since we have not introduced any formal machinery for referring to classes within our theory.[4] The idea of limiting the formal theory to the grounded collections only was first considered by von Neumann (1925) and Zermelo (1930), who proposed axioms designed to achieve it. But neither of these authors had an *argument* that there are no ungrounded collections. Their motive in restricting the theory to sets was simply that they wanted to discuss the possibility of proving categoricity results, and the clear structure of the universe of sets makes such results easier to come by if we restrict ourselves to that domain. Zermelo (1930, p. 31), for instance, said only that the axiom 'has always been satisfied in all practical applications to date, and therefore introduces for the moment no essential restriction of the theory'.

But even if the axiom of foundation made it easier to prove categoricity results, it remained without any application *outside* the theory of collections. I have already remarked on the prominent role the regressive method played in determining the theory mathematicians adopted. So the fact that they showed no inclination to adopt the axiom of foundation is unsurprising. The pattern remained that it was treated solely as a tool for specialists in the theory of collections. Thus as late as 1954 Bourbaki saw no reason to include it in his axiom system, which was explicitly intended to act as a foundation for *mathematics* and not as a basis for metatheoretic study.

Treated as a case study in the history of mathematics this episode thus provides strong evidence for the influence of the regressive method on the practice of axiom selection in the foundations of mathematics. Because the axiom of foundation did not have *mathematical* consequences, mathematicians showed no inclination to adopt it: interest in it was limited to specialists concerned with its metatheoretic consequences.

Matters began to change only when Gödel (1947, p. 519) presented the grounded collections not merely, as Mirimanoff had done, as a sub-universe of the universe of collections but rather as an independently motivated hierarchy which, as he pointed out, 'has never led to any antinomy whatsoever'. Since the 1960s the assumption that every collection is grounded has been adopted enthusiastically by set theorists, and the idea that the *only* coherent conception is the iterative one has become widespread.

But the literature contains very few arguments in favour of the claim that every collection is a set. Most of those who do attempt an argument draw attention to the difficulty of *conceiving* of an ungrounded collection. Suppes (1960, p. 53) simply challenges the reader who doubts this to try to come up

[4]For more on this issue, see appendix C.

with an example. Mayberry (1977, p. 32) suggests that 'anyone who tries ... to form a clear picture of what a non-well-founded collection might be ... will see why extensionality forces us to accept the well-foundedness of the membership relation.' Drake (1974, p. 13) stigmatizes ungrounded collections as 'strange' and says it is 'difficult to give any intuitive meaning' to them. Parsons (1983, p. 296) regards the evidence of the principle as 'more a matter of our not being able to understand how non-well-founded [collections] could be possible rather than a stricter insight that they are *impossible*'.

I am not as pessimistic as Parsons about the prospects for a non-psychologistic argument for the iterative conception. In §3.3 I sketched an argument that the hierarchy is well-founded. It is true that that argument seems to require a premise which I there characterized as internalist — that collections should be representable, or available to reason, or whatever — but that does not reduce the argument to psychologism. And it would, if correct, rule out the possibility Parsons (1983, p. 296) is prepared to countenance that 'someone might conceive a structure very like a "real" $\in$-structure which violated foundation but which might be thought of as a structure of sets in a new sense closely related to the old'.

Even if the argument for well-foundedness that I gave earlier is correct, though, it undeniably depends on an extra premise that goes beyond mere realism — the premise which I branded 'internalist'. It therefore seems prudent, lest I lose readers who do not feel the internalist pull so strongly, not to assume the axiom of foundation. In practice that is no great concession, however, since we shall focus exclusively on grounded collections (i.e. sets) in everything that we do from now on. Readers who believe there are ungrounded collections as well will thus find nothing here with which they can reasonably disagree: the most they are entitled to is a mounting sense of frustration that I am silent about them.

Notes

In 1906*a* Russell canvassed three forms a solution to the paradoxes might take: the no-class theory, limitation of size, and the zigzag theory. It is striking that a century later all of the theories that have been studied in any detail are recognizably descendants of one or other of these. Russell's no-class theory became the theory of types, and the idea that the iterative conception is interpretable as a cumulative version of the theory of types was explained with great clarity by Gödel in a lecture he gave in 1933 (printed in Gödel 1986–2003, vol. III), although the view that it is an independently motivated notion rather than a device to make the theory more susceptible to metamathematical investigation is hard to find in print before Gödel 1947. The analysis of this motivation given here is greatly indebted to Parsons (1977). Wang (1974, ch. 6) and Boolos (1971; 1989) are also central to the modern philosophical literature on the topic. The doctrine of limitation of size (discussed in §13.5) has received rather less philosophical attention, but the cumulatively detailed analysis in Hallett 1984 can be recommended. The principal modern descendant of Russell's zigzag theory — the idea that a property is collectivizing provided that its syntactic expression is not too complex — are Quine's two theories NF and ML. Research into their properties has always been a minority sport: for the current state of knowledge consult Forster 1995. What remains elusive is a proof of the consistency of NF relative to ZF or any of its common strengthenings.

Aczel 1988 is a lucid introduction to non-well-founded set theory. Rieger 2000 and Barwise and Etchemendy 1987 are recent attempts to argue for its philosophical significance. On predicative set theory the standard reference is Wang 1963. The theory of levels which we gave in §3.6 is due to Scott (1974), but his treatment used an extra assumption (the 'axiom of accumulation') which is redundant. The proof of this redundancy which I gave in my earlier book was due to John Derrick, who had been lecturing on the subject at Leeds for some years. The slightly different version I have given here makes use of an idea from Doets 1999.

The relationship between the hierarchy and the law of the excluded middle has been extensively discussed by Dummett (1993). Lear (1977) gives a novel argument, which has in turn been criticized by Paseau (2003).

Chapter 4

The theory of sets

We know quite a lot now about sets, but one thing we cannot yet prove is that there are any. The reason for this is that all the axioms we are committed to so far, the instances of separation, take the form of universal generalizations and are therefore vacuously satisfied if there are no levels to instantiate them. It is time now to rescue our theory from this vacuity. Doing so will require us to enter into some ontological commitments concerning levels.

4.1 How far can you go?

What this amounts to is that we need to determine how many levels there are in the hierarchy. The constructivist, of course, has a criterion by which to settle this: the levels owe their existence to our construction of them in thought, and so to discover how many levels there can be, we need only determine the limits of our capacities for performing such constructions. There may be different answers to this depending on how liberal our conception of the creative subject is, but it is at least clear what the terms of the debate the constructivist must engage in should be.

For the platonist, by contrast, the matter is much more problematic. The difficulty is that everything we have said so far about the platonist's understanding of the dependency relation is negative. The argument has been that the nature of collections constrains them by means of the metaphysical requirements of dependency, but this constraint does nothing in itself to show that there *are* any collections.

The iterative conception is often presented as if it *on its own* delivered the existence of at least a significant number of levels in the hierarchy, but the only version of platonism which has much prospect of justifying this is that which regards it as a limiting case — constructivism without the shackles of time and finiteness. For this reason, perhaps, limiting case platonism is a commonplace of expositions of the iterative conception. It seems to underpin the account in Boolos 1971, for instance. Since I have already disparaged limiting case platonism in §3.2, let us put it to one side. The question that remains is whether any *other* sort of platonism is entitled to the resources to rescue the theory of sets from vacuity. What is needed is plainly a further principle to which our concept of set can be seen to be answerable, one not entailed by what we have said about the concept so far.

The principle, which we shall call the *second principle of plenitude* (after Parsons 1996), states, roughly, that all levels exist which are not ruled out by the metaphysical constraints on the dependency relation already mentioned.

But this *is* rough, and as it stands it is far from unproblematic. We are entitled to an existence principle asserting that everything exists which is not ruled out only if we have a reason to believe that we have stated all relevant constraints, and how are we to know that? Moreover, even if we could be sure that we *have* stated all the constraints on collections implicit in their metaphysical nature, what would be the reason for supposing that the existence of any collections followed? The claim being made here is evidently related to the postulationist's idea that consistency implies existence. It is not, of course, the wholly general claim of the postulationist that the consistency of *any* axiom system implies the existence of mathematical objects with the properties postulated, but it is a particular case of it. We therefore owe an explanation of what it is in the concept *collection* that makes such a principle of ontological plenitude applicable to it. If consistency does not imply existence quite generally, why should it do so here?

And the difficulty we face in coming up with such an explanation is that it is far from clear how to give the principle of plenitude a coherent formulation. The most natural way to express it is by means of a modality: there exist all the collections it is possible for there to be. But if this is not to be vacuous, the modality in question cannot be one according to which mathematical objects exist necessarily. Presumably the modality is to be constrained by the metaphysical restrictions we placed on collections in the last chapter, but then the difficulty is to see how it can be prevented from delivering too much. Whatever levels there are in the hierarchy, why could there not be another one beyond them? The answer must be that for there to be another level beyond those there are would violate the metaphysical constraints, but a great deal more would have to be said to make that response convincing.

Indeed in the crude version we have stated so far the principle of plenitude has a crippling flaw. We cannot simply insist that there are all the sets that are *logically possible*, because however many there are it is logically possible for there to be more. This is no doubt the reason why we often find the principle expressed in terms of other sorts of possibility — conceptual, perhaps, or metaphysical. But to impose a *conceptual* variant of the principle of plenitude seems to import constructivist considerations which the platonist deems inappropriate; and the metaphysical variant is in danger of collapsing into vacuity once more.

4.2 The initial level

These are deep waters, and we shall return to them later, but in the meantime we need to make progress, and to do that we shall have to state axioms which make existence claims. In this section we shall do this only in the most modest way that our theory of levels permits, namely by asserting that there is at least one level (or, equivalently, that there is at least one set).

Temporary axiom. *There is at least one level.*

The reason we call this assumption a temporary axiom is that very shortly (§4.9) we shall add another axiom which entails this one, and at that point we shall be able to withdraw the temporary axiom from service. Since a level is defined to be the accumulation of a history, our temporary axiom asserts that there is at least one history that has an accumulation. We have already noted that $\emptyset$ is a history if it exists, so our temporary axiom amounts to the same as asserting that the collection $\text{acc}(\emptyset) = \{x : x \text{ is an individual}\}$ exists.

Definition. *Let* V_0 *be the earliest level.*

(4.2.1) **Proposition.** $V_0 = \{x : x \textit{ is an individual}\}$.

Proof. Immediate [proposition 3.6.8]. □

(4.2.2) **Corollary.** *If* Φ *is any formula, then*

$$\{x : x \textit{ is an individual and } \Phi(x)\}$$

is a set.

Proof. Immediate. □

The temporary axiom does no more than guarantee that the individuals form a set and does not allow us to prove the existence of any sets that are not sets of individuals. The theory with only it and separation as axioms is therefore probably as fair a representation as a formal theory can be of our ordinary language uses of the term 'set', since these uses do not generally countenance iterative constructions such as sets of sets and the like.

Given the axiom scheme of separation in the form we stated in the last chapter, the existence of *any* set entails the existence of V_0, and so we cannot deny V_0 without reducing our theory of sets to triviality. But even if we had not assumed the existence of a set of all individuals, it would still have been possible to develop a theory of pure levels. We could then have restricted the axiom scheme of separation only to them, and so could have assumed the existence of a hierarchy of sets without thereby committing ourselves to the existence of a set of individuals. This is of some significance because, as

Parsons (1977, p. 359, n. 4) has remarked, whether the existence of a set of individuals is even consistent depends to some extent on the nature of the individuals. The example he quotes is that of the ordinal numbers. We shall introduce ordinals in §11.2 as a particular kind of set, and will demonstrate there the Burali-Forti paradox, which shows that the ordinals do not form a set. This result does not threaten inconsistency in the temporary axiom of our system, of course, since we have ensured quite explicitly that sets, and therefore in particular ordinals as we shall define them, are not individuals and hence are not potentially problematic candidates for membership of V_0. But Parsons' point is that it would be possible to formulate a theory of ordinals quite independent of and prior to the theory of sets. Conceived of in this way ordinals would be individuals and not sets. It would then be natural to worry about whether there could indeed be a set V_0 of all individuals as the temporary axiom insists.

The point is well taken. What it serves to highlight is that individuals are to be conceived of as being, in the language we have been using to describe the hierarchy, conceptually *prior* to sets. If we are to present ordinals as individuals, therefore, it is a requirement that our theory of ordinals should be *independent* of our theory of sets. If so, there would be nothing to prevent the ordinals forming a set. Burali-Forti's contradiction would arise only if we then made the mistake of adding a further principle linking ordinals to sets (e.g. an axiom claiming that we can index the levels of the hierarchy in such a way that every ordinal has a level corresponding to it).

The dependency theorist expresses this point by saying that all the individuals are prior to all the collections. As we noted earlier, though, the priority conception on its own is purely restrictive in effect and does not underpin even the modest ontological commitment expressed by the temporary axiom without the addition of some positive principle.

4.3 The empty set

(4.3.1) **Proposition.** *$\emptyset$ is a set.*

Proof. V_0 exists [temporary axiom], so the set $a = \{x \in V_0 : x \neq x\}$ exists [axiom scheme of separation], and

$$x \in a \Rightarrow x \neq x \Rightarrow \text{contradiction},$$

i.e. a is empty. □

The existence of the empty set is entailed, in the presence of the axiom scheme of separation, by the existence of any set whatever. But, just as in the much older case of the number zero, the existence of the empty set was not at first wholly uncontroversial. What is hard to disentangle now is whether the early suspicions really amounted to an argument that the empty *set* is illegitimate or to a continuing confusion between this and the uncontroversial point noted earlier that there can be no such thing as an empty *fusion*. One important early source on this issue is Dedekind (1888): he regarded what he called the empty *system* as no more than a convenient fiction, but this is scarcely surprising since, as we have already noted, there is other evidence that he meant by 'system' what we now call a fusion. Lewis (1991), whose sympathies are with fusions rather than collections every time, is also reduced to giving a regressive justification for the empty set.

> You'd *better* believe in it, and with the utmost confidence; for then you can believe with equal confidence in its singleton, ... and so on until you have enough modelling clay to make the whole of mathematics. (p. 12)

Of course, for these regressive purposes the empty set does not actually have to be empty — to be, as Lewis puts it (p. 13), 'a little speck of sheer nothingness, a sort of black hole in the fabric of Reality itself ... a special individual with a whiff of nothingness about it.' So Lewis just makes an arbitrary choice — the fusion of all the individuals, as it happens — and lets that serve as the empty set. Of course, this stipulation has the minor disadvantage that if there were nothing at all, there would be no such fusion (since there cannot be a fusion of nothing), but in that case, he says, 'maybe we can let mathematics fall. Just how much security do we really need?'

Thus Lewis at his most fey. But what is rather more puzzling is that Zermelo (1908*b*, p. 263), who was certainly dealing with collections and not fusions, also called the empty set 'improper'; and Gödel (1944, p. 144) was willing at least to tolerate, if not actually to endorse, a similar idea.

One might wonder whether these repetitions of the idea that the empty set is fictitious or has to be constructed arbitrarily show the persistence of intuitions derived from fusions in the conception of aggregation being employed; but this can be no more than a conjecture since neither Zermelo nor Gödel gives in the texts mentioned any *argument* for regarding the empty set as merely fictitious. On the other hand, it is rare to find in the literature any direct reason for believing that the empty set does exist, except for variants of the argument from convenience originally deployed by Dedekind. But notice that *convenience* seems hardly sufficient, even for someone who subscribes to the regressive method, to constitute an argument for the *truth* of the assumption that the empty set exists, since it is so evidently possible (although admittedly inconvenient) to manage without it.

According to the formal development we have adopted here, no collection is an individual, and hence in particular the empty set is not one. This was achieved by taking the notion of an individual as primitive by means of the predicate $\mathrm{U}(x)$. Authors who do not do this (e.g. Fraenkel et al. 1958) are left with the awkwardness that they are unable to distinguish formally between $\emptyset$ and an individual, since individuals, we may suppose, share with $\emptyset$ the property of having no members. Two solutions to this formal problem are possible, neither of them altogether satisfactory.

The first solution, adopted by Quine, is to say that an individual x is not in fact memberless as we had supposed, but to stipulate that it has itself as a member, so that $x = \{x\}$. Individuals then become collections, albeit non-grounded ones, and sets are distinguished from individuals by being collections which do not belong to themselves. This procedure is also suggested by Frege (1893–1903, §18, n. 1), but for the somewhat different (although equally technical) reason that it enables him to settle the truth-conditions of statements identifying an individual with a set. The principal disadvantage of proceeding in this way, however, is that it *is* so obviously just a device: there is no ground whatever for thinking that individuals really do belong to themselves, so why adopt a theory according to which it is true?

The second solution is to say that the empty set is an individual, but one picked at random to fulfil this role (see Fraenkel et al. 1958, p. 24). This is an instance of a procedure which is very common in the foundations of mathematics and which we shall meet again in these pages, namely the procedure of *arbitrary choice*. It arises when we attempt to reduce one theory to another by means of an embedding and discover that there is more than one way of doing it. On some occasions there is an extrinsic reason to prefer one embedding to another; on others, as now, the various embeddings are quite on a par and so we must either make a wholly arbitrary choice or abandon the reduction.

4.4 Cutting things down to size

We mention now a technical device which is sometimes useful in cases where the set $\{x : \Phi(x)\}$ does not exist. The idea, which is due to Scott (1955) and Tarski (1955), is that instead of attempting the doomed task of forming the non-existent set of all those x such that $\Phi(x)$, we restrict ourselves to collecting only such x of earliest possible birthday, i.e. those occurring as low as possible in the hierarchy of levels. This defines a set in every case — and indeed one which may retain enough information about Φ to be useful as a partial representation of it.

Definition. *Suppose that $\Phi(x)$ is a formula. If V is the earliest level such that $(\exists x \in V)\Phi(x)$, then we let $\langle x : \Phi(x)\rangle = \{x \in V : \Phi(x)\}$. If there is no such level, then we let $\langle x : \Phi(x)\rangle = \emptyset$.*

(4.4.1) **Proposition.** *If Φ is a formula and V is a level, then*

$$(\exists y \in V)\Phi(y) \Leftrightarrow \emptyset \neq \langle x : \Phi(x)\rangle \subseteq V.$$

Proof. Suppose first that $y \in V$ and $\Phi(y)$. Then there exist elements z of earliest possible birthday such that $\Phi(z)$, i.e. $\langle x : \Phi(x)\rangle \neq \emptyset$. Moreover, if $z \in \langle x : \Phi(x)\rangle$, then either z is an individual, in which case certainly $z \in V$, or it is a set, in which case $z \subseteq \mathrm{V}(z) \subseteq \mathrm{V}(y) \in V$ and hence again $z \in V$; thus $\langle x : \Phi(x)\rangle \subseteq V$. The converse implication is trivial. □

We shall put this device to use when we define the notion of cardinality in §9.1.

4.5 The axiom of creation

Definition. *The* level above *a level V is the lowest level V' such that $V \in V'$.*

(4.5.1) **Proposition.** *If V' is the level after the level V, then x belongs to V' iff x is either an individual or a subcollection of V.*

Proof. Now $V'' \subseteq V \Rightarrow V'' \in V'$ [proposition 3.6.7] and

$$\begin{aligned} V'' \in V' &\Rightarrow V \notin V'' \\ &\Rightarrow V'' \in V \text{ or } V'' = V \text{ [proposition 3.6.11]} \\ &\Rightarrow V'' \subseteq V \text{ [proposition 3.6.14]}. \end{aligned}$$

Hence

$$V'' \in V' \Leftrightarrow V'' \subseteq V, \tag{1}$$

and so

$$\begin{aligned} x \in V' &\Leftrightarrow x \text{ is an individual or } (\exists V'' \in V')(x \subseteq V'') \text{ [proposition 3.6.8]} \\ &\Leftrightarrow x \text{ is an individual or } (\exists V'' \subseteq V)(x \subseteq V'') \text{ by (1)} \\ &\Leftrightarrow x \text{ is an individual or } x \subseteq V. \end{aligned}$$ □

Axiom of creation. *For each level V there exists a level V' such that $V \in V'$.*

More briefly: there is no highest level. This axiom ensures that for every level V there is a level above it: by the previous proposition this level will be $\mathrm{V}_0 \cup \mathfrak{P}(V)$. We are thus immediately guaranteed the existence of infinitely many levels V_0, $\mathrm{V}_0 \cup \mathfrak{P}(\mathrm{V}_0)$, $\mathrm{V}_0 \cup \mathfrak{P}(\mathfrak{P}(\mathrm{V}_0))$, etc.

(4.5.2) **Lemma.** *If a is a set, then there is a level V such that $a \in V$.*

Proof. If a is a set, there is a level V such that $a \subseteq \mathrm{V}(a) \in V$ [axiom of creation], and so $a \in V$ [proposition 3.6.8]. □

The axiom of creation allows us to extend our inventory of legitimate methods of set formation.

(4.5.3) **Proposition.** *If a is a set, then $\mathfrak{P}(a)$ is a set.*

Proof. If a is a set, there exists V such that $a \in V$ [lemma 4.5.2]. Now

$$(b \in V \text{ and } b \subseteq a) \Leftrightarrow b \subseteq a$$

[proposition 3.6.6], whence $\mathfrak{P}(a) = \{b \in V : b \subseteq a\}$ exists [axiom scheme of separation] and is a set. □

(4.5.4) **Proposition.** *If a is a set, then $\{a\}$ is a set.*

Proof. There is a level V such that $a \in V$ [lemma 4.5.2]. So $\{a\} = \{x \in V : x = a\}$, which exists [axiom scheme of separation]. □

(4.5.5) **Proposition.** *If x and y are either sets or individuals, then $\{x, y\}$ exists.*

Proof. If either x or y is an individual, then its singleton is a set by corollary 4.2.2; if it is a set, the singleton is a set by proposition 4.5.4. Either way, $\{x\} \cup \{y\}$ is then a set by proposition 3.7.10. □

These results obviously give the axiom of creation some regressive support: in the practice of mathematics it is undoubtedly convenient to be able to make free use of such constructions as the power set $\mathfrak{P}(a)$ of a set a. This is in fact the central case: since the level above V is $V_0 \cup \mathfrak{P}(V)$, whether the axiom of creation is true really hinges on whether every set has a power set. But is there a non-regressive argument to think this is so?

The dependency theorist might try to give such an argument on the basis of the principle of plenitude: if a is a set, it is *possible* for there to be a power set of a, and our principle of plenitude will then tell us that there *is* one. Once again, though, there has been a tendency, at least among platonists, to suppose that this key step is so obvious as not to require argument at all. Some who *have* thought about the issue have been reduced to regarding the power set operation as essentially primitive. But to say that it is 'primitive' or a 'given' is evidently only to label the difficulty, not to solve it.

A constructivist will, as we have seen, be likely to adopt a different hierarchy, in which the next level after V is not $V_0 \cup \mathfrak{P}(V)$. If we had adopted such a hierarchy, the axiom of creation and the existence of power sets would come apart: we would have to address the possibility that for every level there is a next level but that this process never exhausts even the subsets of the first level. Something like this idea is expressed by Lusin (1927, pp. 32–3), who suggests that in order to encompass all the subsets of an infinite set in one set, we would have to be able to circumscribe the laws for defining such sets, which he claims is impossible. A somewhat different argument against the existence of the power set of an infinite set is to be found in Mayberry (2000).

4.6 Ordered pairs

The ordered pair (x, y) is supposed to be a single object which codes within it in some way the identities of the two objects x and y. What is required is that the *ordered pair principle*,

$$(x, y) = (z, t) \Rightarrow x = z \text{ and } y = t,$$

should be satisfied for all x, y, z, t. When we are working within the theory of sets there are various technical tricks that enable us to do this. The first satisfactory method is due to Wiener (1914), but here we shall use one discovered by Kuratowski (1921).

Definition. $(x, y) = \{\{x\}, \{x, y\}\}$ *('the* ordered pair *of x and y').*

We shall write (x, y, z) for $((x, y), z)$, (x, y, z, t) for $((x, y, z), t)$, etc. Ordered pairs evidently exist whenever their terms are either sets or individuals [proposition 4.5.5], but before we make use of them we need to demonstrate that they satisfy the ordered pair principle enunciated above.

(4.6.1) **Lemma.** *If x, y, z are sets or individuals, then*

$$\{x, y\} = \{x, z\} \Rightarrow y = z.$$

Proof. Suppose that $\{x, y\} = \{x, z\}$. Then $y \in \{x, y\} = \{x, z\}$, so that either $y = z$ as required or $y = x$: but if $y = x$, then $z \in \{x, z\} = \{x, y\} = \{y\}$, and so $y = z$ in this case as well. □

(4.6.2) **Proposition.** *If x, y, z, t are sets or individuals, then*

$$(x, y) = (z, t) \Rightarrow x = z \text{ and } y = t.$$

Proof. Suppose that $(x, y) = (z, t)$, i.e. $\{\{x\}, \{x, y\}\} = \{\{z\}, \{z, t\}\}$. So either $\{x\} = \{z\}$, in which case $x = z$, or $\{x\} = \{z, t\}$, in which case $x = z = t$. Hence in either case $\{x\} = \{z\}$, so that $\{x, y\} = \{z, t\}$ [lemma 4.6.1] and therefore $y = t$ [lemma 4.6.1 again]. □

If it were our purpose to minimize the number of axioms, the advantage of being able to define ordered pairs would be clear: Whitehead and Russell (1910–13) did not know about such devices and therefore had no choice but to develop two parallel but distinct theories, one of collections and one of relations, and to state most of their axioms twice, once for each case. Nevertheless, when Russell learnt about the possibility of defining ordered pairs from Wiener, he did not express 'any particular approval' (Wiener 1953, p. 191), and he did not bother to mention the manoeuvre at all in the introduction to the 2nd edition of *Principia* (Whitehead and Russell 1927).

One reason to be wary of it is that all the explicit definitions of pairing terms have, because of their arbitrariness, various more or less accidental consequences. This is a case of an issue that has plagued the foundations of mathematics — the problem of *doing too much*. It is sometimes referred to in the literature as Benacerraf's problem, because he famously raised an instance of it in 'What numbers could not be' (1965), but in fact it dates back at least to Dedekind.

The problem typically arises when we try to synthesize in some theory a notion of which we take ourselves already to have a conception independent of that theory: we have to choose some particular way of modelling the notion within our theory, and in doing so we invest the model with extraneous properties which we could not have derived from our prior conception of the notion.

At the formal level one way out of this difficulty is to introduce the problematic entity explicitly by means of a new primitive term governed by appropriate axioms. In the case at hand, this would involve treating '(x, y)' as a primitive term and stating the ordered pair principle as an axiom. If we adopted this practice, the derivation of proposition 4.6.2 would continue to have a metamathematical interest, since it would demonstrate that the theory with the ordered pair axiom included is conservative over the old theory without the axiom, but it would not be needed for the formal development.

There is a difficulty with this strategy, however. It is caused by the fact that among the objects we wish to form ordered pairs of are the sets themselves. If we are serious about the idea that ordered pairs are not sets but distinct entities of which we have an independent conception, we shall therefore have to alter our definition of 'level' to ensure that ordered pairs get included in the hierarchy. One way of doing this would be to replace accumulations with extended accumulations: the *extended accumulation* of a consists of the accumulation of a together with the ordered pairs whose terms belong to the accumulation of a. This manoeuvre would of course complicate the formal development slightly, but if it were adopted, hardly anything in the rest of the book would have to be changed.

The problem of doing too much is one that most mathematicians would regard as irrelevant to their concerns: they are habituated by frequent use to the practice of modelling one theory within another and simply disregarding the extraneous properties which this practice throws up. Moreover, they are undoubtedly aided in this by the fact that the properties at issue *are* extraneous — answers to questions it would never occur to us to ask, lying out among what Quine evocatively calls the 'don't cares'. Kuratowski's trick has certainly proved popular: among those writing after it became well-known, Bourbaki (1954) is the rare exception in not making use of it, instead treating the ordered pair as an extra primitive in his formal system; but even Bourbaki relented eventually and (in the 4th edition of *Théorie des ensembles*) followed what had by then become the universal practice among mathematicians of adopting the trick as a definition of the ordered pair.

As the reader will note, I have chosen to follow this practice here too. I must therefore give up the idea that the collections called 'ordered pairs' here are the genuine article. The reason is that the moral Benacerraf drew — that since their numerical properties do not determine which sets numbers are, numbers are not sets — applies in this case too. If there are such things as ordered pairs in the proper sense — entities governed solely by the ordered pair principle — then for the same reason they are not sets. So the theory of sets does not contain the theory of ordered pairs, but only a convenient surrogate of it. But that is all we need: $\{\{x\}, \{x, y\}\}$ is a single set that codes the identities of the two objects x and y, and it is for that purpose that we use it; as long as we do not confuse it with the *genuine* ordered pair (if such there is), no harm is done. In other words, the ordered pair as it is used here is to be thought of only as a technical tool to be used within the theory of sets and not as genuinely explanatory of whatever prior concept of ordered pair we may have had.

Definition. *If z is an ordered pair, let*

$$\mathrm{dom}(z) = \imath! x(\exists y)(z = (x, y)) \text{ ('the first coordinate of } z\text{');}$$

$$\mathrm{im}(z) = \imath! y(\exists x)(z = (x, y)) \text{ ('the second coordinate of } z\text{').}$$

Thus $\mathrm{dom}(x, y) = x$ and $\mathrm{im}(x, y) = y$.

4.7 Relations

According to the standard conception, relations correspond to binary predicates in the same way that sets correspond to unary ones.

Definition. *A set is called a* relation *if every element of it is an ordered pair.*

If $\Phi(x, y)$ is a formula, then $\{(x, y) : \Phi(x, y)\}$ is a relation provided that it is a set; it is said to be the relation between x and y *defined by* the formula $\Phi(x, y)$. Conversely, any relation r is defined by the formula $(x, y) \in r$, which is customarily written $x \; r \; y$. The set $\{(y, x) : x \; r \; y\}$ is called the *inverse* of r — some authors call it the 'converse' — and it is denoted r^{-1}. The sets $\mathrm{dom}[r] = \{\mathrm{dom}(z) : z \in r\}$ and $\mathrm{im}[r] = \{\mathrm{im}(z) : z \in r\}$ are called the *domain* and *image* of r respectively. If c is a set, we let

$$r[c] = \{y : (\exists x \in c)(x \; r \; y)\}.$$

An element y is said to be r*-minimal* if there is no x such that $x \; r \; y$. If r and s are relations, then we let $r \circ s$ denote the relation

$$\{(x, z) : (\exists y)(x \; s \; y \text{ and } y \; r \; z)\}.$$

The order of r and s in the definition of $r \circ s$ is not what one might expect: it is dictated by the widespread practice, which we shall follow here, of writing function symbols on the *left* of their arguments (see below). If $r \subseteq s$, we say that s is an *extension* of r and that r is a *restriction* of s.

Definition. $a \times b = \{z : \text{dom}(z) \in a \text{ and } \text{im}(z) \in b\}$ *('the cartesian product of a and b').*

The members of $a \times b$ are thus the ordered pairs whose first coordinate is in a and whose second coordinate is in b.

(4.7.1) **Proposition.** *If a and b are sets, then $a \times b$ is a set.*

Proof. Every ordered pair with a first coordinate in a and a second coordinate in b is a member of $\mathfrak{P}(\mathfrak{P}(a \cup b))$. The result follows [propositions 3.7.10 and 4.5.3]. □

Definition. *A relation which is a subset of $a \times b$ is said to be a relation* between *a and b. A relation between a set a and itself (i.e. a subset of $a \times a$) is said to be a relation* on *a.*

For example, the formula $x = y$ defines on any set a a relation which is often called the *diagonal* of a. If c is a subset of a, then the restriction $r \cap (c \times c)$ of r is a relation on c which is denoted r_c.

We conclude this section by mentioning another technical device that is occasionally useful. It depends on picking once for all two definite objects. For the trick to work, all that matters is that the chosen objects are distinct: once we have got round to defining the natural numbers 0 and 1 (which will be in §5.4), we might as well use them for this purpose, so we shall state the definition as if those are the ones we have chosen, but nothing of importance hinges on the matter.

Definition. $a \uplus b = (a \times \{0\}) \cup (b \times \{1\})$ *('disjoint union').*

The purpose of this definition is to tag each member of a with the label 0 and each member of b with the label 1, so that they retain their distinctness when we form the union.

Exercises

1. (a) Show that $\{x\} \times \{x\} = \{\{\{x\}\}\}$.

   (b) Suppose that a is a set. Show that $a \times a = a \Leftrightarrow a = \varnothing$. [If a is non-empty, consider an element of a of earliest possible birthday.]

   (c) Give an example of sets a, b, c, d such that $a \times b = c \times d$ but $a \neq c$ and $b \neq d$.

2. Suppose that r, s and t are relations.

   (a) Show that $(r \circ s) \circ t = r \circ (s \circ t)$.

   (b) Show that $(r \circ s)^{-1} = s^{-1} \circ r^{-1}$.

4.8 Functions

A relation f between a and b is said to be *functional* if for every $x \in a$ there is exactly one $y \in b$ such that $x \ f \ y$. Functional relations between a and b are more usually called *functions from* a *to* b. If $\tau(x)$ is a term, then the set $\{(x, \tau(x)) : x \in a\}$ (if it exists) is a function which is said to be *defined by* the term $\tau(x)$; it is denoted $(\tau(x))_{x\in a}$ or '$x \mapsto \tau(x)$ $(x \in a)$'. or, if the domain is clear from the context, simply $(\tau(x))$ or '$x \mapsto \tau(x)$'.[1] Conversely, if f is a function from a to b, it is defined by the term $\imath!y(x \ f \ y)$, which is denoted $f(x)$ and called the *value* of f for the *argument* x.

Another terminology which we shall sometimes use is to call a function a *family*, its domain the *indexing set* and its image the *range* of the family: if the members of the range all have some property F, we call it a *family of* Fs. When we are using this terminology, it is usual to express the value of f for the argument x as f_x.

One function that crops up often enough to be worth singling out is the function id_a from a set a to itself given by the assignment $x \mapsto x$ $(x \in a)$: it is sometimes known as the *identity function* on a, although it is in fact the same thing as the diagonal relation on a.

The set of all functions from a to b is denoted ${}^a b$. Somewhat more generally, if $(b_x)_{x\in a}$ is a family of sets, we write $\prod_{x\in a} b_x$ for the set of all the functions f from a to $\bigcup_{x\in a} b_x$ such that $f(x) \in b_x$ for all $x \in a$; so ${}^a b = \prod_{x\in a} b$.

The fact that most authors use $f(x)$ to denote the value of the function f at x rather than $(x)f$ or $x|f$ is a historical accident with nothing except tradition to commend it. Dedekind used the notation $x|f$ in an early draft of *Was sind und was sollen die Zahlen?* (see Dugac 1976, app. LVI), but changed this to $f(x)$ in the published version.

(4.8.1) **Proposition.** *If* $(f_i)_{i\in I}$ *is a family of functions such that* $f_i \cup f_j$ *is a function for all* $i, j \in I$, *then* $\bigcup_{i\in I} f_i$ *is a function.*

Proof. Suppose that $(x, y), (x, z) \in \bigcup_{i\in I} f_i$. So there exist $i, j \in I$ such that $(x, y) \in f_i$ and $(x, z) \in f_j$. So (x, y) and (x, z) belong to $f_i \cup f_j$, which is a function by hypothesis, and therefore $y = z$. □

If f is a function from a to b and $c \subseteq a$, then the restriction $f \cap (c \times b)$ of f is a function from c to b which is denoted $f|c$. If $f|c = g|c$, i.e. if $f(x) = g(x)$ for all $x \in c$, we say that f and g *agree on* c.

A function f from a to b is said to be *one-to-one* if for each $y \in b$ there is at most one $x \in a$ such that $y = f(x)$; it is said to be a function *onto* b if for each $y \in b$ there is at least one $x \in a$ such that $y = f(x)$. It is said to be a *one-to-one*

[1] The term $(\tau(x))_{x\in a}$ does not depend on x; in this expression, the letter x is being used as a dummy variable and could be replaced by any other variable which does not already occur in the expression.

correspondence between a and b if it is a one-to-one function from a onto b: this is so just in case for each $y \in b$ there is exactly one $x \in a$ such that $y = f(x)$, or equivalently if the inverse relation f^{-1} is a function from b to a. We shall say that a and b are *equinumerous* if there is a one-to-one correspondence between them.

Exercises

1. If f is a one-to-one function from a to b and g is a one-to-one function from b to c, show that $g \circ f$ is a one-to-one function from a to c.

2. If f is a function from a onto b and g is a function from b onto c, show that $g \circ f$ is a function from a onto c.

3. If f is a function from a to b, show that f^{-1} is a function from $\mathrm{im}[f]$ onto a iff f is one-to-one.

4.9 The axiom of infinity

There is a lowest level, and for each level there is another level above it. It follows from this that there are infinitely many levels, but not that there is any level with infinitely many levels below it — with, that is to say, an infinite history. For that we need another axiom.

Definition. *A* limit level *is a level that is neither the initial level nor the level above any other level.*

(4.9.1) **Proposition.** *If V is a level other than the lowest, then it is a limit level iff* $(\forall x \in V)(\exists V' \in V)(x \in V')$.

Necessity. Suppose that V is a limit level and $x \in V$. Then x is either a set or an individual [proposition 3.7.2]. If x is an individual, $x \in V_0 \in V$. If x is not an individual, on the other hand, $x \subseteq V' \in V$ [proposition 3.6.8] and so if V'' is the level above V, then $x \in V'' \in V$.

Sufficiency. If V is not a limit level, then it is the level above some level V', and $V' \in V$ but there is no level V'' such that $V' \in V'' \in V$. □

Axiom of infinity. *There is at least one limit level.*

To explain why we have called this the axiom of infinity, we need to introduce a definition of infinity due to Dedekind.

Definition. *A set is* infinite *if it is equinumerous with a proper subcollection of itself.*

Definition. *The lowest limit level is denoted* V_ω.

(4.9.2) **Proposition.** *The history of* V_ω *is infinite.*

Proof. The history of V_ω is the set of levels belonging to it. Consider the function which maps each such level to the one above it: this function is one-to-one, but the initial level V_0 is not in its image. The history of V_ω is therefore infinite. □

Our intuitions about infinite collections are undoubtedly more nebulous and more capable of arousing controversy than those about finite ones. Indeed many of the properties of infinite collections are paradoxical at first sight. When Cantor wrote his works on set theory, an abhorrence of actually infinite collections had been commonplace since Aristotle: they were, for example, scrupulously avoided by Euclid (c.300 B.C.). There were certainly a few heretics — Galileo and Bolzano, for example — but Aristotle's views had been widely accepted.

It was just at the time when infinitesimals had been successfully expunged from analysis by Weierstrass (Cantor's teacher) — not to be rehabilitated until the 1950s — that Cantor himself tamed the paradoxes of the actual infinite in his work on cardinal and ordinal arithmetic. However, he had to brook considerable opposition and devoted lengthy passages in his published work to the defence of his views against what he called the *horror infiniti*, 'a kind of shortsightedness which destroys the possibility of seeing the actual infinite' (1886, p. 230).

Perhaps the most influential of those whom Cantor (1886, pp. 225–6) took to be opponents of the actual infinite was Gauss, who had protested in a frequently quoted letter to Schumacher 'against treating infinite magnitudes as something completed, which is never admissible in mathematics' (Gauss 1860–65, vol. II, p. 269) As so often, however, it pays to examine the quotation in context. In the letter to which Gauss was responding Schumacher had tried to prove that space is Euclidean by an ingenious but fallacious argument involving the construction of a large semi-circle. If the semi-circle is kept fixed and the triangle is made small, Schumacher's argument does indeed show that the sum of the angles of the triangle will tend to 180°. What Gauss quite correctly objected to, though, was Schumacher's use of the opposite procedure — keeping the triangle fixed and letting the radius of the semi-circle tend to infinity — to show that the sum of the angles of the triangle is 180°.

But even if Gauss was not in this famous quotation opposing the notion of an actually infinite *set*, there were certainly many others who did. Indeed it was not uncommon at first to regard the paradoxes of set theory as paradoxes of the *infinite*. Some years after Cantor's work Poincaré (1906, p. 316) could still assert that 'there is no actual infinite; the Cantorians forgot this and fell into contradiction'. On the issue of bare consistency, however, Cantor's views did eventually prevail: hardly anyone would now try to argue that the existence of infinite collections is *inconsistent*: the modern finitist is more likely to make the much weaker claim that there is no reason to suppose they exist.

That is not to say, of course, that we can *prove* the logical consistency of the axiom of infinity. We might indeed be able to supply an argument for this: however, it would not be a proof *ab initio*, but only relative to some other infinitary theory for which we might have some independent ground (e.g. Euclidean geometry). Even if that were possible, though, it would heavily qualify our theory's suitability to take on the foundational role for which we were grooming it: in the case just quoted, for example, the theory would obviously be ruled out as a candidate to act as a foundation for Euclidean geometry. More importantly, though, it would not even meet the case since, as we have already seen, there seems to be no reason to think the mere logical possibility of a set implies its actual existence, and indeed it is dubious whether the supposition that it does imply it is even coherent.

Modern mathematical practice makes use of infinite collections everywhere, at least if we take its surface grammar at face value: not only does the construction of proxies for the standard objects of mathematics in set theory that we shall outline in part II need there to be at least one infinite set in order for it to get started, but even independently of that construction mathematicians quite routinely use infinite sets in reasoning about these standard objects. So if we believe that what mathematicians say is true, there is nowadays a regressive justification for assuming an axiom of infinity, i.e. an axiom asserting the existence of at least one infinite collection. But there are two distinct ways in which we could have done this, depending on whether we supposed the infinite collections to exist because there are infinitely many individuals or because there are levels infinitely far up the hierarchy. If we had opted for the first of these, our axiom would have been as follows.

Axiom of infinity$_1$. V_0 *is infinite.*

The axiom of infinity which we have asserted here is of the second sort. (In the next section we shall suggest an informal way of picturing the hierarchy according to which the first of these suppositions makes the hierarchy infinitely wide and the second makes it infinitely high.)

The requirement of mathematics is sets from which to construct the relevant objects, and we do not care much which sets they are. So the regressive method does not directly give us a ground for preferring one sort of axiom of infinity over the other. In fact, it makes remarkably little difference to anything that follows which of these two axioms we assume: for mathematical purposes it would be sufficient to assume the following axiom, which is entailed by each of them.

Axiom of infinity$_2$. *There exists an infinite set.*

One route that has been tried, most famously by Dedekind but also by Russell (1903, p. 357) and others, is to give a direct (i.e. non-regressive) argument for the existence of infinitely many individuals. Such an argument would not be attractive to set theorists since the individuals are (by definition) not sets, so for mathematics to depend on them in this manner would be for it to depend on facts extraneous to it. Philosophers, whose subject habituates them to seeking any port in a storm, might not be so dismissive. Dedekind's argument is as follows:

> My own realm of thoughts, i.e. the totality S of all things which can be objects of my thought, is infinite. For if s signifies an element of S, then the thought s' that s can be an object of my thought, is itself an element of S. If we regard this as an image $\phi(s)$ of the element s, then the mapping ϕ of S thus determined has the property that the image S' is part of S; and S' is certainly a proper part of S, because there are elements in S (e.g. my own ego) which are different from such a thought s' and therefore are not contained in S'. Finally it is clear that if a, b are different elements of S, their images a', b' are also different, and that therefore the mapping ϕ is one-to-one. Hence S is infinite, which was to be proved. (Dedekind 1888, no. 66)

Of course, this does not purport to be a proof that the empirical world must be infinite, but only that the realm of thought must. If it were to turn out that mathematics depends for its truth on an inevitable feature of that realm, that would certainly be a radical conclusion, for it would at a stroke convert our philosophy of mathematics into idealism, but it is far from clear that this would be absurd.

Note, though, that even if it is sound, Dedekind's proof shows the existence only of infinitely many objects, not of a set to which they all belong. But the axioms we have stated *already* demonstrate that, for we can prove the existence of, for instance, the sets

$$\emptyset, \{\emptyset\}, \{\{\emptyset\}\}, \ldots$$

Dedekind's proof is therefore of use only to a dependency theorist who, invoking Aristotle's celebrated distinction between potential and actual infinity, regards the relation of dependency that holds between the levels as showing that they form only a potential infinity, not an actual one, and for that reason hesitates to assert that there is a set V_ω containing all of them. For such a person Dedekind's proof is not redundant, since it attempts to demonstrate that the individuals are actually, not just potentially, infinite in number, and therefore gives a stronger ground for believing that there is a set containing all of them.

In any case, we have chosen here not to follow the course of supposing that there are infinitely many individuals. Our reason is that to have done so would have limited the applicability of our theory: if we assumed that there are infinitely many individuals, the resulting theory would be inconsistent in those cases where the theory T to which set theory is being added had only finite models. That would contravene our (surely reasonable) desire to provide a theory of sets that can, at the very least, be adjoined to *any* theory T without introducing inconsistency.

Instead, therefore, we have asserted that there is a level V_ω which has an infinite history — a level, in other words, which depends on infinitely many lower levels. The constructivist's route to establishing this principle is relatively clear and depends solely on the coherence of the notion of a supertask, as already discussed. Platonists, on the other hand, seem to have frustratingly little they can say by way of justification for this form of infinity axiom.

4.10 Structures

The general framework we set up in §2.2 consisted of a prior theory T, to which we added a predicate $\mathrm{U}(x)$ to mean 'x is an individual', before relativizing the axioms of T to U. Let $\mathsf{Z}[T]$ be the theory obtained by adding to this framework the axioms of creation and infinity together with all the instances of the separation scheme. (Although it will not figure prominently in this book, we should also mention the corresponding second-order theory $\mathsf{Z2}[T]$, in which the separation scheme is replaced by the second-order separation principle.) Two special cases are worthy of note. If we start from the theory **null** which has no axioms at all, so that nothing whatever is assumed about the individuals, the theory Z[**null**] thus obtained is usually denoted simply ZU. If we start from the theory **empty** whose only model is empty, we get a theory Z[**empty**] which is usually denoted simply Z. When **empty** is relativized to U, it asserts that there are no individuals, so another way of arriving at Z would be to obtain it from ZU by adding the axiom of purity of §3.8.

But treating its objects as individuals is not the only way in which a theory can be embedded in set theory.

Definition. *An ordered pair* (A, r) *is called* a structure *if* r *is a relation on* A.

We shall sometimes refer to A as the *carrier set* of the structure and to (A, r) as the result of *endowing* it with the relation r. There is a widely used convention, which we shall occasionally make use of, that (A, r) may be denoted simply by A if the identity of the relation r is clear from the context.

Consider now a formal language in which the only non-logical symbol is a binary relation symbol R. If (A, r) is a structure, in the sense just defined, we can interpret any sentence in the language as making a set-theoretic claim about (A, r). If the language is first-order, we do this by replacing $x \; R \; y$ with $(x, y) \in r$ and relativizing all the quantifiers in the sentence to A. If the language is second-order, then because our theory of sets is first-order, this procedure will leave us with second-order quantifiers relativized to A which

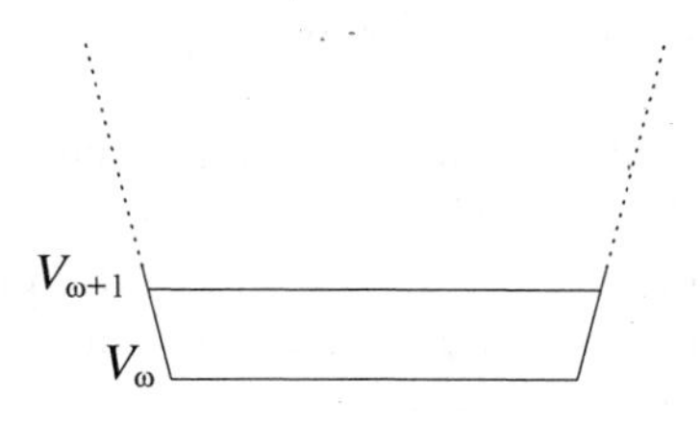

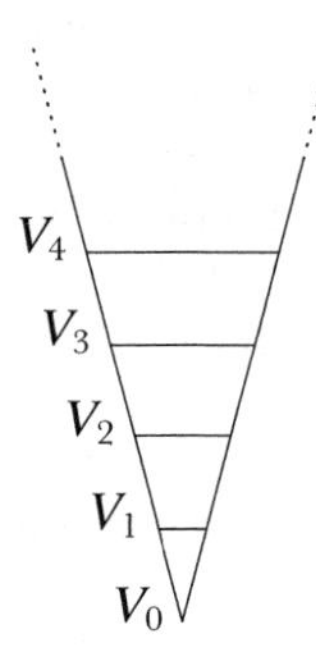

A picture of Z

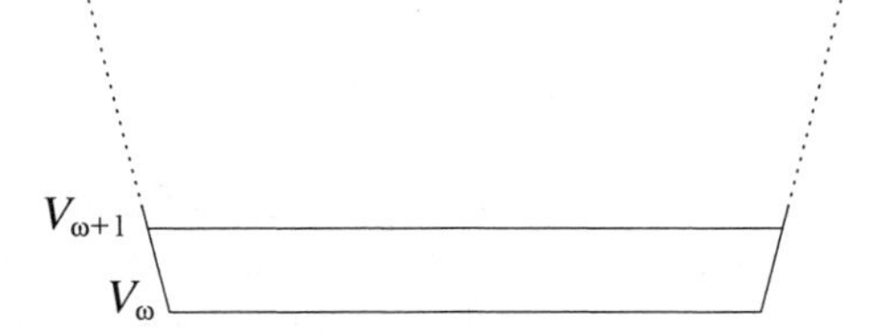

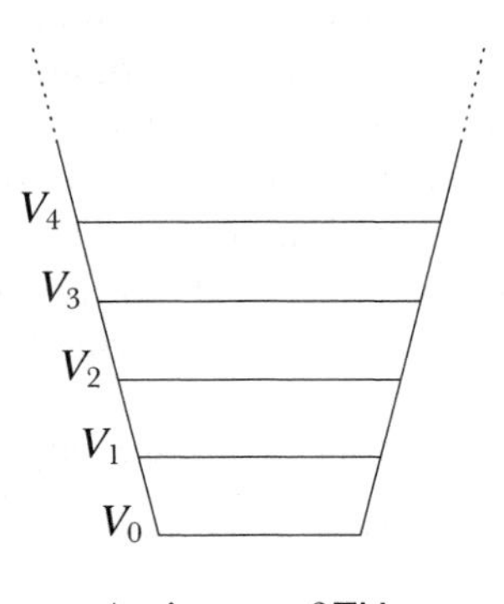

A picture of ZU

we still need to interpret: we do this by interpreting them as ranging over all the subsets of A. The result of this process of interpretation is that we can talk of a sentence in the language as being *true* in a particular structure (A, r).[2]

If all the sentences of T are true in (A, r), we say that it is a *set-theoretic model* of T. This method of interpreting a theory in a structure is indeed a generalization of the method considered earlier. For the proposal then was that starting from a theory T we add a predicate $\mathrm{U}(x)$ to mean 'x is an individual' and relativize the axioms of T to U. If we do this, we can form a structure by taking its domain to be the set V_0 of all the individuals; interpret each binary relation symbol R in the language of T by the relation on V_0 defined by the formula $x\ R\ y$ (and similarly for relation symbols of other arities); and interpret each unary function symbol f by the function on V_0 defined by the term $f(x)$ (and similarly for function symbols of other arities). The result will be a set-theoretic model of T since the relativizations to U of all the axioms of T have been adopted as axioms of $\mathsf{ZU}[T]$.

Definition. *An* isomorphism *between structures (A, r) and (A', r') is a one-to-one correspondence f between the sets A and A' such that*

$$(\forall x, y \in A)(x\ r\ y \Leftrightarrow f(x)\ r'\ f(y)). \tag{2}$$

Two structures are said to be *equivalent* if just the same sentences are true in each of them, and *elementarily equivalent* if just the same first-order sentences are true in each of them. Isomorphic structures are evidently equivalent, and hence in particular elementarily equivalent.

If the notion of a set-theoretic model generalizes the ideas of §2.2 by including them as the special case in which the domain of the model consists of individuals, it is natural to ask under what circumstances it is a *genuine* generalization. If, in other words, we have a theory T which has a set-theoretic model (A, r), under what circumstances is there a model whose domain consists of individuals? It is easy to see that this is an issue that hinges solely on how *many* individuals there are. For if the domain A is equinumerous with any other set A' (whether a set of individuals or not), we can use (2) as a *definition* of a relation r' on A' which will make (A, r) and (A, r') isomorphic. And as isomorphic structures are equivalent, if (A, r) is a model of T, (A', r') will be one too.[3]

From a metatheoretic perspective we can also proceed in the other direction and, given some set-theoretic structures, ask whether there is a theory whose

[2]The case we have dealt with here is that of a language with a single binary relation symbol, but it can easily be generalized to more elaborate languages, and we shall occasionally use the vocabulary we have just introduced in this generalized meaning, so that for instance the ordered triple $(\boldsymbol{\omega}, \mathrm{s}, 0)$ will be referred to as a 'structure' in §5.5.

[3]This simple observation is the nub of a telling objection made by Newman (1928) to Russell's (1927) causal theory of perception.

set-theoretic models are just those structures. As an instance of this, consider the following properties.

Definition. *A relation r on A is said to be:*

reflexive *on A if* $(\forall x \in A)(x\ r\ x)$;

irreflexive *on A if* $(\forall x \in A)(\text{not}\ x\ r\ x)$;

transitive *if* $(\forall x, y, z)((x\ r\ y \text{ and } y\ r\ z) \Rightarrow x\ r\ z)$;

symmetric *if* $(\forall x, y)(x\ r\ y \Rightarrow y\ r\ x)$;

antisymmetric *if* $(\forall x, y)((x\ r\ y \text{ and } y\ r\ x) \Rightarrow x = y)$.

Each of these classes of structure evidently corresponds to a first-order axiom in the language of the binary relation symbol '*R*'. Thus the reflexive structures, for example, are just the set-theoretic models of the first-order axiom

$$(\forall x)\, x\ R\ x.$$

But not all the classes of structure we might be interested in can be characterized by first-order axioms in this way. Model theory is a rich source of results on what is possible in this regard. For instance, logicians have been especially interested in theories that are *categorical*, which means that any two set-theoretic models are isomorphic to each other, but it is an easy consequence of the compactness theorem for first-order logic that no first-order theory with an infinite set-theoretic model can be categorical.

Notes

What we have given here is only an outline of the elementary theory of sets, functions and relations. It is described in greater detail in many, many textbooks: one that is quite close in spirit to ours is Tourlakis 2003, vol. II.

The issue concerning the validity of the axiom of creation, which amounts to the question whether every set has a power set, has not always received as much attention as it deserves. Some sceptical thoughts on this are to be found in Hobson 1921 and Lusin 1927.

The argument about the status of the infinite has a long and tangled history, the best introduction to which is A. W. Moore 1990. Waterhouse 1979 is a clear account of the correspondence between Gauss and Schumacher.

Conclusion to Part I

Although we shall discuss other set-theoretic axioms later in the book, we shall not adopt them as part of our system, which will be ZU. This is sufficiently important that it deserves to be repeated with emphasis.

We shall assume the axioms of ZU *throughout the remainder of this book.*

ZU — the theory whose axioms are creation (§4.5), infinity (§4.9) and all the instances of separation (§3.5) that can be stated in the language of the theory — will, that is to say, be our *default theory*. When we state a theorem, therefore, what we are claiming is that the statement in question is provable in ZU. Of course, it follows at once *a fortiori* that it is also a theorem of Z[T] for any theory T, and in particular a theorem of Z.

As I have said, we shall encounter other set-theoretic axioms later. One warning in the meantime, though: what plays the role of the default theory in other textbooks is most often a theory of sets equivalent either to ZF, which is obtained from Z by adding a stronger axiom of infinity ensuring the existence of many more levels in the hierarchy, or to ZFC, which is obtained from ZF by adding the axiom of choice as well.[1] But anything that is a theorem of ZU is also a theorem of these stronger theories, so what is said here remains valid in these other contexts too.

I have already commented at some length on two other respects in which our default theory is weaker than standard formulations in other books: it allows for there to be both individuals and ungrounded collections. Of these two permissions the latter is the less significant. I have argued that platonists with internalist sympathies might well be willing to grant that all collections are grounded. In recognition of this we shall adopt the convention from now on that our quantifiers are restricted to range over sets and individuals only.

So if the theory I have chosen leaves room for ungrounded collections, it is only because I do not want to pick a fight unnecessarily. That it leaves room for individuals, on the other hand, is central to ensuring its applicability. We deduce directly from Ramsey's theorem in combinatorics, for instance, that

[1] For explanations of these theories see part IV; for a discussion of the rather different sort of axiomatization of ZF used in most other books see appendix A.

at a party attended by ten people there must be either four acquaintances or three strangers. But if our theory of sets is pure, we cannot do this, at any rate not directly: we are only entitled to apply the theorem to Ø and sets formed from it by iterations of the power-set operation; and it begins to seem miraculous that mathematics applies to the world at all.

If for some reason we were determined to study only the pure theory, all would not be lost quite yet, it is true: we could try to repair the damage later by adding appropriate bridging principles connecting the pure sets of our theory to the denizens of the real world that we want eventually to be able to count. But it is very hard to see what the point would be of proceeding in this fashion.

Although our theory is non-standard in the two respects just mentioned, we have sided with the majority in two others, the adoption of classical logic and the acceptance of impredicative definitions. Denial of these two principles has long been a badge of constructivism, but it is worth noting that in fact they are independent of each other, not just technically but philosophically: semi-intuitionists such as Poincaré argued against impredicative sets without querying classical logic, while others (e.g. Lear 1977) argued for intuitionistic logic in set theory without bringing impredicative sets into question.

It is worth noting, too, that if it were only the proof-theoretic strength of the resulting system that concerned us, only the elimination of impredicative sets would represent a significant restriction. This is because there are methods (such as the so-called negative translation) for re-interpreting the classical connectives in any sentence intuitionistically so as to preserve provability (from which, of course, we may deduce that the classical system is consistent if the intuitionistic one is). In the case of arithmetic, for which the result was proved by Gödel (1933), the translation is especially simple because the atomic sentences (i.e. numerical equations) are intuitionistically decidable, and hence can be left intact. In the case of set theory we have to translate the atomic formula $x \in y$ into its double negation, and the treatment of extensionality is also rather delicate. This lessens somewhat the naturalness of the translation, and it is not as clear in this case as in the arithmetical one that its existence creates problems for arguments that aim to reject excluded middle on grounds of manifestability (see Potter 1998).

Nonetheless, any translation, whether natural or not, gives us a relative consistency result. Impredicativity, by contrast, makes an enormous difference to the formal strength of the first-order system we are studying, and it does so independently of whether the underlying logic is classical. There are various ways of explaining the reason for this, but one is that impredicativity allows for elaborate feedback loops between different levels in the hierarchy, greatly enriching its structure. This is immediately clear if we imagine the difficulty involved in constructing from scratch a model of the theory. We can use a set at one level V to define a set at a higher level, but then we may have to go back to V and construct further sets forced on us because they can be defined by instantiating quantifiers in a term of the form $\{x \in V : \Phi\}$. In the predicative theory, by contrast, there are no such feedback loops, and which sets we include in some level of a model depends only on the preceding levels. The consequences of the richness of structure which the impredicative form of the separation scheme permits will be a constant theme of this book. In particular, the whole theory of infinite cardinals which we develop in part III would be impossible without it.

Part II

Numbers

Introduction to Part II

In the first part of this book we have developed a theory with wide application. It systematizes our talk in what might be called the *extensional mode* — talk, that is to say, that depends only on which objects have a property rather than on how the property is presented to us. The use of the extensional mode is by no means confined to mathematics — it is a commonplace of almost every sphere of discourse that comes to mind — but in mathematics it is of especial importance. Mathematicians are persistently inclined, once they have proved two properties extensionally equivalent, to ignore the difference and treat them thereafter as one. They therefore find a particular utility in a language which shares this blindness. Set theory thus gives us a way to systematize manners of talking and patterns of reasoning that are already widespread in mathematics.

But what we have introduced is not simply a language: the axioms of the theory carry existential commitments significant enough for their truth to be a matter of some controversy. If my goldfish Bubble exists, set theory asserts that Bubble's singleton exists too; and it is by no means clear that this is something to which the belief in Bubble of itself commits me. This lack of ontological innocence has caused set theory to be pressed into the service of a different and more ambitious project which aims not merely to express certain ways of grouping objects that are common in mathematics but to supply the objects themselves.

The principal locus of this latter project has traditionally been the part of mathematics that is usually called either 'calculus' or 'analysis' — that is, roughly, the study of real-valued functions of a real variable, together with the corresponding case for the complex numbers. The idea of studying such functions is certainly not new — Archimedes, for example, used methods that recognizably belong to the calculus to prove results on the areas of various figures — but the subject owes its modern importance to the extraordinarily fruitful developments which began with the discoveries of Newton and Leibniz in the late 17th century. For over a hundred years, however, the methods that mathematicians used to prove the theorems of the calculus made apparently ineliminable use of diagrams. If it is indeed ineliminable, this use of diagrams has consequences for both the nature of mathematical knowledge and its certainty, since our knowledge of the theorems of the calculus will be of the same

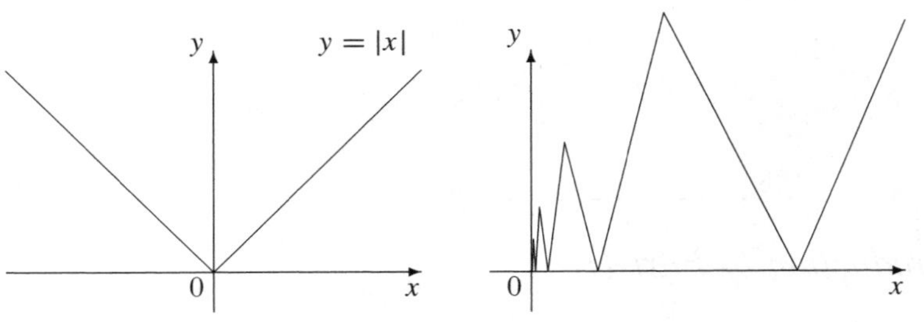

nature as our knowledge of the diagrams and no more certain than it.

Of these two issues — the nature of knowledge and its certainty — the former is a philosophical, the latter a practical one. Inevitably, then, it was overwhelmingly the latter that motivated mathematicians, not the former. Berkeley in *The Analyst* (1734) famously mocked the unclarity and lack of rigour to be found in the arguments used by early expositors of the calculus, but his criticisms had little effect on the practice of mathematicians. What eventually came to concern them were cases where appeal to diagrams was shown to be capable of leading them into error. The examples that gave rise to this concern are well known. Here I shall mention one of the most striking, which arises when we investigate the relationship between the two most important properties that fall within the scope of the calculus, continuity and differentiability. A function f of a real variable is said to be *continuous* at an argument a if $f(x)$ tends to $f(a)$ as x tends to a; it is said to be *differentiable* at a if $\frac{f(x)-f(a)}{x-a}$ tends to a limit as x tends to a, the limit being called the *derivative* of f at a and denoted $f'(a)$. Every differentiable function is continuous, but it is easy to give examples where the converse fails. The modulus function $x \mapsto |x|$, for instance, is continuous but not differentiable at 0. If we call the set of points at which a continuous function fails to be differentiable its *exceptional set*, it is then natural to wonder what sorts of sets can be exceptional in this sense. It is not hard to see by extending the example of the modulus function that every finite set of real numbers is exceptional. And by constructing a function which performs a damped oscillation we can see that *some* infinite sets, such as $\{1, 1/2, 1/4, \dots\}$, are exceptional. But intuitions derived from diagrams suggest (or at any rate suggested to some 19th century analysts) that there should be some sense in which all exceptional sets are small. If, for example, we imagine a particle traversing the graph of a continuous function, then an exceptional point will be one at which the particle abruptly changes its direction in a discontinuous manner. This is an instance of a familiar notion from kinematics: it is what is supposed occurs in an elastic collision. But could a particle change direction discontinuously at *every* moment? It seems clear that this is impossible. For this reason, or reasons similar to it, analysts supplied various properties of smallness which, they conjectured, all exceptional sets must have. One consequence these conjectures had in common was that the whole real line could not be an exceptional set: there could not,

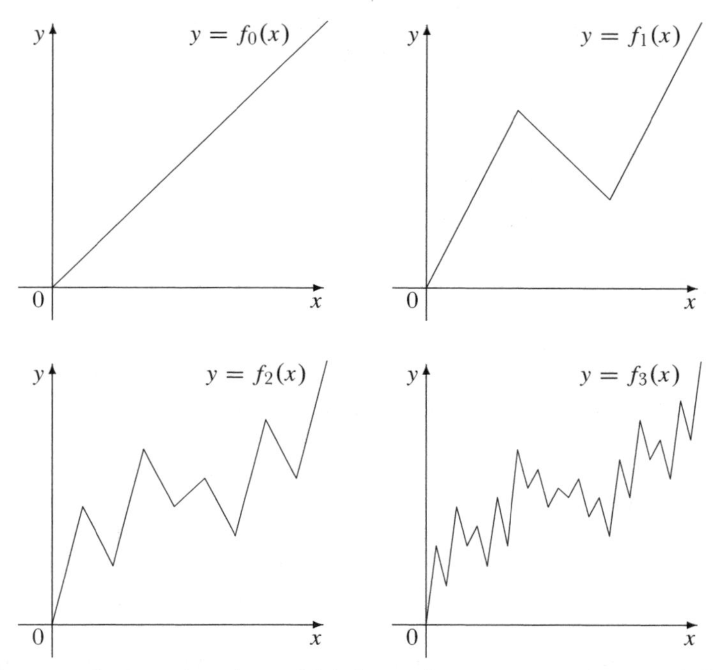

in other words, be a function which is continuous everywhere and differentiable nowhere.

But there can. Consider the sequence (f_n) of functions defined on the unit interval as follows. First let $f_0(x) = x$. Then, when f_n has been defined, let

$$f_{n+1}\left(\frac{k}{3^n}\right) = f_n\left(\frac{k}{3^n}\right)$$

$$f_{n+1}\left(\frac{k}{3^n} + \frac{1}{3^{n+1}}\right) = f_n\left(\frac{k}{3^n} + \frac{2}{3^{n+1}}\right)$$

$$f_{n+1}\left(\frac{k}{3^n} + \frac{2}{3^{n+1}}\right) = f_n\left(\frac{k}{3^n} + \frac{1}{3^{n+1}}\right)$$

and complete the definition by requiring that f_{n+1} should be linear in each of the 3^{n+1} intervals $[\frac{k}{3^{n+1}}, \frac{k+1}{3^{n+1}}]$ $(0 \leqslant k < 3^{n+1})$. It can be shown that these functions converge to a limit function f which is continuous throughout the unit interval but not differentiable anywhere in it.

The first example of a continuous nowhere differentiable function was discovered by Bolzano in 1840, but he did not publish it, and the phenomenon did not influence the mathematical community until Weierstrass discovered another example, which he published in 1872. From the pragmatic perspective that most mathematicians adopt, what this surprising example and others like it demonstrate is the need for rigour. The research programme to which they thus gave birth was prosecuted in the second half of the 19th century, most prominently by Weierstrass himself, who in lecture courses he gave at the University of Berlin gradually developed the rigorous treatment of analysis that was being sought.

What this rigorous treatment did was to remove from proofs the appeal to what is obvious from the diagram, and replace it with symbolic reasoning from axioms. By the end of this period it had thus become possible to prove all the results of the calculus solely on the assumption that the real numbers form what is called a complete ordered field.[1]

This systematization of the calculus thus shifted the epistemic focus away from what we can glean from diagrams and onto two issues which could now be seen to be distinct: the status of the symbolic reasoning used in diagram-free proofs; and the status of the initial assumption presupposed by that reasoning, namely that the real numbers form a complete ordered field.

The first of these issues was addressed above all by Frege, whose *Begriffsschrift* (1879) provided not only a notation for expressing arguments involving multiple generality but a codification of the rules of proof used in such arguments. Frege's work was ignored, however: it was left to Peano to invent a more typographically tractable notation, and to others such as Russell and Hilbert to turn the logic of multiple generality into a manageable tool for the codification of mathematical reasoning. This logic came to be seen by some, indeed, as not only a tool but a canon: whether an argument could be symbolized in the formal system came to be seen as a criterion of its very correctness. It was Gödel's first incompleteness theorem that put paid to this inflated ambition by showing that no formal system codifies all the types of argument that we are entitled to regard as correct.

It is also worth adding the further caveat that even if a type of argument *can* be codified within the currently accepted formal system (which in the present work is, as already noted, taken to be the classical first-order predicate calculus with identity), this does not answer the epistemological question of how we know that the argument is valid: it merely postpones that question by palming it off on the underlying logic. In particular, the fact that proofs in analysis proceed by modes of reasoning formalizable in the accepted formal system does not in itself show that Kant was wrong when he talked of continued appeal to intuition in mathematics: this further conclusion will follow only if there is a route independent of intuition to the correctness of the formal system.

Nevertheless, it took some time for this further problem to emerge as a central issue in the justification of mathematical analysis. What concerned the more foundationally minded mathematicians much more pressingly was the

[1] This expression will be defined in chapter 8. It means that, in addition to the ordinary algebraic properties satisfied by the order relation and the operations of addition and multiplication, the real numbers have the completeness property: every non-empty bounded set of real numbers has a least upper bound.

second of the two parts into which Weierstrass' work had allowed the question to be split, namely that of justifying his assumption that the real numbers form a complete ordered field.

Once again this relates both to the certainty and to the character of the truths of analysis. It relates to their certainty because the conception of the real numbers to which Weierstrass' axiomatization is answerable is not the only one that is available. To see why, we need to say a little more about the detail of his treatment. The characteristic form of argument that is deployed in the calculus seems *prima facie* to involve the idea of a real number quantity *tending* to a limit. The method Weierstrass used for representing such arguments replaces this locution with quantified variables. The key point to note, of course, is that the word 'variable' here is now to be understood in the sense in which it is used in logic, and there is no longer any requirement to conceive of a quantity as actually *varying*. This is what is usually known as the *epsilon-delta* method, so called because in textbook expositions it is traditional for epsilon and delta to be used as the variables. Thus Weierstrass treated the expression

$$f(x) \text{ tends to } f(a) \text{ as } x \text{ tends to } a$$

as *meaning* that

$$(\forall \varepsilon > 0)(\exists \delta > 0)(\forall x \in \mathbf{R})(|x - a| < \delta \Rightarrow |f(x) - f(a)| < \varepsilon).$$

The epsilon-delta method is thus what finally allowed mathematicians to banish the practice of glossing the definition of continuity as

$$f(x) \text{ is infinitesimally close to } f(a) \text{ when } x \text{ is infinitesimally close to } a.$$

This talk of infinitesimals had been one of the primary targets of Berkeley's ridicule in *The Analyst*; but however much they tried to avoid it, mathematicians had not until Weierstrass seen how to eliminate it completely. Now they could, and did. For 80 years after Weierstrass it was a commonplace taught to undergraduates that talk of infinitesimals was at best a non-rigorous heuristic for proof-searching and a quaint throwback to an earlier and more primitive era, at worst a dangerous invitation to serious logical error.

But then, in the 1960s, the subject now known as *non-standard* analysis grew up, principally at the hands of Abraham Robinson. What he showed was that there is a conception of the continuum — let us call it the *non-standard* conception — according to which there *are* infinitesimal quantities lying in it; and that this conception is just as consistent as the standard Weierstrassian one.

The development of non-standard analysis has not in fact broken the stranglehold of classical analysis to any significant extent, but that seems to be a matter of taste and practical utility rather than of necessity. Moreover, the development of non-standard analysis has led to some revisions in pre-Weierstrassian history. The most striking of these concerns a proof Cauchy gave in his *Cours d'Analyse* (1821) that if a sequence of continuous functions converges pointwise to a limit function f, then the limit function is also continuous. When it is represented in epsilon-delta form, Cauchy's proof is exposed as fallacious, since it then involves an illicit inversion of the order of two of the quantifiers occurring in it. Indeed it is easy to come up with examples showing that Cauchy's result cannot be true. In order to make it correct, we need to replace the reference to pointwise convergence with the stronger notion of *uniform* convergence. The difference between the two lies purely in the order of the quantifiers in their epsilon-delta definitions; and it is lack of attention to just this that Cauchy stands accused of.

Thus the standard post-Weierstrassian view. But there is something very suspect about it. The counterexamples one has to construct to show that Cauchy's result is fallacious are extremely simple. It is enough to note, for instance, that as n tends to infinity,

$$x^n \text{ tends to } \begin{cases} 0 & \text{if } 0 \leqslant x < 1 \\ 1 & \text{if } x = 1 \end{cases}$$

and the limit function is obviously discontinuous at $x = 1$. It beggars belief that a mathematician of Cauchy's ability would not have noticed this.

A possible explanation for this puzzle has been provided by Lakatos (1978). He seized on the striking fact that there is a way of interpreting Cauchy's result as *true* about functions defined not on the standard but on the non-standard continuum, and used this to argue that Cauchy was guilty of no logical fallacy but was simply talking about a different subject from Weierstrass. The suggestion, then, is that the situation in analysis is analogous to that in the study of aggregation which we described in part I. In both cases mathematicians were working with an inchoate blend of properties of two distinct (but related) concepts. The point to note here is the common role played in each narrative by the process of axiomatization, which serves to clarify concepts by enunciating their properties.

But even when the concept of the continuum that is in play has been clarified sufficiently, a question still remains as to whether there is anything that answers to it. According to Kant, our knowledge of the truths of analysis is derived ultimately from our knowledge of the spatio-temporal structure of experience. A Weierstrassian axiomatization of analysis does nothing to challenge this account unless it can be shown how it is possible to construct objects satisfying the axioms without recourse to that same spatio-temporal structure. The research programme that attempts to do this has dominated work in the foundations of mathematics ever since Weierstrass.

The first step in executing this programme is nowadays known as *arithmetization* and consists in the construction of a complete ordered field within the theory of sets, taking the natural numbers as given. This step was carried out by Cantor (1872), Dedekind (1872), Heine (1872), Méray (1872), and others, using several quite distinct methods.

But once again this postpones the Kantian problem rather than solving it. We still need to ground the knowledge of the natural numbers that these constructions presuppose. This next step, which might be called the *set-theorization* of arithmetic, is therefore to construct the natural numbers in pure set theory. Once again this can be done in quite distinct ways. The first construction to be carried out in full detail is contained in Dedekind's *Was sind und was sollen die Zahlen?* (1888), but Frege's *Die Grundlagen der Arithmetik* (1884) contains a sketch from which another workable treatment can be derived.

As is so often the case in foundational matters, however, the logical direction is the opposite of the direction of discovery, and we shall here follow the former rather than the latter. Thus we begin this part of the book with Dedekind's construction of the natural numbers and end with the real numbers. If we chain these together with the Weierstrassian account of the calculus, we obtain an embedding of real analysis in the theory of sets. We shall leave to the end a discussion of the significance of this embedding.

Chapter 5

Arithmetic

We have to show that the axioms of ZU guarantee the existence of structures with the familiar properties we expect the natural, rational and real numbers to have. In this chapter we make a start on this project by considering the natural numbers.

5.1 Closure

We focus first on the problem of identifying a structure which has all the right *algebraic* properties for the natural numbers. (We defer consideration of their order properties to the next chapter.) Dedekind's original (1888) treatment of this problem has scarcely been bettered since, and so we shall follow it quite closely here. The key to Dedekind's treatment is something that he called a 'chain' (*Kette*) but is nowadays more usually called a *closure*.

Definition. *Suppose that r is a relation on a set A. A subset B of A is said to be r-*closed *if $r[B] \subseteq B$.*

As an example, consider the set of all adherents of a religion which bans mixed marriages: this set is supposed to be closed under the relation of marriage since anyone married to an adherent of the religion is also an adherent.

Consider now the problem of finding the smallest closed set containing a given set. If our starting set is not closed, we might not get a closed set simply by adding all the relatives of its members, since some of them might have other relatives not yet included. To get closure, we need to add in all their relatives, all *their* relatives, and so on *ad infinitum*. But what does 'and so on' mean here? We might be tempted to gloss it as 'for any finite number of iterations', but that would, as Dedekind remarked in a letter to Keferstein, 'contain the most pernicious and obvious kind of vicious circle' (van Heijenoort 1967, pp. 100–01), since it is our intention to use this notion to *define* what we mean by 'finite'. Dedekind's solution to this problem was to see that there is another, quite different way of specifying what is meant by the closure of a set. Instead of working from the inside out, adding more and more things until we get a closed set, we can operate in the other direction, taking the intersection of all the closed sets which contain our starting set.

Definition. *Suppose that r is a relation on a set A. The intersection of all the r-closed subsets of A which contain B will be denoted $\mathrm{Cl}_r(B)$ and called the r-*closure *of B.*

What makes this definition work is that any intersection of r-closed sets is r-closed. It follows from this that $\mathrm{Cl}_r(B)$, which is by definition contained in every r-closed set containing B, is itself r-closed, and hence must be the smallest r-closed set containing B, as we wanted. Evidently, then, B is r-closed iff $\mathrm{Cl}_r(B) = B$. This is all utterly trivial, but it is the key to the treatment that follows.

(5.1.1) **Proposition.** *Suppose that r is a relation on A and $B \subseteq A$.*

(a) $\mathrm{Cl}_r(B) = B \cup \mathrm{Cl}_r(r[B])$;

(b) $\mathrm{Cl}_r(r[B]) = r[\mathrm{Cl}_r(B)]$.

Proof of (a). Clearly $B \subseteq \mathrm{Cl}_r(B)$. Moreover $r[B] \subseteq \mathrm{Cl}_r(B)$, and so $\mathrm{Cl}_r(r[B]) \subseteq \mathrm{Cl}_r(B)$. Hence $B \cup \mathrm{Cl}_r(r[B]) \subseteq \mathrm{Cl}_r(B)$. Now $r[\mathrm{Cl}_r(r[B])] \subseteq \mathrm{Cl}_r(r[B])$ and $r[B] \subseteq \mathrm{Cl}_r(r[B])$, so that

$$\begin{aligned} r[B \cup \mathrm{Cl}_r(r[B])] &= r[B] \cup r[\mathrm{Cl}_r(r[B])] \\ &\subseteq \mathrm{Cl}_r(r[B]) \\ &\subseteq B \cup \mathrm{Cl}_r(r[B]), \end{aligned}$$

and therefore $B \cup \mathrm{Cl}_r(r[B])$ is r-closed. But $B \subseteq B \cup \mathrm{Cl}_r(r[B])$, and so $\mathrm{Cl}_r(B) \subseteq B \cup \mathrm{Cl}_r(r[B])$.

Proof of (b). $\mathrm{Cl}_r(B)$ is r-closed, hence so is $r[\mathrm{Cl}_r(B)]$. But $B \subseteq \mathrm{Cl}_r(B)$, so that $r[B] \subseteq r[\mathrm{Cl}_r(B)]$, and therefore $\mathrm{Cl}_r(r[B]) \subseteq r[\mathrm{Cl}_r(B)]$. Now $r[B] \subseteq \mathrm{Cl}_r(r[B])$ and $r[\mathrm{Cl}_r(r[B])] \subseteq \mathrm{Cl}_r(r[B])$, and so

$$\begin{aligned} r[\mathrm{Cl}_r(B)] &= r[B \cup \mathrm{Cl}_r(r[B])] \text{ by (a)} \\ &= r[B] \cup r[\mathrm{Cl}_r(r[B])] \\ &\subseteq \mathrm{Cl}_r(r[B]). \end{aligned}$$

□

If $x \in A$, we shall sometimes write $\mathrm{Cl}_r(x)$ instead of the lengthier $\mathrm{Cl}_r(\{x\})$.

5.2 Definition of natural numbers

With Dedekind's theory of chains at our disposal it is now a simple matter to state the basic properties on which our development of the natural numbers will be based. What Dedekind took as fundamental was the function which takes each natural number to its successor: the properties he required this function to have are that no two numbers have the same successor (i.e. the successor function is one-to-one), that zero is not the successor of any natural number, and that every number is in the closure of zero with respect to the successor function. We shall call a structure which has these properties a 'Dedekind algebra' in his honour.

Definition. A Dedekind algebra *is a structure* (A, f) *such that* f *is a one-to-one function from* A *to itself and* $A = \mathrm{Cl}_f(a)$ *for some* $a \in A \setminus f[A]$.

It is obvious that if (A, f) is a Dedekind algebra, then A is is infinite. By using the technique of closures, we can easily show, conversely, that any infinite set contains a Dedekind algebra.

(5.2.1) **Theorem.** *There exists a (pure) Dedekind algebra.*

Proof. Suppose that B is any (pure) infinite set. (The existence of such sets is guaranteed by the axiom of infinity.) Then there is by hypothesis a one-to-one function g from B into itself that is not onto. So B has an element a which is not in the image of g. If we let $A = \mathrm{Cl}_g(a)$ and $f = g|A$, it is evident that (A, f) is a Dedekind algebra. □

This theorem is needed so that we can justify the following definition.

Definition. *We choose once for all some (pure) Dedekind algebra (of lowest possible birthday) and denote it* $(\boldsymbol{\omega}, \mathrm{s})$. *The elements of* $\boldsymbol{\omega}$ *are called* natural numbers*: the function* s *is called the* successor function *on* $\boldsymbol{\omega}$.

The parenthetical stipulation that the Dedekind algebra we choose should be pure and of lowest possible birthday is of no great mathematical import, but it does at least prevent Julius Caesar from being a natural number, an outcome of notorious concern to Frege (1884). Whether a set is the carrier set of a Dedekind algebra is solely a matter of cardinality, and the lowest level at which a pure set of the appropriate size occurs is the first limit level. So the parenthetical stipulation has the effect of fixing the level at which the set of natural numbers occurs in the hierarchy. Nonetheless, I should stress that these stipulations are still very far from determining even the carrier set uniquely, let alone the successor function: these choices are wholly conventional.

Mathematicians are inclined to avoid such vagueness by fixing on one and ignoring any rivals. One popular choice is to make $\boldsymbol{\omega}$ the closure of $\emptyset$ under the operation $a \mapsto a \cup \{a\}$. But defining the natural numbers in this way makes many more things true about them than we want: it becomes the case that $\emptyset \in 2$, for instance. This is of course another instance of the problem of doing too much that we met in §4.6 when we defined the ordered pair. The way out we took then was to give an explicit definition but make clear that we did not claim to be defining the notion of ordered pair of which we had a prior grasp. We could evidently do the same here: use the standard explicit definition just mentioned but make clear that what we have defined are not *really* natural numbers (since *of course* the empty set is not a member of the number two). For purely mathematical purposes that might do well enough (which is presumably why mathematicians do it), but it would scarcely help us in the project we are engaged in here. What allowed us to adopt Kuratowski's ersatz ordered pairs with a tolerably clear conscience was that we did not pretend our aim in this book was to provide an account of the genuine article. But when it comes to numbers, things stand differently. We *are* concerned here with a foundational project: we aim to assess, and hence should try to present the best version of, the proposal that set theory could serve as a foundation for arithmetic, in the sense of explicitly supplying its content, or at any rate of explaining our knowledge of it.

So what else can we do? One possibility would be to add a new primitive, 's', to our language, define $\boldsymbol{\omega}$ to be its domain, and add the stipulation that $(\boldsymbol{\omega}, \mathrm{s})$ is a Dedekind algebra to our system as an extra axiom. This procedure would get round the problem of doing too much, but was famously stigmatized by Russell for the opposite vice of doing too little and hence having 'the advantages of theft over honest toil' (1919, p. 71). Certainly if we planned to use the device of postulation as a means of avoiding the tedious business of proving theorem 5.2.1, Russell's jibe would be entirely fair. But what if we agreed to adopt the method of postulation *only* in cases where a proof is available that the procedure is conservative? In that case there would no longer be a sense of having got something for nothing, since we should still have to supply our proof of the theorem, slightly repackaged now in metalinguistic form as a proof of the conservativeness of the postulation. The problem is rather that we still owe an explanation of why a proof of conservativeness is enough. We need to explain, that is to say, how the new primitive we introduce gains a meaning if it is conservative (and, equally, how it fails to gain a meaning if not).

The formal development that follows is neutral between these options, of course. All we shall need in proofs is that $(\boldsymbol{\omega}, \mathrm{s})$ is a Dedekind algebra, and it will not matter which device was used to achieve this.

By definition there is an element $0 \in \boldsymbol{\omega} \setminus \mathrm{s}[\boldsymbol{\omega}]$ such that $\boldsymbol{\omega} = \mathrm{Cl}_{\mathrm{s}}(0)$. Moreover,

$$\begin{aligned} \boldsymbol{\omega} &= \mathrm{Cl}_{\mathrm{s}}(0) \\ &= \{0\} \cup \mathrm{Cl}_{\mathrm{s}}(\mathrm{s}(0)) \text{ [proposition 5.1.1(a)]} \\ &= \{0\} \cup \mathrm{s}[\mathrm{Cl}_{\mathrm{s}}(0)] \text{ [proposition 5.1.1(b)]} \\ &= \{0\} \cup \mathrm{s}[\boldsymbol{\omega}], \end{aligned}$$

and so if $0' \in \boldsymbol{\omega} \setminus \mathrm{s}[\boldsymbol{\omega}]$, then $0 = 0'$. In other words, 0 is the unique s-minimal element of ω.

Let us also temporarily define $n + 1 = s(n)$. This is a special case of a more general notation we shall be introducing in §5.4. Using it, the defining properties of the set ω of natural numbers take the following more familiar form:

(1) $m + 1 = n + 1 \Rightarrow m = n$ for all $m, n \in \omega$.

(2) $0 \neq n + 1$ for any $n \in \omega$.

(3) If $0 \in A \subseteq \omega$ and if $(\forall n)(n \in A \Rightarrow n + 1 \in A)$, then $A = \omega$.

It is a remarkable fact that all the arithmetical properties of the natural numbers can be derived from such a small number of assumptions. The last of them is standardly known as the *principle of mathematical induction*. For convenience we shall restate it in a slightly different form.

(5.2.2) **Simple induction scheme.** *If $\Phi(n)$ is a formula,*

$$(\Phi(0) \text{ and } (\forall n \in \omega)(\Phi(n) \Rightarrow \Phi(n + 1))) \Rightarrow (\forall n \in \omega)\Phi(n).$$

Proof. Let $A = \{n \in \omega : \Phi(n)\}$ and apply (3) above. □

It is worth emphasizing that we are licensed by this scheme to substitute for Φ *any* formula in the language of set theory, even one that ineliminably involves reference to non-arithmetical entities such as real numbers or sets of higher level. We shall discuss the effects of this licence at various points in what follows.

The elementary arithmetic of the natural numbers has of course been known for thousands of years — Euclid (c. 300 B.C.) describes elementary number theory (divisibility, prime numbers, etc.) in some detail — but the centrality of induction did not emerge until much more recently. Indeed induction seems hardly to have been used explicitly as a method of proof until the mid-17th century (Pascal 1665); what older texts contain instead are much vaguer instructions to repeat an argument 'as many times as required' or illustrations of general reduction methods in particular cases.

5.3 Recursion

A family whose indexing set is ω (or, what is the same thing, a function whose domain is ω) is often called an *infinite sequence* (or sometimes just a *sequence*). There is a standard method of defining a sequence by what is known as *recursion*: this consists in defining the initial term x_0 of the sequence and then defining x_{n+1} in terms of x_n. This method is very widely used in mathematics: often it is signposted by the use of temporal language. One might, for instance, find a mathematician saying

Start by letting $x_0 = a$; then once x_n has been defined, let $x_{n+1} = f(x_n)$.

But what justifies this method of definition? It plainly has something to do with the principle of mathematical induction, but it is not quite the same thing, since induction is a method of *proof*, recursion a method of *definition*. The justification for such definitions is supplied by the following theorem.

(5.3.1) **Simple recursion principle (Dedekind 1888).** *If A is a set, f is a function from A to itself, and a is a member of A, then there exists exactly one sequence (x_n) in A such that $x_0 = a$ and $x_{n+1} = f(x_n)$ for all $n \in \boldsymbol{\omega}$.*

Existence. We can define a function $s \times f$ on $\boldsymbol{\omega} \times A$ by

$$(\mathrm{s} \times f)(n, x) = (\mathrm{s}(n), f(x)).$$

Let $g = \mathrm{Cl}_{\mathrm{s}\times f}((0, a)) \subseteq \boldsymbol{\omega} \times A$ and let

$$B = \{n \in \boldsymbol{\omega} : \text{there is exactly one } x \in A \text{ such that } n \; g \; x\}.$$

Then it is a straightforward exercise to show that $0 \in B$ and that B is s-closed. Hence by induction $B = \boldsymbol{\omega}$; i.e. g is a function from $\boldsymbol{\omega}$ to A. It follows at once that it is the sequence we require.

Uniqueness. Suppose g, g' are functions from $\boldsymbol{\omega}$ to A such that $g(0) = g'(0)$, $g(n + 1) = f(g(n))$ and $g'(n + 1) = f(g'(n))$ for all $n \in \boldsymbol{\omega}$. If $C = \{n \in \boldsymbol{\omega} : g(n) = g'(n)\}$, then C is s-closed and $0 \in C$, so that $C = \boldsymbol{\omega}$ by induction. Hence $g = g'$. □

The sequence (x_n) is said to be *defined by recursion* from the equations

$$x_0 = a$$
$$x_{n+1} = f(x_n).$$

(5.3.2) **Corollary.** *Every Dedekind algebra is isomorphic to $(\boldsymbol{\omega}, \mathrm{s})$.*

Proof. Suppose that (A, f) is a Dedekind algebra. The same argument which showed in the proof of theorem 5.3.1 that $g = \mathrm{Cl}_{\mathrm{s}\times f}((0, a))$ is a function from $\boldsymbol{\omega}$ to A shows that its inverse is a function from A to $\boldsymbol{\omega}$. Since $g(\mathrm{s}(n)) = f(g(n))$ for all n, it follows that g is an isomorphism between $(\boldsymbol{\omega}, \mathrm{s})$ and (A, f). □

Occasionally we may wish to define a sequence (x_n) in such a way that the definition of x_{n+1} in terms of x_n depends on the value of n. The simple recursion principle as we have stated it does not justify this, but an easy extension of it does.

(5.3.3) **Simple recursion principle with a parameter.** *If A is a set, (f_n) is a sequence of functions from A to A, and a is a member of A, then there exists a unique sequence (x_n) in A such that $x_0 = a$ and $x_{n+1} = f_n(x_n)$ for all $n \in \boldsymbol{\omega}$.*

Proof. If we define a function h from $\boldsymbol{\omega} \times A$ to $\boldsymbol{\omega} \times A$ by

$$h(n, x) = (n + 1,\ f_n(x)),$$

then by simple recursion we can obtain a sequence (m_n, x_n) such that

$$\begin{aligned}(m_0, x_0) &= (0, a)\\(m_{n+1}, x_{n+1}) &= h(m_n, x_n).\end{aligned}$$

But this is equivalent to specifying that

$$\begin{aligned}m_n &= n\\x_0 &= a\\x_{n+1} &= f_n(x_n).\end{aligned}$$

The result follows. □

As an example of definition by recursion, suppose that r is a relation on a set A. The relation r^n, called the nth *iterate* of r on A, is defined recursively by the equations

$$r^0 = \mathrm{id}_A, \quad r^{n+1} = r \circ r^n.$$

Note that if we apply this notation to the successor function itself we can prove by induction that

$$\mathrm{s}^n(0) = n \text{ for all } n. \tag{1}$$

For

$$\mathrm{s}^0(0) = \mathrm{id}_{\boldsymbol{\omega}}(0) = 0,$$

and if $\mathrm{s}^n(0) = n$, then

$$\mathrm{s}^{n+1}(0) = \mathrm{s}(\mathrm{s}^n(0)) = \mathrm{s}(n) = n + 1.$$

In words rather than symbols, this says that the number n is (as we would expect) the result of applying the successor function n times to 0.

The recursion principle is one of Dedekind's most impressive achievements, principally because it was not obvious until he proved it that there was really anything substantial to prove. For it is tempting to suppose that definition by recursion can be justified by a much more straightforward inductive argument as follows.

We wish to show that a sequence (x_n) can be defined in such a way that

$$x_0 = a$$
$$x_{n+1} = f(x_n).$$

Now plainly x_0 can be defined. And if x_n can be defined, then so can x_{n+1}. Hence by induction x_n can be defined for all $n \in \omega$.

It was Dedekind's achievement to see that this argument is fallacious. The fallacy is that it involves an application of induction to the formula 'x_n can be defined'. But in advance of knowing that the *whole* sequence (x_n) can be defined (which is what we are trying to prove), we are not in a position to know what it means to say that 'x_n can be defined' for any particular n.

The error, although perhaps obvious enough when it is pointed out, is subtle enough to have fooled some able mathematicians, even ones writing after Dedekind. Landau made it in the first draft of his *Grundlagen der Analysis* and had to have his mistake pointed out to him by a colleague (see Landau 1930, p. ix). And Peano seems to have been quite oblivious to the issue: in 1889 he blithely defines addition and multiplication recursively without giving any justification, and even as late as 1921 he could mention such examples in the course of a discussion of the nature of definitions in mathematics and yet remain silent about the problem of proving the existence of the functions in question.

Exercises

1. Show that there is no sequence (A_n) of sets such that $A_{n+1} \in A_n$ for all $n \in \omega$.
2. Fill in the details in the proof of Dedekind's simple recursion principle.
3. (a) If r and s are relations such that $r \circ s = s \circ r$, show that $r^n \circ s^m = s^m \circ r^n$ for all $m, n \in \omega$ and that $(r \circ s)^n = r^n \circ s^n$ for all $n \in \omega$.

   (b) If r is a relation, show that $(r^m)^n = (r^n)^m$ for all $m, n \in \omega$.
4. Fill in the details in the proof of corollary 5.3.2.

5.4 Arithmetic

Now that we have Dedekind's recursion principle, it is a straightforward matter to give a recursive definition of the familiar algebraic operation of addition on ω.

Definition. *The* addition *function* $(m, n) \mapsto m+n$ *from* $\omega \times \omega$ *to* ω *is defined by the recursion equations*

$$m + 0 = m \tag{2}$$
$$m + \mathrm{s}(n) = \mathrm{s}(m + n). \tag{3}$$

If we also define $1 = \mathrm{s}(0)$, this definition gives us

$$\mathrm{s}(n) = \mathrm{s}(n + 0) = n + \mathrm{s}(0) = n + 1$$

(thus coinciding, as promised, with the notation we introduced in §5.2), and (3) attains the more familiar form

$$m + (n + 1) = (m + n) + 1. \tag{4}$$

Note also that if r is a relation on a set A, then

$$r^{m+n} = r^n \circ r^m. \tag{5}$$

For

$$r^{m+0} = r^m = \mathrm{id}_A \circ r^m = r^0 \circ r^m;$$

and if $r^{m+n} = r^n \circ r^m$, then

$$\begin{aligned} r^{m+(n+1)} = r^{(m+n)+1} = r \circ r^{m+n} = r \circ (r^n \circ r^m) \\ = (r \circ r^n) \circ r^m = r^{n+1} \circ r^m, \end{aligned}$$

whence the result. In the case where r is the successor function, this gives by (1)

$$m + n = \mathrm{s}^{m+n}(0) = \mathrm{s}^n(\mathrm{s}^m(0)).$$

This last equation has the appearance at first sight of being usable as an *explicit* definition of addition, but of course it does not circumvent the appeal to Dedekind's recursion principle, as that was needed to justify the definition of the iterate s^n in the first place. (Incidentally, readers of the *Tractatus* will notice a formal similarity between (5) and the explicit definition of addition proposed there.)

All the familiar properties of addition of natural numbers follow from (2) and (3). For example, if we define $2 = 1 + 1$, $3 = 2 + 1$ and $4 = 3 + 1$, it is reassuring to discover that

$$2 + 2 = 2 + (1 + 1) = (2 + 1) + 1 = 3 + 1 = 4.$$

General properties of addition are proved by induction. To prove the associative law

$$k + (m + n) = (k + m) + n,$$

for example, we use induction on n, noting first that

$$k + (m + 0) = k + m = (k + m) + 0 \text{ by (2)},$$

so that the result is true for 0. And if it is true for n, then

$$\begin{aligned} k + (m + (n+1)) &= k + ((m+n)+1) \text{ by (3)} \\ &= (k + (m+n)) + 1 \text{ by (3)} \\ &= ((k+m)+n) + 1 \text{ by the induction hypothesis} \\ &= (k+m) + (n+1) \text{ by (3)}, \end{aligned}$$

so that it is also true for $n + 1$. It follows by induction that the result is true for all n.

Definition. *The* multiplication *function* $(m, n) \mapsto mn$ *from* $\boldsymbol{\omega} \times \boldsymbol{\omega}$ *to* $\boldsymbol{\omega}$ *is defined by the recursion equations*

$$m0 = 0 \tag{6}$$

$$m(n+1) = mn + m. \tag{7}$$

Note that if r is a relation on a set A,

$$r^{mn} = (r^m)^n. \tag{8}$$

For

$$r^{m0} = r^0 = \mathrm{id}_A = (r^m)^0,$$

and if $r^{mn} = (r^m)^n$, then

$$r^{m(n+1)} = r^{mn+m} = r^m \circ r^{mn} = r^m \circ (r^m)^n = (r^m)^{n+1},$$

whence the result. Once again, readers of the *Tractatus* will notice how this could be used as a definition if the method of recursion were presupposed. In the case where r is the successor function, (8) gives

$$mn = \mathrm{s}^{mn}(0) = (\mathrm{s}^m)^n(0).$$

Definition. *The* exponentiation *function* $(m, n) \mapsto m^n$ *is defined by the recursion equations*

$$m^0 = 1 \tag{9}$$

$$m^{n+1} = m^n m. \tag{10}$$

Various other arithmetical functions can be defined by the same method. For instance, we can define a function $n \mapsto 2_n$ by the equations

$$2_0 = 1$$

$$2_{n+1} = 2^{2_n}.$$

This function arises naturally in complexity theory: it provides a compact notation for expressing very large numbers, since

$$2_n = 2^{2^{2^{\cdot^{\cdot^{\cdot^{2}}}}}} \Big\} n \text{ exponents}$$

is enormous even for quite small values of n. Another example is the *factorial* function $n \mapsto n!$, defined by the recursion equations

$$0! = 1$$
$$(n+1)! = n!(n+1).$$

This is noteworthy because it is our first example of a definition by recursion involving a parameter.

Now that we have defined all the arithmetical operations, it would be entirely a matter of routine to derive their elementary properties from the definitions, and for that reason we shall omit the details. The proofs were first carried out by Grassmann (1861).

Exercises

1. Prove the following results by induction on n:
   (a) $1 + n = n + 1$.
   (b) $m + n = n + m$.

2. (a) If the function f from A to itself is one-to-one, prove that the nth iterate f^n is also one-to-one for all $n \in \boldsymbol{\omega}$.
   (b) Prove that $k + n = k + m \Rightarrow n = m$.

5.5 Peano arithmetic

In this chapter we have sketched how to embed arithmetic in the theory of sets. One thing worth stressing about this development is how few assumptions we need to make in order to ground it. Another is that the assumptions in question can be stated as a self-standing theory quite independent of the theory of sets. In order to state this theory, we need a language which has a unary function symbol S, but we shall also (for convenience — we could eliminate it if we wanted) suppose that we have a primitive constant 0. The axioms are then as follows:

$$(\forall x)(\forall y)(Sx = Sy \Rightarrow x = y)$$

$$(\forall x)Sx \neq 0$$

$$(\forall X)((X(0) \text{ and } (\forall x)(X(x) \Rightarrow X(Sx))) \Rightarrow (\forall x)Xx).$$

This theory is called *second-order Peano arithmetic* and denoted **PA2**. Expressed in these terms our definition in §5.2 amounts to the stipulation that $(\boldsymbol{\omega}, \mathrm{s}, 0)$ should be a model of **PA2**, and corollary 5.3.2 is the statement that **PA2** is categorical.

If we intended an independent development of this theory, however, we might be troubled by the fact that the third axiom, the induction axiom, is second-order. If we were determined that our theory should be first-order, we would be forced to replace it with its first-order surrogate,

$$(\Phi(0) \text{ and } (\forall x)(\Phi(x) \Rightarrow \Phi(Sx))) \Rightarrow (\forall x)\Phi(x)$$

for any formula Φ. What results from substituting this first-order scheme for the second-order induction axiom is known as the *elementary theory of the successor function*. It is used as a standard example in textbooks on model theory, which often include the proof that it is complete, i.e. that there is nothing to be said in a first-order language about the successor function that does not follow from these axioms. It follows immediately from this completeness result that the elementary theory of the successor function is decidable. Textbooks also contain the proof that the theory is not finitely axiomatizable.

But the particular axiom scheme we have given is by no means inevitable. We could, for instance, have substituted for it the axioms

$$(\forall x)(\exists y)(x = 0 \text{ or } x = Sy)$$

$$(\forall x)\underbrace{SS\ldots S}_{n \text{ terms}}x \neq x \quad \text{for every } n \geqslant 1.$$

The fact that this substitution is possible should arouse the suspicion that the elementary theory of the successor function is actually rather weak. And so it proves. The trouble is that in the theory as now constituted we are asserting only those instances of induction that can be expressed in a language containing 0 and S, and it turns out that hardly anything of mathematical interest *can* be so expressed. In particular, the recursive definitions of addition given in §5.4 cannot be reproduced in this first-order theory.

The solution is to add addition as a new primitive function and its recursion equations as axioms.

$$(\forall x)\, x + 0 = x$$
$$(\forall x)(\forall y)\, x + (y + S0) = (x + y) + S0.$$

This theory is known as *Presburger arithmetic* after the person who showed that it is complete (Presburger 1930). It is still not finitely axiomatizable, but although significantly stronger than the elementary theory of the successor function, it once again turns out to be too weak to capture arithmetical practice — too weak, in fact, to define multiplication.

So we need to add multiplication as a primitive function too, together with its recursion equations.

$$(\forall x)\, x0 = 0;$$
$$(\forall x)(\forall y)\, x(y + S0) = xy + x.$$

Now, finally, we are in business. The first-order theory that results from this extension, which is called *Peano arithmetic* or PA, is distinctly non-trivial.

Nevertheless, it follows easily from the compactness theorem that PA is not categorical, and hence that there are true second-order sentences (such as the induction axiom itself) which it does not entail. Not only that, but there are even true sentences in the first-order language of arithmetic itself which PA does not entail (Gödel 1931). On the other hand, it is surprising, after our cautionary experience with addition and multiplication, to discover that we do not need to posit an endless succession of new primitives, one for exponentiation, one for the factorial function, and one for every other primitive recursive function: once we have addition and multiplication, all the other primitive recursive functions are definable in PA (Gödel 1931). It is above all this remarkable stability that makes PA seem to be something more than a merely accidental fragment of the second-order theory PA2.

What gives Peano arithmetic this (relative) strength is the complexity generated by the *interplay* between addition and multiplication in the induction scheme. As we have just noted, the elementary theory of addition is complete and therefore weaker than PA. *Skolem arithmetic*, the elementary theory of multiplication on its own (i.e. without + or S), is also complete (Skolem 1931) and even finitely axiomatizable (Cegielski 1981), so it is again significantly weaker than PA.

PA itself, by contrast, suffices to derive a large amount of highly non-trivial mathematics. Consider, for example, the part of arithmetic known as number theory. This begins with the definition of divisibility: the formula $(\exists r \in \omega)(n = mr)$ is written $m \mid n$ and read 'm divides n'. If m and n are natural numbers, then the least natural number they both divide is called their *least common multiple* and denoted $\mathrm{lcm}(m, n)$; the greatest number which divides both of them is called their *greatest common divisor* and denoted $\gcd(m, n)$. We say that m and n are *coprime* if $\gcd(m, n) = 1$. As an illustration of how results in elementary number theory can be established, let us show now that for any $m, n \in \omega$ we have

$$mn = \mathrm{lcm}(m, n)\gcd(m, n). \tag{11}$$

This is trivial if $\gcd(m, n) = 0$, since then $m = n = 0$. So suppose $\gcd(m, n) \neq 0$. Then certainly $\gcd(m, n)|n$, and so

$$m = \frac{m}{\gcd(m, n)}\gcd(m, n) \mid \frac{mn}{\gcd(m, n)}.$$

But equally

$$n \mid \frac{mn}{\gcd(m, n)},$$

whence

$$\operatorname{lcm}(m, n) \mid \frac{mn}{\gcd(m, n)}.$$

Now there exist $r, s \in \omega$ such that $\operatorname{lcm}(m, n) = mr = ns$. Hence $\frac{m}{s} = \frac{n}{r}$. But $\frac{m}{s} \mid m$ and $\frac{n}{r} \mid n$, so $\frac{m}{s} \mid \gcd(m, n)$, whence

$$\frac{mn}{\gcd(m, n)} = \frac{ns}{\gcd(m, n)} \frac{m}{s} \mid ns = \operatorname{lcm}(m, n).$$

Equation (11) now follows.

The purpose of rehearsing this proof here is to make the point that the reasoning involved is entirely first-order, and hence that equation (11) is not merely true but provable in the first-order theory PA. It is important to realize, though, that there are no general guarantees. Whenever we are confronted with a proof of an arithmetical proposition, there will be a question whether the properties the induction principle is applied to in the course of the proof can be defined using first-order arithmetical formulae. If they cannot, the result may fail to be a theorem of PA even though it is provable set-theoretically. Consider, for example, the sentence Con(PA) which formalizes the consistency of PA. This is not provable in PA, but it must have a set-theoretic proof, since we have proved in set theory that PA has a model.

Notes

The set-theoretic construction of the natural numbers we have given here is due to Dedekind (1888). The following year Peano (1889) converted it into an axiom system now known almost universally as Peano's axioms. In chapter 1 we described, even if we did not entirely explain, the rise of first-order logic as the supposed framework for rigorous mathematics. This made it natural for PA, the first-order fragment of Peano's axioms, to be the object of intensive study. For the results of this study consult Kaye 1991. What is much more recent is the suggestion advanced by Isaacson (1987; 1992) that the theoretical stability of PA is not an accident but reflects a correspondingly stable notion of arithmetical truth. The limitative results concerning subsystems of PA which we have mentioned here are explained further in Smoryński 1991.

The derivation of the basic properties of the natural numbers from Peano's axioms is laid out in detail by Henkin, Smith, Varineau and Walsh (1962). These properties are not the subject of active study nowadays, of course, but it is striking how similar they are in form to substantial problems in number theory such as Goldbach's conjecture or the twin-prime conjecture. The classical foundations of number theory are elegantly described by Hardy and Wright (1938). In the present context, however, it is of particular interest to focus on the branch of the subject known as analytic number theory, in which non-arithmetical methods are used to prove arithmetical theorems. One of the most famous instances is Dirichlet's theorem that if r and s are coprime, there are infinitely many prime numbers of the form $rn + s$. Dirichlet's (1837) proof of this uses elegant methods in complex function theory: for a modern presentation see Serre 1973. It is at first sight very surprising that analytic methods should be usable in this way, and it is natural to wonder whether they are eliminable. In this case they are: an elementary proof of Dirichlet's theorem was discovered by Selberg (1949). Nathanson 2000 is a good source of information on this issue.

Chapter 6

Counting

In this chapter we study those properties of numbers that depend not on the algebraic operations but on the order in which they are arranged. We shall start by setting up the terminology for talking about and classifying relations of order. Beyond that our objective will be to complete the unfinished business of the last chapter by defining an ordering on the natural numbers and showing how this ordering enables us to use the natural numbers in counting finite sets.

6.1 Order relations

Definition. *A relation* $\leqslant$ *on a set* A *is called a* weak partial ordering *if it is transitive, antisymmetric and reflexive on* A.

Definition. *A relation* $<$ *on a set* A *is called a* strict partial ordering *if it is transitive and irreflexive.*

There is a close relationship between weak partial orderings and strict partial orderings: if $\leqslant$ is a weak partial ordering on A, then the relation on A defined between x and y by the formula

$$x \leqslant y \text{ and } x \neq y$$

is a strict partial ordering on A; and if $<$ is a strict partial ordering on A, then the relation on A defined between x and y by the formula

$$x < y \text{ or } x = y$$

is a weak partial ordering on A. Moreover, these two operations are inverses of one another. What this amounts to is that it is a matter of indifference whether it is the weak partial ordering or its associated strict partial ordering that we specify. We shall therefore describe as a *partially ordered set* a structure in which the relation is *either* a weak partial ordering *or* a strict partial ordering. There is no confusing the two because a relation cannot be both a weak *and* a strict partial ordering (except in the trivial case of the empty relation on the empty set).

It is the *structure*, i.e. the ordered pair, that is called a partially ordered set. But mathematicians routinely talk of the carrier set A alone as if it were itself the structure. This is not strictly correct, since the set plainly does not encode any information about the identity of the partial ordering. It works only because in most of the cases one actually deals with it is obvious from the context which relation on A is intended. As an example of this, if the structure $(A, \leqslant)$ is a partially ordered set and $B \subseteq A$, then in any reference to B as a partially ordered set without further qualification it will invariably be $(B, \leqslant_B)$, i.e. B with the partial ordering it inherits from A by restriction, that is meant. The inverse of $\leqslant$ is of course denoted $\geqslant$, and its associated strict partial ordering is denoted $>$.

There is a rather large amount of terminology that needs to be introduced at this point. So suppose that $(A, \leqslant)$ is a partially ordered set and $B \subseteq A$. Two elements a and b of A are said to be *comparable* if either $a \leqslant b$ or $b \leqslant a$. An element $a \in A$ is called a *lower bound* [resp. a *strict lower bound*] for B if for every $x \in B$ we have $a \leqslant x$ [resp. $a < x$]. A lower bound for B which belongs to B is unique if it exists: it is then called the *least* element of B and denoted $\min B$. If an element b of B is $<$-minimal, i.e. if there does not exist $x \in B$ such that $x < b$, we say simply that it is *minimal* in B. A least element of B must be minimal in B; the converse is not true in general. [Strict] lower bounds and least and minimal elements with respect to the inverse partial ordering are called respectively [strict] *upper bounds* and greatest and maximal elements; the greatest element of B (if it exists) is denoted $\max B$. B is said to be *bounded* in A if it has both an upper and a lower bound in A. The least and greatest elements of A are sometimes denoted $\bot$ ('bottom') and $\top$ ('top') respectively. The least upper bound [resp. greatest lower bound] of B (if it exists) is denoted $\sup B$ [resp. $\inf B$].

A subset of A is $\leqslant$-closed in A iff it is $<$-closed, and in that case we say that it is *final* in A; we say that it is *coinitial* in A if its $\leqslant$-closure in A is A. Dually, a subset of A is *initial* in A if it is $\geqslant$-closed or equivalently if it is $>$-closed; it is *cofinal* in A if its $\geqslant$-closure is A.

A subset B of A is *convex* if whenever $x \leqslant y \leqslant z$ and $x, z \in B$ it follows that $y \in B$: this is the case iff B is the intersection of an initial and a final subset of A. We say that B is *dense* in A if whenever $x < z$ in A there exists $y \in B$ such that $x < y < z$.

It is worth singling out the case of the partial ordering on a collection of sets $\mathcal{A}$ defined by $A \subseteq B$; when we wish to regard $\mathcal{A}$ as a partially ordered set in this way, we refer to it as '$\mathcal{A}$ partially ordered by inclusion'. One important case of this is the power set $\mathfrak{P}(A)$. Its least element with respect to inclusion is $\varnothing$, and its greatest element is A. The least upper bound of a set $\mathcal{B} \subseteq \mathfrak{P}(A)$ is $\bigcup \mathcal{B}$, and the greatest lower bound is $\bigcap \mathcal{B}$ provided that $\mathcal{B} \neq \varnothing$; the set $\bigcap \varnothing$ does not exist since everything would have to belong to it; the greatest lower bound of $\varnothing$ in $\mathfrak{P}(A)$, on the other hand, is just $\mathfrak{P}(A)$ itself.

Another important case, with a vocabulary of its own, is that of the set $\mathrm{Po}(A)$ of all the partial orderings on A. The least element of $\mathrm{Po}(A)$ is the trivial partial ordering in which no element is less than any other: it is called the *total unordering* on A, and its associated weak partial ordering is just the diagonal relation on A. At the other extreme are the maximal elements of $\mathrm{Po}(A)$, i.e. the partial orderings on A which cannot be extended while remaining partial orderings. These are called *total orderings*, or sometimes just *orderings*, of A.

(6.1.1) **Proposition.** *A partial ordering on a set A is total iff any two elements of A are comparable.*

Necessity. Suppose $a, b \in A$ are not comparable with respect to $<$, i.e. $a \nleqslant b$ and $b \nleqslant a$. Define $x <' y$ iff *either* $x < y$ *or* $x \leqslant a$ and $b \leqslant y$. Now if $x <' y <' z$ then either (1) $x < y \leqslant a$ and $b \leqslant z$, or (2) $x \leqslant a$ and $b \leqslant y < z$, or (3) $x < y < z$; but in each case $x <' z$. So $<'$ is transitive. And it is obviously irreflexive. So it is a partial ordering on A which properly extends $<$, and hence $<$ is not total.

Sufficiency. Obvious. □

A collection $\mathcal{A}$ is called a *chain* if it is totally ordered by inclusion, i.e. if for all $A, B \in \mathcal{A}$ we have either $A \subseteq B$ or $B \subseteq A$.

Definition. *Suppose that $(A, \leqslant)$ and $(B, \leqslant)$ are partially ordered sets. A function f from A to B is said to be* increasing *if*

$$x \leqslant y \Rightarrow f(x) \leqslant f(y) \text{ for all } x, y \in A,$$

and strictly increasing *if*

$$x < y \Rightarrow f(x) < f(y) \text{ for all } x, y \in A.$$

[Strictly] increasing functions from A to B with the inverse ordering are called [strictly] decreasing.

Any one-to-one increasing function is strictly increasing, and any strictly increasing function is obviously increasing, but a strictly increasing function need not be one-to-one. (Consider a constant function defined on a totally unordered set.) Fortunately, though, a function is an isomorphism between the weak partially ordered sets $(A, \leqslant)$ and $(B, \leqslant)$ iff it is an isomorphism between the associated strictly partially ordered sets $(A, <)$ and $(B, <)$, so we can continue to be vague about that distinction when talking about isomorphisms between partially ordered sets as well.

Exercises

1. (a) If a partially ordered set $(A, \leqslant)$ has a least element, show that it is the unique minimal element of A.

(b) Is it true, conversely, that if A has a unique minimal element, then it is necessarily the least element of A?

2. Show that B is dense in the ordered set A iff $a = \sup\{x \in B : x < a\}$ for all $a \in A$.

3. If $(A, \leqslant)$ is totally unordered and $(B, \leqslant)$ is partially ordered, show that every function from A to B is strictly increasing. Provide an example of a strictly increasing one-to-one correspondence which is not an isomorphism.

6.2 The ancestral

If we have a relation which is not itself one of order, it may nevertheless be possible to extend it in such a way that the extended relation is an ordering. We shall now investigate the conditions under which this can occur.

Definition. *If r is a relation on A, the intersection of the transitive relations on A containing r is called the* strict ancestral *of r and denoted $r^{\mathbf{t}}$; the intersection of the reflexive transitive relations on A containing r is called the* weak ancestral *of r in A and denoted $r^{\mathbf{T}}$.*

The terminology arises from the case in which r is the relation of parenthood on the set of all humans, since then $r^{\mathbf{t}}$ is literally the ancestral relation.

It is easy to see that any intersection of [reflexive] transitive relations is also a [reflexive] transitive relation. It follows that the ancestral [resp. strict ancestral] of r is the smallest reflexive transitive relation [resp. transitive relation] on A containing r. The strict ancestral is consequently called the *transitive closure* by some authors, but I shall avoid this terminology here because it is widely used by set theorists for a different (though related) concept.

(6.2.1) **Proposition.** *Suppose that r is a relation on a set A.*

(a) $r^{\mathbf{T}} = r^{\mathbf{t}} \cup \mathrm{id}_A$.

(b) $r \circ r^{\mathbf{T}} = r^{\mathbf{t}} = r^{\mathbf{T}} \circ r$.

Proof of (a). Note first that $r^{\mathbf{t}} \subseteq r^{\mathbf{T}}$ and that a relation on A is reflexive iff it contains id_A, so that $\mathrm{id}_A \subseteq r^{\mathbf{T}}$. Thus $r \subseteq r^{\mathbf{t}} \cup \mathrm{id}_A \subseteq r^{\mathbf{T}}$. But $r^{\mathbf{t}} \cup \mathrm{id}_A$ is transitive and reflexive, so $r^{\mathbf{t}} \cup \mathrm{id}_A = r^{\mathbf{T}}$.

Proof of (b). Observe that $r \subseteq r \circ r^{\mathbf{T}}$ since $r^{\mathbf{T}}$ is reflexive, and that $r \circ r^{\mathbf{T}}$ is transitive, so that $r^{\mathbf{t}} \subseteq r \circ r^{\mathbf{T}}$. Moreover, if $x\ (r \circ r^{\mathbf{T}})\ z$, then there exists $y \in A$ such that $x\ r^{\mathbf{T}}\ y$ and $y\ r\ z$: we then have either $y\ r^{\mathbf{t}}\ z$ or $y = z$ by part (a), and in either case $x\ r^{\mathbf{t}}\ z$. In other words $r \circ r^{\mathbf{T}} \subseteq r^{\mathbf{t}}$, whence $r \circ r^{\mathbf{T}} = r^{\mathbf{t}}$. A very similar argument shows that $r^{\mathbf{t}} = r^{\mathbf{T}} \circ r$. □

(6.2.2) **Proposition.** *Suppose that r is a relation on a set A.*

(a) $r^{\mathbf{T}}[B] = \mathrm{Cl}_r(B)$ *for all* $B \subseteq A$.

(b) $r^{\mathbf{t}}[B] = r[\mathrm{Cl}_r(B)]$ *for all* $B \subseteq A$.

Proof of (a). Suppose first that $x \in r^{\mathbf{T}}[B]$ and $x\ r\ y$. Then there exists $a \in B$ such that $a\ r^{\mathbf{T}}\ x$. Also $x\ r^{\mathbf{T}}\ y$, and so $a\ r^{\mathbf{T}}\ y$ by transitivity, i.e. $y \in r^{\mathbf{T}}[B]$. This shows that $r^{\mathbf{T}}[B]$ is r-closed, whence $\mathrm{Cl}_r(B) \subseteq r^{\mathbf{T}}[B]$ since evidently $B \subseteq r^{\mathbf{T}}[B]$ by the reflexivity of $r^{\mathbf{T}}$. Now observe that the relation on A defined by the formula '$y \in \mathrm{Cl}_r(x)$' is reflexive and transitive and contains r: consequently $r^{\mathbf{T}}$ is contained in it. So if $y \in r^{\mathbf{T}}[B]$, then there exists $x \in B$ such that $x\ r^{\mathbf{T}}\ y$, hence $y \in \mathrm{Cl}_r(x) \subseteq \mathrm{Cl}_r(B)$. Therefore $r^{\mathbf{T}}[B] \subseteq \mathrm{Cl}_r(B)$, and so $r^{\mathbf{T}}[B] = \mathrm{Cl}_r(B)$.

Proof of (b).
$$\begin{aligned} r^{\mathbf{t}}[B] &= (r \circ r^{\mathbf{T}})[B] \text{ [proposition 6.2.1(a)]} \\ &= r[r^{\mathbf{T}}[B]] \\ &= r[\mathrm{Cl}_r(B)] \text{ by part (a).} \end{aligned}$$
□

(6.2.3) **Corollary.** *Suppose that r is a relation on a set A and $x, y \in A$.*

(a) $x\ r^{\mathbf{T}}\ y \Leftrightarrow \mathrm{Cl}_r(y) \subseteq \mathrm{Cl}_r(x)$.

(b) $x\ r^{\mathbf{t}}\ y \Leftrightarrow \mathrm{Cl}_r(y) \subseteq r[\mathrm{Cl}_r(x)]$.

Proof of (a).
$$\begin{aligned} x\ r^{\mathbf{T}}\ y &\Leftrightarrow y \in r^{\mathbf{T}}[x] \\ &\Leftrightarrow y \in \mathrm{Cl}_r(x) \text{ [proposition 6.2.2(a)]} \\ &\Leftrightarrow \mathrm{Cl}_r(y) \subseteq \mathrm{Cl}_r(x). \end{aligned}$$

Proof of (b).
$$\begin{aligned} x\ r^{\mathbf{t}}\ y &\Leftrightarrow y \in r^{\mathbf{t}}[x] \\ &\Leftrightarrow y \in r[\mathrm{Cl}_r(x)] \text{ [proposition 6.2.2(b)]} \\ &\Leftrightarrow \mathrm{Cl}_r(y) \subseteq r[\mathrm{Cl}_r(x)] \text{ [proposition 5.1.1(b)].} \end{aligned}$$
□

(6.2.4) **Proposition.** *Suppose that r is a relation on A.*

(a) $r^{\mathbf{t}} = \bigcup_{n \in \omega \setminus \{0\}} r^n$.

(b) $r^{\mathbf{T}} = \bigcup_{n \in \omega} r^n$.

Proof of (a). Let $r' = \bigcup_{n \in \omega \setminus \{0\}} r^n$. It is easy to show by induction on n that $r^n \subseteq r^{\mathbf{t}}$ for all $n \in \omega \setminus \{0\}$; consequently $r' \subseteq r^{\mathbf{t}}$. And if $x\ r^n\ y$ and $y\ r^m\ z$, then $x(r^n \circ r^m)z$ and so $x\ r^{n+m}\ z$, from which it follows that r' is transitive, so that $r^{\mathbf{t}} \subseteq r'$. Thus $r^{\mathbf{t}} = r'$ as required.

Proof of (b).
$$\begin{aligned} r^{\mathbf{T}} &= r^{\mathbf{t}} \cup \mathrm{id}_A \text{ [proposition 6.2.1(a)]} \\ &= r' \cup r^0 \text{ by part (a)} \\ &= \bigcup_{n \in \omega} r^n. \end{aligned}$$
□

(6.2.5) **Corollary.** *$r^{\mathbf{t}}$ is a strict partial ordering on A iff there do not exist $x \in A$ and $n \in \boldsymbol{\omega} \setminus \{0\}$ such that $x\ r^n\ x$.*

Proof. This follows immediately from proposition 6.2.4(a). □

Exercises

1. Check that the r-closed subsets of a structure (A, r) are the closed sets of a *topology* on A, in other words that:

 $\emptyset$ and A are r-closed;

 B,C r-closed $\Rightarrow B \cup C$ r-closed;

 B_i r-closed for all $i \in I \Rightarrow \bigcap_{i \in I} B_i$ r-closed.

2. (a) Show that the smallest reflexive relation on A containing r (sometimes called the *reflexive closure* of r) is $r \cup \mathrm{id}_A$.

 (b) Show that the smallest symmetric relation on A containing r (sometimes called the *symmetric closure* of r) is $r \cup r^{-1}$.

3. Show that a subset of A is r-closed iff it is $r^{\mathbf{T}}$-closed.

4. If r is a relation on A, show that $r^{\mathbf{t}} = \bigcap\{s \subseteq A \times A : r \cup (r \circ s) \subseteq s\}$.

6.3 The ordering of the natural numbers

(6.3.1) **Lemma.** *$\mathrm{s}^{\mathbf{t}}$ is a strict partial ordering on $\boldsymbol{\omega}$, and $\mathrm{s}^{\mathbf{T}}$ is the associated partial ordering.*

Proof. The only part which is not obvious is that $\mathrm{s}^{\mathbf{t}}$ is irreflexive. But

$$\begin{aligned} 0\ \mathrm{s}^{\mathbf{t}}\ 0 &\Rightarrow 0 \in \mathrm{s}[\mathrm{Cl}_{\mathrm{s}}(0)] = \mathrm{s}[\boldsymbol{\omega}] \text{ [proposition 6.2.2(b)]} \\ &\Rightarrow \text{contradiction,} \end{aligned}$$

and

$$\begin{aligned} \mathrm{s}(n)\ \mathrm{s}^{\mathbf{t}}\ \mathrm{s}(n) &\Rightarrow \mathrm{s}(n) \in \mathrm{s}[\mathrm{Cl}_{\mathrm{s}}(\mathrm{s}(n))] \text{ [proposition 6.2.2(b)]} \\ &\Rightarrow n \in \mathrm{Cl}_{\mathrm{s}}(\mathrm{s}(n)) = \mathrm{s}[\mathrm{Cl}_{\mathrm{s}}(n)] \text{ [proposition 5.1.1(b)]} \\ &\Rightarrow n\ \mathrm{s}^{\mathbf{t}}\ n \text{ [proposition 6.2.2(b)].} \end{aligned}$$

So it follows by induction that $\mathrm{s}^{\mathbf{t}}$ is an irreflexive relation on $\boldsymbol{\omega}$. □

Thus $\mathrm{s}^{\mathbf{t}}$ is a strict partial ordering on $\boldsymbol{\omega}$ containing s. In fact it is the only one (see exercise 4 below). So if we want to have $n < \mathrm{s}(n)$ for all $n \in \boldsymbol{\omega}$, then we have no choice but to adopt the following definition.

Definition. *We write $<$ for $\mathrm{s}^{\mathbf{t}}$ and $\leqslant$ for $\mathrm{s}^{\mathbf{T}}$.*

Whenever we refer to $\boldsymbol{\omega}$ as a partially ordered set without qualification, it will be with respect to this partial ordering.

(6.3.2) **Least element principle.** *Every non-empty subset of* $\boldsymbol{\omega}$ *has a least element.*

Proof. Suppose on the contrary that B is a subset of $\boldsymbol{\omega}$ which does not have a least element. Now

$$\boldsymbol{\omega} = \mathrm{Cl}_s(0) = \mathrm{s}^{\mathbf{T}}[0] \text{ [proposition 6.2.2(a)]}$$

and so 0 is certainly a lower bound for B. And if n is a lower bound for B, then $n \notin B$ (since otherwise n would be the least element of B, contrary to hypothesis), so that

$$B \subseteq \mathrm{s}^{\mathbf{t}}[n] = \mathrm{s}^{\mathbf{T}}[\mathrm{s}(n)] \text{ [proposition 6.2.1(b)]}$$

and hence $\mathrm{s}(n)$ is a lower bound for B. Consequently, by induction every element of $\boldsymbol{\omega}$ is a lower bound for B. So if $n \in B$, then in particular $\mathrm{s}(n)$ is a lower bound for B and so $\mathrm{s}(n) \leqslant n < \mathrm{s}(n)$. Contradiction. □

(6.3.3) **Corollary.** $\boldsymbol{\omega}$ *is totally ordered.*

Proof. If $m, n \in \boldsymbol{\omega}$, then $\{m, n\}$ has a least element [least element principle]; this least element is either m, in which case $m \leqslant n$, or n, in which case $n \leqslant m$. □

Definition. $\boldsymbol{n} = \{m \in \boldsymbol{\omega} : m < n\}$.

(6.3.4) **Corollary.** *If* $A \subseteq \boldsymbol{\omega}$ *is such that* $\boldsymbol{n} \subseteq A \Rightarrow n \in A$ *for every* $n \in \boldsymbol{\omega}$, *then* $A = \boldsymbol{\omega}$.

Proof. Suppose $A \neq \boldsymbol{\omega}$. So $\boldsymbol{\omega} \setminus A$ is non-empty and therefore [least element principle] has a least element n. Hence $\boldsymbol{n} \subseteq A$, and so by hypothesis $n \in A$. Contradiction. □

(6.3.5) **General induction scheme.** *If* $\Phi(n)$ *is a formula such that*

$$(\forall n \in \boldsymbol{\omega})((\forall m < n)\Phi(m) \Rightarrow \Phi(n)),$$

then $(\forall n \in \boldsymbol{\omega})\Phi(n)$.

Proof. Let $A = \{n \in \boldsymbol{\omega} : \Phi(n)\}$ and apply corollary 6.3.4. □

A proof that utilizes this scheme is sometimes called a 'proof by infinite descent'.

(6.3.6) **General recursion principle with a parameter.** *If for each $n \in \omega$ we have a function s_n from $\mathfrak{P}(A)$ to A, then there exists exactly one function g from ω to A such that $g(n) = s_n(g[\boldsymbol{n}])$ for all $n \in \omega$.*

Proof. Exercise 6. □

The use of the general recursion principle in the course of a proof is often signposted by temporal terminology such as:

Once $g(r)$ has been defined for all $r < n$, let $g(n) = s_n(g[\boldsymbol{n}])$.

Definition. *A family indexed by* $\boldsymbol{n}$ *for some natural number n is called a* finite sequence *(or sometimes a* string*) of length n. The set $\bigcup_{n \in \omega} {}^{\boldsymbol{n}}A$ of all strings in A is denoted* String(A).

Exercises

1. (a) Show that $m < n \Leftrightarrow \mathrm{s}(m) \leqslant n$ for all $m, n \in \omega$.
   (b) Prove by induction on n that $q < r \Rightarrow n + q < n + r$ for all $n, q, r \in \omega$.

2. Prove that if $m \leqslant n$, then there exists a unique element $n - m \in \omega$ such that $n = m + (n - m)$. [*Existence.* Consider the least element of $\{r \in \omega : m + r \geqslant n\}$.]

3. Let n be a natural number.
   (a) Prove that $n + 1$ is the unique successor of n.
   (b) Show that if $n \neq 0$, then $n - 1$ is the unique predecessor of n.

4. Show that $<$ is the only strict partial ordering on ω with the property that $n < \mathrm{s}(n)$ for all $n \in \omega$.

5. Show that the proper initial subsets of ω are precisely the sets $\boldsymbol{n}$ for $n \in \omega$.

6. Prove the general recursion principle.

7. If r is a relation on A, show that $(r \cup \mathrm{id}_A)^n = \bigcup_{m=0}^{n} r^m$ for all $n \in \omega$.

6.4 Counting finite sets

We now take $\boldsymbol{n} = \{0, 1, 2, \ldots, n - 1\}$ as a prototype and define what it is for a set to have n members by reference to it.

Definition. *We say that a set* has n members *if it is equinumerous with* $\boldsymbol{n}$.

Note that for this definition to be possible we needed to have available the *order* structure of the natural numbers (so as to be able to pick out $\boldsymbol{n}$), even though no reference is made in the definition to any ordering of the set being counted.

(6.4.1) **Theorem.** *There is no natural number n such that* $\boldsymbol{n}$ *is infinite.*

Proof. $\mathbf{0}$ is obviously not infinite. Suppose if possible that $\boldsymbol{n}$ is not infinite but $\boldsymbol{n}+\mathbf{1}$ is. So $\boldsymbol{n}+\mathbf{1}$ is equinumerous with a proper subset B of itself. There are three cases to consider.

(1) $n \in B$. So $B \setminus \{n\}$ is a proper subset of $\boldsymbol{n}$ which is equinumerous with it, contradicting the assumption that $\boldsymbol{n}$ is not infinite.

(2) B is a proper subset of $\boldsymbol{n}$. So $\boldsymbol{n}$ is equinumerous with a proper subset of itself, which once again is impossible since by hypothesis it is not infinite.

(3) $B = \boldsymbol{n}$. So $\boldsymbol{n}$ is equinumerous with $\boldsymbol{n}+\mathbf{1}$, which is impossible since the latter is infinite and the former is not.

Thus in every case the assumption is contradictory. The result follows by induction. □

Definition. *We say that a set is* finite *if it has n members for some $n \in \boldsymbol{\omega}$.*

Of course $\boldsymbol{n}$ itself is finite since it trivially has n members; in particular, $\varnothing$ is finite. Less trivially, if there is a function from A onto B and A is finite, then B is finite. It follows from this that every subset of a finite set is finite.

(6.4.2) **Corollary.** *No set is both finite and infinite.*

Proof. This follows at once from the theorem. □

We shall examine later (§9.4) whether it is possible for a set to be neither finite *nor* infinite.

(6.4.3) **Corollary.** *For each finite set there is exactly one natural number n such that it has n members.*

Proof. If not, then there are natural numbers m and n such that $m < n$ but $\boldsymbol{m}$ and $\boldsymbol{n}$ are equinumerous, in which case $\boldsymbol{n}$ is equinumerous with a proper subset of itself and is therefore infinite, contrary to the theorem. □

It is this corollary that gives us finally the technical resources to use the natural numbers as unequivocal measures of the sizes of finite sets.

Definition. *If A is finite, then the unique number n such that A has n members is called the* number of members *of A.*

We write $\mathfrak{F}(A)$ for the set of all finite subsets of A and $\mathfrak{F}_n(A)$ for the set of all n-element subsets of A, so that $\mathfrak{F}(A) = \bigcup_{n\in\boldsymbol{\omega}} \mathfrak{F}_n(A)$.

(6.4.4) **Proposition.** *If there exist natural numbers n such that $\mathfrak{F}_{n+1}(A) = \varnothing$, then A is finite, and the least such n is the number of elements in A.*

Proof. Suppose that n is the least natural number such that $\mathfrak{F}_{n+1}(A) = \emptyset$. Then A has an n-element subset B (since $n < n + 1$). But if B were a proper subset of A, there would be an element b in $A \setminus B$, and $B \cup \{b\}$ would be a subset of A with $n + 1$ elements, contrary to hypothesis. It follows that $A = B$, and hence A has n elements as required. □

(6.4.5) **Theorem.** *Every non-empty finite partially ordered set has a maximal element.*

Proof. Plainly every partially ordered set with 1 member has a maximal element. So suppose now that every partially ordered set with n members has a maximal element and let $(A, \leqslant)$ be a partially ordered set with $n + 1$ members. If $a \in A$, then $A \setminus \{a\}$ has n members and therefore by the induction hypothesis has a maximal element b: if $b \leqslant a$, then a is a maximal element of A, and if $b \nleqslant a$, then b is a maximal element of A. The result follows by induction. □

We have already characterized fully (up to isomorphism) the structure of $(\boldsymbol{\omega}, \mathrm{s})$ [corollary 5.3.2]; we shall now do the same for the ordered set $(\boldsymbol{\omega}, \leqslant)$.

(6.4.6) **Theorem.** *A partially ordered set is isomorphic to* $(\boldsymbol{\omega}, \leqslant)$ *iff it is non-finite, totally ordered, and every proper initial subset of it is finite.*

Necessity. We have already shown that $(\boldsymbol{\omega}, \leqslant)$ is a totally ordered set and that it is not finite. Moreover, if B is a proper initial subset of $\boldsymbol{\omega}$, then there exists $n \in \boldsymbol{\omega} \setminus B$, so that $B \subseteq \boldsymbol{n}$, and hence B is finite. The result follows.

Sufficiency. Suppose that $(A, \leqslant)$ has the properties described in the theorem. Let us first observe that if a subset B of A is non-empty, then it has a least element: for if $b \in B$, then either $\{x \in B : x < b\}$ is empty, in which case b is the least element of B, or it is finite and non-empty, in which case it has a least element [theorem 6.4.5] which is evidently also the least element of B. Now by the general recursion principle we can define a function f from $\boldsymbol{\omega}$ to A as follows: when $f(m)$ has been defined for all $m < n$, let $f(n)$ be the least element of $A \setminus f[\boldsymbol{n}]$ (which is non-empty since A is non-finite). Now f is evidently strictly increasing; moreover, $f[\boldsymbol{\omega}]$ is an initial subset of A which is not finite and must therefore be A itself. Since $\boldsymbol{\omega}$ is totally ordered, it follows that f is an isomorphism between $(\boldsymbol{\omega}, \leqslant)$ and $(A, \leqslant)$. □

Exercises

1. Show that if $(A, \leqslant)$ is a partially ordered set, then these three assertions are equivalent:

   (i) $(A, \leqslant)$ is isomorphic to $(\boldsymbol{n}, \leqslant)$ for some $n \in \boldsymbol{\omega}$;

   (ii) A is finite and $\leqslant$ is a total ordering;

(iii) Every non-empty subset of A has a greatest and a least element.

2. Show that A is finite iff there exists a function f on A such that the only f-closed subsets of A are $\emptyset$ and A itself.

3. Suppose that $(A, \leqslant)$ is a totally ordered set. Show that any sequence $(a_n)_{n\in\omega}$ in A has a monotonic subsequence. (We say that $(a_{n_r})_{r\in\omega}$ is a *subsequence* of (a_n) if (n_r) is a strictly increasing sequence in ω.) [Let $B = \{n \in \omega : a_n < a_r \text{ for all } r > n\}$ and consider two cases according to whether or not B is finite.]

4. (a) If r is the relation on $\mathfrak{P}(A)$ defined between X and Y by the formula 'there exists $a \in A$ such that $Y = X \cup \{a\}$', show that $\mathfrak{F}(A) = \mathrm{Cl}_r(\emptyset)$.

(b) Show that $X, Y \in \mathfrak{F}(A) \Rightarrow X \cup Y \in \mathfrak{F}(A)$.

(c) Show that if A is finite then $\mathfrak{P}(A)$ is finite.

6.5 Counting infinite sets

Definition. *A set A is said to be* countable *if either $A = \emptyset$ or there exists a sequence whose range is A.*

Every finite set is countable, whereas ω is countable but not finite (since if it were finite it would have a greatest element [theorem 6.4.5]). If there is a function from A onto B and A is countable, then B is countable. It follows that every subset of a countable set is countable.

Definition. *A set A is said to be* countably infinite *if it is both countable and infinite. A is said to be* uncountably infinite *if it is infinite but not countable.*

The set of natural numbers is countably infinite; hence so is any set equinumerous with it. In fact any countably infinite set is equinumerous with ω. For if A is an infinite set enumerated by a sequence $(x_n)_{n\in\omega}$ and we recursively define y_n to be the element x_m of $A \setminus \{y_r : r < n\}$ with m chosen as small as possible, the sequence $(y_n)_{n\in\omega}$ enumerates A without repetitions.

(6.5.1) **Proposition.** *$\omega \times \omega$ is countably infinite.*

Proof. It can be shown that the function $(m, n) \mapsto 2^n(2m + 1) - 1$ is a one-to-one correspondence between $\omega \times \omega$ and ω. (The quickest way to see this is to note that by dividing successively by 2 we can express any natural number except 0 uniquely in the form of a power of 2 multiplied by an odd number.) □

It follows by an easy induction, of course, that ω^n is countably infinite for all $n \in \omega$.

(6.5.2) **Proposition.** *$\mathfrak{F}(\omega)$ is countably infinite.*

Proof. The function $A \mapsto \sum_{n\in A} 2^n$ is a one-to-one correspondence between $\mathfrak{F}(\omega)$ and ω as required. □

(6.5.3) **Proposition.** String($\boldsymbol{\omega}$) *is countably infinite.*

Proof. Consider the function which maps any finite string $(n_1, n_2, \ldots, n_k)$ of natural numbers to the single natural number

$$p_1^{n_1} p_2^{n_2} \ldots p_k^{n_k+1} - 1,$$

where $p_1, p_2, \ldots$ is a listing of the prime numbers in increasing order of size. It is a standard fact of elementary number theory (often known as the *fundamental theorem of arithmetic*) that this is a one-to-one correspondence between String($\boldsymbol{\omega}$) and $\boldsymbol{\omega}$. (The '-1' at the end of the expression is there so as to include 0; the '$+1$' in the last exponent is there to deal with the problem of non-uniqueness caused by trailing zeros in the input string.) □

Of course we could also extend these results to show that $\mathfrak{F}(A)$ and String(A) are countable if A is, and that $A \times B$ is countable if A and B are.

(6.5.4) **Proposition.** $\mathfrak{P}(\boldsymbol{\omega})$ *is uncountably infinite.*

Proof. It is easy to see that $\mathfrak{P}(\boldsymbol{\omega})$ is infinite. Suppose if possible that it is countable, so that there is a sequence $(A_n)_{n\in\boldsymbol{\omega}}$ which enumerates all the subsets of $\boldsymbol{\omega}$. Let $A = \{n \in \boldsymbol{\omega} : n \notin A_n\}$. Then by hypothesis $A = A_n$ for some $n \in \boldsymbol{\omega}$. But then

$$n \in A \Leftrightarrow n \notin A_n \Leftrightarrow n \in A.$$

Contradiction. □

6.6 Skolem's paradox

We cautioned earlier about the dangers of reflexively applying results proved in a theory such as ZU to that theory itself. One such danger is exemplified by the following two theorems.

Completeness theorem. *Every consistent first-order theory has a set-theoretic model.*

Löwenheim/Skolem theorem. *Every structure is elementarily equivalent to a countable structure.*

These two theorems of logic have been used to mount the following argument against a realist understanding of set theory. Suppose that I am a realist. I believe that the axioms of ZU are true and hence consistent. I therefore believe that the set-theoretic formalization of the claim that they are consistent is true, even though it is not provable from the axioms.[1] Because of the two theorems

[1]It is at this point that the realist parts company with the formalist: formalists might perhaps have a non-mathematical belief that ZU is consistent, not because it is true but on inductive grounds, for instance; however, they would not take that as a reason to believe the set-theoretic claim obtained by formalizing this belief.

just quoted, therefore, I ought also to believe that **ZU** has a countable set-theoretic model (and even a model whose domain is $\boldsymbol{\omega}$). Every theorem of **ZU** is true in this model: in particular, therefore, it is true in the model that $\mathfrak{P}(\boldsymbol{\omega})$ is uncountable. Nevertheless, to repeat, the model is countable. This is known as *Skolem's paradox*.

But it is a paradox only in the sense that it is initially surprising, not that it is a contradiction, for its resolution is straightforward. What it means for a set to be uncountable is expressed by a fairly complicated logical expression involving $\in$; what it means for a set to be uncountable *in the model* is obtained by replacing $\in$ with whatever relation is the interpretation of '$\in$' in the model, and then relativizing all the quantifiers to the domain of the model. The resulting sentence will not express at all the same claim as that the set in question is actually uncountable. So Skolem's paradox, at least in the form we have been considering here, is certainly very far from being a formal contradiction. When taken at face value, indeed, it is hard even to find it especially surprising.

But consider now what happens if we temporarily adopt a metalinguistic perspective from which to study our own set-theoretic language, the language we have been using in the bulk of this book so far, which is now regarded as the object language. By running through the above argument once more, we discover from this new perspective that set theory has a countable model. This is disconcerting because it prompts the thought that this countable model might be *our* model: the *intended* model could turn out — behind our backs — to be countable.

Skolem's paradox has therefore been used by various authors as a springboard to an argument that there are no uncountable sets. Thus Kaufmann (1930, ch. 5), for example, used it to question the coherence of impredicatively specified sets; and Wright (1985) has essayed a direct argument that we cannot grasp the notion of an arbitrary subset of a given infinite set such as the set of natural numbers.[2] But we need to be careful. For the conclusion we have reached in the metalanguage is that the object language theory **ZU** has a countable model. If we try to re-express this in the object language, we get not a contradictory thought but no thought at all. There is no formula in the object language expressing what we mean when we say in the metalanguage that a set is countable.

Nonetheless, we are not yet quite ready to conclude that the theorem is without philosophical consequences. Putnam (1980), for instance, has used it to mount a challenge on the 'moderate realist position which seeks to preserve the centrality of the classical notions of truth and reference without postulat-

[2]Even if Wright's argument works, it is not clear that it should worry the platonist, at least in the form in which Wright presents it, since it appeals to constructivist assumptions which the platonist need feel under no pressure to accept (see Clark 1993*b*).

ing nonnatural mental powers'. In general, if we treat the meanings of the logical vocabulary as fixed, but the meanings of the non-logical vocabulary as completely open, then even a complete set of axioms does not pin down our intended meaning fully. However, if that is all we want to establish, we can do it by means of a simple permutation argument (Putnam 1981), which shows that if there is such a thing as an 'ideal theory', it cannot implicitly define its own reference relation; and this permutation argument is applicable to a second-order theory just as much as a first-order one. The Löwenheim/Skolem theorem does indeed show that in the case of a first-order theory this relativity is somewhat more radical: if only the first-order logical vocabulary is treated as fixed, we do not even pin down the cardinality of the domain of reference. What is harder to see is why the second, more radical relativism should be more troubling than the first.

Notes

The ancestral is defined and its elementary properties are proved in Frege's *Begriffsschrift* (1879), but it was Dedekind's rediscovery of the notion in (1888) which led to its popularization.

The use of natural numbers to measure the size of finite sets is evidently central to any understanding of their applicability. It remains controversial whether the ordinal use is primary (see Dummett 1991, p. 293). In Dedekind's treatment the principle which justifies using numbers for counting was derived as a theorem. Frege, by contrast, proposed to move in the opposite direction, taking this theorem as his starting point and deriving the properties of natural numbers from it. This project stalled when he unwisely used an inconsistent theory to justify the procedure, but interest in it was renewed by Wright (1983). Although most philosophers remain unconvinced, the detailed working out of this programme (see Hale and Wright 2001) has contributed significantly to our understanding of the logicist project.

Chapter 7

Lines

The task of this chapter is to characterize the ordering properties of the rational and real lines and to prove within ZU the existence of sets which satisfy these characterizations. The question of how points on these lines can be added or multiplied, and how they can be used to *measure* quantities, will be set on one side to be dealt with in the following chapter.

We start with some terminology. A subset of an ordered set $(A, \leqslant)$ is called an *open interval* if it is of one of the following four forms:

$\{x \in A : a < x\}$ for some a in A;

$\{x \in A : x < b\}$ for some b in A;

$\{x \in A : a < x < b\}$ for some $a < b$ in A;

the whole of A.

Definition. *A totally ordered set in which every open interval is non-empty is called a* line.

Otherwise stated, a line is an ordered set which is dense in itself and has no greatest or least element. It follows [theorem 6.4.5] that a line cannot be finite. The elements of a line are usually called *points*. We say that a subset B of a line is *open* if it is a union of open intervals.

A subset B of a line is said to be *closed* if its complement is open. A *limit point* of B is a point a such that every open interval containing it contains at least one point of B other than a. The set of limit points of B is called the *derived set* of B and denoted B'. Evidently B is closed iff it contains all its limit points, i.e. iff $B' \subseteq B$. We say that B is *perfect* if $B' = B$, i.e. if it is a closed set in which every point is a limit point.

7.1 The rational line

(7.1.1) **Theorem.** *There exists a (pure) countable line.*

Proof. Let $\mathcal{Q} = \mathfrak{F}(\boldsymbol{\omega}) \setminus \{\emptyset\}$ and define an ordering on $\mathcal{Q}$ by writing $A < B$ iff $A \neq B$ and the least element in one of A and B but not in both is in A. It is easy to verify that $\mathcal{Q}$ with this ordering is a line. Moreover, $\mathfrak{F}(\boldsymbol{\omega})$ is countable [proposition 6.5.2], and hence $\mathcal{Q}$ is countable too. □

Definition. *Some (pure) countable line (of lowest possible birthday) is denoted* $(\mathbb{Q}, \leqslant)$ *and called the* rational line.

The properties just mentioned are sufficient to characterize the rational line up to isomorphism.

(7.1.2) **Theorem (Cantor 1895).** *Every countable line is isomorphic to the rational line.*

Proof. Let $(A, \leqslant)$ be a countable line and let (a_n) and (b_n) be sequences whose images are A and $\mathbb{Q}$ respectively. We construct a function f from A to $\mathbb{Q}$ recursively: once $f(a_r)$ has been defined for all $r < m$, let $f(a_m)$ be the element b_n of least possible index such that it bears the same order relation to $f(a_0), \ldots, f(a_{m-1})$ as a_m bears to $a_0, \ldots, a_{m-1}$. (Such an element always exists because the ordering on $\mathbb{Q}$ is dense and has no greatest or least element.) Now it is clear that the resulting function f from A to $\mathbb{Q}$ is strictly increasing (and therefore one-to-one); we will be finished if we can show that its image is $\mathbb{Q}$. So suppose not. There therefore exists an element $b_n \in \mathbb{Q} \setminus f[A]$ of least possible index. If $r < n$, then $b_r \in f[A]$, and so we can let m_r be the least natural number such that $f(a_{m_r}) = b_r$. If m is the least natural number greater than $m_0, \ldots, m_{n-1}$ such that a_m bears the same order relation to $a_{m_0}, \ldots, a_{m_{n-1}}$ as b_n bears to $f(a_{m_0}), \ldots, f(a_{m_{n-1}})$, then it is clear that $f(a_m) = b_n$. Contradiction. □

The specification we have just given of the rational *line* is not a specification of the set of rational *numbers*, since it omits any reference to their algebraic (ordered field) structure. We shall postpone discussion of this algebraic structure to the next chapter, where we shall see explicit examples showing that it is not determined uniquely by the ordering.

Exercise

Let $(A, \leqslant)$ be a countable partially ordered set.

(a) Prove that there exists an increasing one-to-one function from $(A, \leqslant)$ into $(\mathbb{Q}, \leqslant)$.

(b) Deduce that the partial ordering of A can be extended to a total ordering.

7.2 Completeness

Geometry was traditionally conceived of as the study of those constructions which are possible using only a straightedge (to draw straight lines) and a pair of compasses (to draw circles). Straight lines, as this sort of geometry conceives of them, are 'gapless' in the sense that between any two points there is another, so when thought of as ordered sets they are indeed lines according to the definition of the last section. But there is another (and, as it turns out, more constraining) sort of gaplessness that we might expect a geometrical line to possess. One way of expressing it is by means of the idea that if a function is thought of as describing a path that a particle might traverse, it cannot pass through two points without passing through all the points between them. This (or at any rate a plausible formalization of it) is known as the *intermediate value property*.

Definition. *Suppose that $(A, \leqslant)$ and $(B, \leqslant)$ are ordered sets. A function f from A to B is said to have the* intermediate value property *if whenever b lies between $f(a_1)$ and $f(a_2)$ in B there exists a between a_1 and a_2 in A such that $f(a) = b$.*

Now we plainly would not expect *any* function whatever to have the intermediate value property, but it has generally been thought plausible — 'intuitively obvious' is how it is usually expressed — that every function of a real variable which is continuous in the sense explained in the Introduction must have the intermediate value property.

Definition. *Suppose that $(A, \leqslant)$ and $(B, \leqslant)$ are ordered sets and f is a function from A to B. We say that $f(x)$* tends *to b as x tends to a if for every open interval J containing b there is an open interval I containing a such that $f[I] \subseteq J$. We say that f is* continuous *at a if $f(x)$ tends to $f(a)$ as x tends to a. And we say that f is* continuous *if it is continuous at every element of A.*

What we want to study, then, is the condition a line must satisfy if every function that is continuous in this sense is to have the intermediate value property. The key property here turns out to be the one on which all rigorous undergraduate calculus courses are based, namely the assumption that every non-empty set of real numbers which has an upper bound has a least upper bound.

(7.2.1) **Lemma.** *If $(A, \leqslant)$ is a partially ordered set, these two assertions are equivalent:*

(i) Every non-empty subset of A which has an upper bound in A has a supremum;

(ii) Every non-empty subset of A which has a lower bound in A has an infimum.

Proof. Suppose that B is a non-empty subset of A and let

$$C = \{x \in A : x \text{ is a lower bound for } B\}.$$

By hypothesis $a = \sup C$ exists. We shall show that $a = \inf B$. To do this, note first that if $x \in C$ and $y \in B$, then $x \leqslant y$: consequently y is an upper bound for C and so $a \leqslant y$. In other words, a is a lower bound for B. And if x is another lower bound for B, then $x \in C$ and so $x \leqslant a$. Thus a is the greatest lower bound for B, i.e. $a = \inf B$. This proves that (i) $\Rightarrow$ (ii); the proof that (ii) $\Rightarrow$ (i) is similar. □

Definition. *A line is said to be* complete *if it satisfies the equivalent conditions of lemma 7.2.1 above.*

(7.2.2) **Theorem.** *Suppose that* $(A, \leqslant)$ *and* $(B, \leqslant)$ *are lines. Then* $(A, \leqslant)$ *is complete iff every continuous function from* A *to* B *has the intermediate value property.*

Necessity. Suppose that f is a continuous function from A to B and that b lies between $f(a_1)$ and $f(a_2)$. Let $C = \{x \in A : f(x) < b\}$ and let $a = \sup C$.

Now if $f(a) < b$, then by the continuity of f there exists $c > a$ such that $f(x) < b$ whenever x lies between c and a. But then each such x is in C, and therefore a is not an upper bound for C. Contradiction.

And if $f(a) > b$, then by the continuity of f there exists $c < a$ such that $f(x) > b$ whenever x lies between c and a. But then no such x is in C, and therefore a cannot be the supremum of C. Contradiction.

The only remaining possibility is that $f(a) = b$ as required.

Sufficiency. Suppose that C is a subset of A that is bounded above but does not have a supremum in A. If we let $C' = \{x \in A : (\exists y \in C)\, x \leqslant y\}$, then C' does not have a supremum in A either. Choose two elements b and b' of B and define a function f from A to B by letting

$$f(x) = \begin{cases} b' & \text{if } x \in C' \\ b & \text{otherwise.} \end{cases}$$

Then it is easy to show that f is continuous. Yet f cannot have the intermediate value property, since B is a line and so has elements between b and b', which are the only values taken by f. Contradiction. □

Exercise

(a) Show that the intersection of a non-empty set of partial orderings is also a partial ordering.

(b) Deduce that $\text{Po}(A)$ is complete with respect to inclusion.

7.3 The real line

If our aim is to construct a complete line, it would evidently be prudent to check first that the *rational* line constructed in the last section is not already complete.

(7.3.1) **Proposition.** *$\mathbb{Q}$ is not complete.*

Proof. Consider once more the explicit model $\mathcal{Q}$ of the rational line constructed in the last section, and suppose for a contradiction that it is complete. In this model consider the two sequences

$$\{0\}, \{0, 2\}, \{0, 2, 4\}, \{0, 2, 4, 6\}, \ldots$$

$$\{0, 1\}, \{0, 2, 3\}, \{0, 2, 4, 5\}, \{0, 2, 4, 6, 7\}, \ldots$$

It is easy to check that every element of the first sequence is greater than every element of the second sequence. It follows that there is a member A of $\mathcal{Q}$ such that

$$\{0\} > \{0, 2\} > \{0, 2, 4\} > \{0, 2, 4, 6\} > \ldots A > \ldots \{0, 2, 4, 5\} > \{0, 2, 3\} > \{0, 1\}.$$

Then $A < \{0\}$, and so $0 \in A$. But $\{0, 1\} < A$, and so $1 \notin A$. Also $A < \{0, 2\}$, and so $2 \in A$. Then $A < \{0, 2, 3\}$, and so $3 \notin A$. And so on. Thus we get eventually that $\{0, 2, 4, 6, \ldots\} \subseteq A$, and so A is infinite, contradicting the definition of $\mathcal{Q}$, which consists only of the non-empty finite subsets of $\boldsymbol{\omega}$. It follows that $\mathcal{Q}$ is not complete. But $\mathcal{Q}$ is isomorphic to $\mathbb{Q}$ [Cantor's theorem 7.1.2], and so $\mathbb{Q}$ cannot be complete either. □

(7.3.2) **Corollary.** *Any complete line is uncountable.*

Proof. Any countable line is isomorphic to $\mathbb{Q}$ [theorem 7.1.2] and hence not complete. □

If no countable line is complete, the next best thing is to try to achieve completeness by a minimal extension of such a line, i.e. by constructing a line which, although not countable itself, nevertheless has a countable line as a dense subset.

Definition. *A complete line which has a countable dense subset is called a* continuum.

Mathematicians have often conceived of space and time as supplying us with instances of continua in this sense. Whether they are justified in this is a matter we shall return to later, but in any case what we have to do here is to construct a continuum independently of such considerations, i.e. without recourse to spatial or temporal intuition.

(7.3.3) **Theorem.** *There exists a continuum.*

Proof. Let $\mathcal{R}$ be the set of all non-empty proper initial subsets of $\mathbb{Q}$ which have no greatest element. It is evidently totally ordered by inclusion and has no least or greatest element. Moreover, if $\mathcal{B}$ is a chain in $\mathcal{R}$, all the elements of which are contained in some element of $\mathcal{R}$, then it is easy to verify that $\bigcup \mathcal{B} \in \mathcal{R}$; it follows that $\mathcal{R}$ is complete. Finally $\{\{x : x < a\} : a \in \mathbb{Q}\}$ is a countable subset of $\mathcal{R}$, since $\mathbb{Q}$ is countable; and it is dense in $\mathcal{R}$, since if $A, B \in \mathcal{R}$ and $A \subset B$, then there exists $x \in B \setminus A$ and so $A \subset \{x : x < a\} \subset B$. We have therefore shown that $\mathcal{R}$ is a complete line which has a countable dense subset, i.e. a continuum. □

Definition. *We choose some (pure) continuum (of lowest possible birthday): it is denoted* $(\mathbb{R}, \leqslant)$ *and called the* real line.

Intermediate value theorem. *Every continuous function from* $\mathbb{R}$ *to itself has the intermediate value property.*

Proof. Immediate [theorem 7.2.2]. □

(7.3.4) **Proposition.** $\mathbb{R}$ *is not countable.*

Proof. This follows at once from corollary 7.3.2. □

We have already come across one example of an uncountable set, namely $\mathfrak{P}(\omega)$. The real line, though, was Cantor's first example, and so his discovery of it in 1873 may be said to mark the birth of the modern theory of cardinality.

Definition. *Suppose that* $(A, \leqslant)$ *and* $(B, \leqslant)$ *are partially ordered sets. A function* f *from* A *to* B *is said to be* normal *if for every set* C *which has a supremum in* A *the set* $f[C]$ *has a supremum in* B *and* $f(\sup C) = \sup f[C]$; f *is said to be* strictly normal *if in addition it is one-to-one.*

Every [strictly] normal function is in particular [strictly] increasing.

(7.3.5) **Proposition.** *Suppose that* B *is a dense subset of a line* $(A, \leqslant)$. *Then every normal function* f *from* B *to* $\mathbb{R}$ *has exactly one normal extension* $\overline{f}$ *from* A *to* $\mathbb{R}$.

Uniqueness. If $a \in A$, then in order that $\overline{f}$ should be normal, we must have

$$\overline{f}(a) = \sup_{x \in B, x < a} \overline{f}(x) = \sup_{x \in B, x < a} f(x),$$

whence the uniqueness of $\overline{f}$.

Existence. If we define

$$\overline{f}(a) = \sup_{x \in B, x < a} f(x)$$

for all $a \in A$ as this suggests, then for $a \in B$ we have

$$\overline{f}(a) = \sup_{x \in B, x < a} f(x) = f(a)$$

since f is normal. And if C is a subset of A with supremum c, then

$$\sup \overline{f}[C] = \sup_{a \in C} \sup_{y \in B, y < a} f(y) = \sup_{y \in B, y < c} f(y) = \overline{f}(c).$$

This shows that $\overline{f}$ is normal. □

We shall now show that the properties we have used to define $(\mathbb{R}, \leqslant)$ — that it is complete and that it has a countable dense subset — are sufficient to determine it up to isomorphism.

(7.3.6) **Lemma.** *If $(A, \leqslant)$ is a line, and B is a dense subset of A which is complete with respect to the inherited ordering, then $B = A$.*

Proof. Suppose that $a \in A$ and let $C = \{x \in B : x < a\}$. C is bounded above in B and hence has a least upper bound in B. Because B is dense in A, this least upper bound must be a, from which it follows that $a \in B$. □

(7.3.7) **Theorem.** *Every continuum is isomorphic to the real line.*

Proof. Suppose that $(A, \leqslant)$ is a continuum and B is a countable dense subset of A. Now $\mathbb{R}$ has by definition a countable dense subset, and there is therefore an isomorphism f from B onto this subset [theorem 7.1.2]. This isomorphism is certainly strictly normal and therefore has a normal extension $\overline{f}$, which is a strictly increasing function from A onto a subset $\overline{f}[A]$ of $\mathbb{R}$ [proposition 7.3.5]. This subset is dense in $\mathbb{R}$, and it is complete because it is isomorphic to A; hence it must be the whole of $\mathbb{R}$ [lemma 7.3.6]. □

We have shown, then, that our definition of a continuum characterizes uniquely a certain sort of order structure. What is less clear, and receives less attention in the literature than it deserves, is whether either space or time supplies us with models of this structure. Neither part of the definition — neither the completeness condition nor the existence of a countable dense subset — is as obvious as is sometimes supposed. As we shall see in the next chapter, matters may be somewhat different once we make the assumption that the operations of addition and multiplication (or even just addition) can be performed, but in the meantime let us consider what can be said without that assumption.

Dedekind (1872, §3) thought it likely that 'everyone will at once grant the truth' of the completeness axiom, but he went on straightaway to grant that space might not in fact satisfy it. In that case, he says,

there would be nothing to prevent us, in case we so desired, from filling up its gaps in thought and thus making it continuous; this filling up would consist in a creation of new points and would have to be carried out in accordance with the above principle.

He does not go into the question of whether it is possible to complete time in the same manner. The Kantian resonance is evident: we can represent space to ourselves as continuous whether or not it really is. Something similar is to be found in Poincaré's view that space is not revealed to us as a continuum directly by the senses. It is necessary, he says, 'that by an active operation of the mind we agree to consider two states of consciousness as identical by *disregarding* their differences' (1913, ch. III, §3). Many other authors have simply taken it as intuitively obvious that this is how space is. The weakness of this is that the completeness principle makes a claim about *arbitrary* subsets of the real line, and it is far from clear that our intuitions about these are reliable: see §15.7 for a theorem about arbitrary subsets of the plane which it takes some mathematical sophistication not to regard as an obvious falsehood.

If we are hesitant about arguing for completeness directly, an alternative might be to argue that the intermediate value theorem — that every continuous function on the real line has the intermediate value property — is an obvious truth. For we have seen that this theorem holds only if the real line is complete. But our intuitive conception of a continuous function is perhaps of one that can be drawn without lifting the pen from the paper, or something of that sort; and it seems plausible that any such function will be at least piecewise differentiable (which, after all, is why the example of a continuous, nowhere differentiable function strikes many people as counter-intuitive when they first come across it). So the definition of continuity already involves a considerable extension of our intuitive conception, and it is unclear why we should regard the intermediate value theorem as intuitively correct when applied to these non-intuitive functions. After all, to the untutored eye the intermediate value property seems to have as good a title to be regarded as an explication of the intuitive concept of continuity as does the definition that has actually been adopted. And of course if 'continuous' had been defined by means of the intermediate value property, we could not have used the intermediate value theorem as a reason for believing that the real line of our intuition is complete.

The other part of our definition, the requirement that the real line should have a countable dense subset, is even harder to motivate intuitively without reference to addition. Indeed it is not even clear that we have *any* direct intuitions that bear directly on the cardinality of the real line beyond the bare fact that it is infinite. Hodges (1998, p. 3) has remarked (in the course of an interesting discussion of amateur criticisms of the proof) that when we come to Cantor's result, 'all intuition fails us. Until Cantor first proved his theorem ... nothing like its conclusion was in anybody's mind's eye. And even now we accept it because it is proved, not for any other reason.' If this is right, then the prospects for an intuitive argument to show that the continuum has a countable dense subset without appealing to its metric properties seem slim.

Exercises

1. If $(A, \leqslant)$ is a totally ordered set which has a countable dense subset, show that it can be embedded in $(\mathbb{R}, \leqslant)$.

2. Give an example of a dense ordered set without a greatest or a least element which is equinumerous with, but not isomorphic to, the real line. [*Hint.* Remove one point from $\mathbb{R}$.]

7.4 Souslin lines

(7.4.1) **Proposition.** *Every pairwise disjoint set of open intervals in $\mathbb{R}$ is countable.*

Proof. Suppose that $\mathcal{A}$ is a set of pairwise disjoint open intervals and $\{q_n : n \in \boldsymbol{\omega}\}$ is a countable dense subset of $\mathbb{R}$. If $\mathcal{A}$ is empty, it is trivially countable. If not, it contains at least one member A_0. For each $n \in \boldsymbol{\omega}$ there is no more than one $A \in \mathcal{A}$ such that $q_n \in A$: if there is one, let it be $f(n)$; if there is none, let $f(n) = A_0$. This evidently defines a function f from $\boldsymbol{\omega}$ onto $\mathcal{A}$. □

So $(\mathbb{R}, \leqslant)$ is a complete line in which every pairwise disjoint set of open intervals is countable. Souslin (1920) speculated that this might constitute an alternative *characterization*, so that any such line would be a continuum.

Definition. A Souslin line *is a complete line with no countable dense subset in which nonetheless every pairwise disjoint set of open intervals is countable.*

Souslin's hypothesis may thus be expressed as the claim that there are no Souslin lines. As it turns out, though, this hypothesis is independent of the formal system we have been using, and even independent of ZFC (Jech 1967; Solovay and Tennenbaum 1971). It is in fact our first explicit example of a genuinely mathematical statement independent of our theory: we shall see several more examples in part IV.

(7.4.2) **Lemma.** *Every non-empty convex subset of $\mathbb{R}$ is an open interval.*

Proof. Let U be a non-empty convex open subset of $\mathbb{R}$. There are four cases to consider, depending on whether U is bounded above or below in $\mathbb{R}$.

(1) U is bounded above but not below. If we let $b = \sup U$, it is easy to show that $U = \{x \in \mathbb{R} : x < b\}$.

(2) U is bounded below but not above. In the same way we can let $a = \inf U$ and show that $U = \{x \in \mathbb{R} : a < x\}$.

(3) U is bounded both above and below. Let $a = \inf U$ and $b = \sup U$, so that $U = \{x \in \mathbb{R} : a < x < b\}$.

(4) U is not bounded above or below. So $U = \mathbb{R}$.

Thus in each case U is an open interval. □

(7.4.3) **Lemma.** *Any open subset of a line is uniquely expressible as the union of a collection of pairwise disjoint non-empty convex open sets.*

Proof. Let U be an open subset of a line and define $x \sim y$ if there is an open interval contained in U to which both x and y belong. It is easy to see that this is an equivalence relation on U and that each of the equivalence classes is a non-empty convex open set. □

(7.4.4) **Proposition.** *Every open subset of $\mathbb{R}$ is uniquely expressible as the union of a countable collection of pairwise disjoint open intervals.*

Proof. If U is an open subset of $\mathbf{R}$, it is expressible as the union of a collection of pairwise disjoint non-empty convex open sets [lemma 7.4.3]. These sets are all open intervals, and hence the collection is countable [proposition 7.4.1]. □

7.5 The Baire line

Definition. *A line obtainable from a continuum by removing a countable dense subset is called an* irrational line.

The work we have done already gives us an easy construction of an irrational line, of course: the real line has a dense subset isomorphic to the rational line; what is left if we remove it is evidently an irrational line. The direct geometrical significance of an irrational line such as this is presumably much less than that of the real line. So irrational lines would not be of any great interest to us if it were not for a surprising connection with the theory of games which arises out of a quite different construction of an irrational line. For this we take the set ${}^{\omega}\omega$ of all sequences of natural numbers, ordered by letting $x < y$ if $x \neq y$ and the least n such that $x(n) \neq y(n)$ is such that

$$x(n) < y(n) \text{ if } n \text{ is even,}$$
$$x(n) > y(n) \text{ if } n \text{ is odd.}$$

We shall call the resulting ordered set the *Baire line*. A subset of the Baire line is closed iff it is the set of all the paths through some tree.

(7.5.1) **Theorem.** *The Baire line is an irrational line.*

The proof of this theorem will not be given here: it can be found in various places, e.g. Truss 1997, ch. 10.

(7.5.2) **Corollary.** *Every irrational line is isomorphic to the Baire line.*

Proof. The work we did in §7.3 shows immediately that any two irrational lines are isomorphic. □

The point of interest for us here is that the Baire line gives us a natural way to encode and study the theory of games between two players. To see this, consider a game for two players such as draughts or chess. Each play of such a game consists of a sequence of moves by the players alternately: if all the possible moves are labelled by natural numbers, a *play* can be represented as a sequence x of natural numbers, i.e. as a member of the Baire line: the even members $x(0), x(2), x(4), \ldots$ of the sequence enumerate the first player's moves; and the odd members $x(1), x(3), x(5), \ldots$ enumerate the second player's moves. What constitutes winning varies depending on the game that is being played, of course; but let us denote by A the set of plays which are wins for the first player. We shall make no assumptions at all about A at this stage, so *every* subset A of the Baire line ${}^{\omega}\omega$ constitutes the first-person wins of some game. The only simplifying assumption we make is that there are no draws, so every play which does not belong to A is automatically a win for the second player. The game in which winning is defined in this manner is called the game *on* A.

A *strategy* for the first player is a function σ from $\bigcup_{n\in\omega} {}^{2n}\omega$ to ω, i.e. a function which for any string of $2n$ natural numbers as input generates a single natural number as output. It should be thought of as telling the first player which move to make at each stage of the play on the basis of the previous moves: a play x *conforms* to the strategy σ just in case

$$x(2n) = \sigma(x(0), x(1), x(2), \ldots, x(2n-1)) \text{ for all } n.$$

We write $\sigma * t$ for the play which results if the first player follows strategy σ against a second player whose moves are enumerated by t; and we say that σ is a *winning* strategy if $\sigma * t$ is a win for the first player for every sequence t. Similarly, a winning strategy for the second player is a function τ from $\bigcup_{n\in\omega} {}^{2n+1}\omega$ to ω such that every play x which conforms to it, in the sense that

$$x(2n+1) = \tau(x(0), x(1), \ldots, x(2n)) \text{ for all } n,$$

is a win for the second player; and we write $s * \tau$ for the game in which the first player's moves are enumerated by s and the second player follows the strategy τ in response.

Let us say that a game is *determined* if one or other player has a winning strategy. The question that naturally presents itself is to settle which games are determined. There are certainly special cases in which this problem can be solved quite easily.

(7.5.3) **Proposition.** *The game on any set which is countable or has a countable complement is determined.*

Proof. Suppose that the sequence $x_0, x_1, x_2, \ldots$ enumerates the plays which constitute wins for the first player. The second player then has a strategy in which the nth move is $x_n(2n+1)+1$: this wins the game for the second player because it diagonalizes the play out of the range which are wins for the first player. □

(7.5.4) **Proposition (Gale and Stewart 1953).** *The game on any closed or open subset of the Baire line is determined.*

Proof. Suppose first that A is closed. As we noted above, A must then be the set of all the infinite paths through some tree. So if the second player does not have a winning strategy, then the first player has a winning strategy, which amounts simply to not making a mistake (i.e. making a move which goes outside the tree). If A is open, then its complement is closed, and so this case reduces to the previous one with the roles of the two players reversed. □

These results give us a large stock of determined games, but they are far from exhausting what there is to be said about which games are determined. We shall return to this issue in part IV, where it will lead us to consider issues concerning further axioms taking us beyond the default theory.

Notes

The notion of a partially ordered set is ubiquitous in mathematics. However, the theory of such structures typically delivers significant results only when the general notion is restricted in various ways. The theory of lines, which results from one such restriction, was studied extensively by Cantor, Hausdorff and others.

The Baire line is so named because it was Baire (1909) who first recommended using it as a tool for descriptive set theory.

Chapter 8

Real numbers

We noted in the Introduction to this part of the book that Weierstrass based his rigorous presentation of the calculus on the assumption that the real numbers form a complete ordered field — that is to say, a continuum on which are defined operations of addition and multiplication satisfying the familiar algebraic laws. Our primary goal in this chapter will be the construction of such a set in our default theory **ZU**. We went some way towards this in the last chapter when we showed how to construct a continuum. But an ordering is not enough structure for the purpose of measurement: addition and multiplication cannot be defined in terms of order structure alone. To see why, just consider a number line made out of elastic: stretching the elastic in a non-uniform manner will distort the *algebraic* relations between the numbers but leave their order relations intact. So if we are to carry out our programme we must start afresh: our strategy will be to use the set $\boldsymbol{\omega}$ of natural numbers to construct in turn sets $\mathbf{Z}$ (of integers), $\mathbf{Q}$ (of rational numbers) and $\mathbf{R}$ (of real numbers) endowed with an algebraic structure as well as an ordering.

8.1 Equivalence relations

Our constructions of the first two of these sets will make use of a standard technique of set construction known as the 'method of equivalence classes'.

Definition. *A relation on a set A is called an* equivalence relation *if it is transitive, reflexive and symmetric.*

The smallest equivalence relation on a set A is the diagonal relation defined by equality $x = y$.

Definition. *The* equivalence classes *of an equivalence relation s on A are the sets $s[a] = \{x \in A : a\ s\ x\}$. The set $\{s[a] : a \in A\}$ of all the equivalence classes is called the* quotient *of A by s and written A/s.*

Equivalence classes are, of course, not classes at all (at least not if we resist the temptation to identify sets with classes), but the terminology is so standard that it would be silly to deviate from it by calling them 'equivalence sets' as perhaps we strictly should. At any rate, what is important about them is that if s is an equivalence relation on A then

$$(\forall a, b \in A)(a\ s\ b \Leftrightarrow s[a] = s[b]).$$

Definition. *A collection $\mathcal{B}$ of subsets of a set A is a* partition *of A if each element of A belongs to exactly one element of $\mathcal{B}$.*

Another way of putting this is that a partition of a set A is a pairwise disjoint collection whose union is A. The connection between partitions and equivalence relations is provided by the following theorem.

(8.1.1) **Proposition.** *If A is a set, then the function $s \mapsto A/s$ is a one-to-one correspondence between the equivalence relations on A and the partitions of A; its inverse function maps a partition $\mathcal{B}$ to the equivalence relation defined on A by the formula $(\exists B \in \mathcal{B})(x, y \in B)$.*

Proof. Exercise 3. □

What this tells us is that in order to specify an equivalence relation, it is a matter of indifference whether we define the relation itself or the corresponding partition of the domain into equivalence classes.

Exercises

1. Show that a reflexive relation s on A is an equivalence relation iff

$$(x\ s\ y \text{ and } z\ s\ y \text{ and } z\ s\ t) \Rightarrow x\ s\ t.$$

2. Show that the intersection of a family of equivalence relations on A is also an equivalence relation on A but that the union of two equivalence relations on A need not be.

3. Prove proposition 8.1.1.

4. If r is a relation on A and $s = (r \cup r^{-1})^{\mathbf{T}}$, show that s is the smallest equivalence relation on A containing r and that r/s is a partial ordering on A/s.

5. If f is a function from A to A and s is an equivalence relation on A, find a necessary and sufficient condition on s for f/s to be a function from A/s to A/s.

8.2 Integral numbers

Our task in this section is to construct a set that mimics the integral numbers, both positive and negative. The guiding idea is that every integer can be written in the form $m - n$ with $m, n \in \boldsymbol{\omega}$. This representation is not unique, of course, but since

$$m - n = m' - n' \Leftrightarrow m + n' = m' + n,$$

we shall wish to treat (m, n) and (m', n') as representing the same integer if $m + n' = m' + n$. So let us write this last relation as $(m, n) \sim (m', n')$: it is obviously an equivalence relation, so we can let $\mathbf{Z} = (\boldsymbol{\omega} \times \boldsymbol{\omega})/\sim$. The definitions of addition, multiplication and the ordering relation can easily be worked out by informal calculations. For instance, the calculation

$$(m - n) + (m' - n') = (m + m') - (n + n')$$

suggests that, writing $[m, n]$ for the equivalence class of the ordered pair (m, n) with respect to $\sim$, we should define

$$[m, n] + [m', n'] = [m + m', n + n'].$$

Similarly, we want

$$(m - n)(m' - n') = (mm' + nn') - (mn' + nm'),$$

and so we define

$$[m, n][m', n'] = [mm' + nn', mn' + nm'].$$

And in order for it to be the case that

$$m - n \leqslant m' - n' \Leftrightarrow m + n' \leqslant n + m',$$

we need to define

$$[m, n] \leqslant [m', n'] \Leftrightarrow m + n' \leqslant n + m'.$$

Before going any further we ought first to check these definitions for consistency, i.e. show that they are independent of the particular ordered pairs chosen to represent the equivalence classes. But this is in each case easy. Then we need to embed the natural numbers in this construction by associating each natural number n with the integer $n - 0$, i.e. by letting $n_{\mathbf{Z}} = [n, 0]$. Once all this is done, it is a straightforward matter to check that the standard algebraic properties of the integers are satisfied by our constructions, and thus to prove the following theorem.

(8.2.1) **Theorem.** *The set* $\mathbf{Z}$ *has defined on it*

(*O*) *a relation* $\leqslant$,

(*A*) *an operation* $(j, k) \mapsto j + k$,

(*M*) *an operation* $(j, k) \mapsto jk$,

(*I*) *a function* $n \mapsto n_{\mathbf{Z}}$ *from* $\boldsymbol{\omega}$ *to* $\mathbf{Z}$

They have these properties:

(A1) $(\forall j,k,l \in \mathbf{Z})\ j+(k+l)=(j+k)+l$;

(A2) $(\forall k \in \mathbf{Z})\ k+0_{\mathbf{Z}}=k$;

(A3) $(\forall k \in \mathbf{Z})(\exists k' \in \mathbf{Z})\ k+k'=0_{\mathbf{Z}}$;

(A4) $(\forall j,k \in \mathbf{Z})\ j+k=k+j$;

(M1) $(\forall j,k,l \in \mathbf{Z})\ (jk)l=(jk)l$;

(M2) $(\forall k \in \mathbf{Z})\ k1_{\mathbf{Z}}=k$;

(M3) $(\forall j,k \in \mathbf{Z})\ jk=kj$;

(M4) $(\forall j,k \in \mathbf{Z})(jk=0_{\mathbf{Z}} \Rightarrow j=0_{\mathbf{Z}} \text{ or } k=0_{\mathbf{Z}})$;

(AM) $(\forall j,k,l \in \mathbf{Z})\ j(k+l)=jk+jl$;

(O) $\leqslant$ *is a total ordering on* $\mathbf{Z}$;

(OA) $(\forall j,k,l \in \mathbf{Z})(j \leqslant k \Rightarrow j+l \leqslant j+l)$;

(OM) $(\forall j,k,l \in \mathbf{Z})(j \leqslant k \text{ and } l \geqslant 0 \Rightarrow jl \leqslant kl)$;

(IA) $(\forall m,n \in \boldsymbol{\omega})\ (m+n)_{\mathbf{Z}}=m_{\mathbf{Z}}+n_{\mathbf{Z}}$;

(IM) $(\forall m,n \in \boldsymbol{\omega})\ (mn)_{\mathbf{Z}}=m_{\mathbf{Z}}n_{\mathbf{Z}}$;

(IO) $(\forall m,n \in \boldsymbol{\omega})\ (m \leqslant n \Leftrightarrow m_{\mathbf{Z}} \leqslant n_{\mathbf{Z}})$.

(8.2.2) **Proposition.** $\mathbf{Z}$ *is countable.*

Proof. We already know that $\boldsymbol{\omega} \times \boldsymbol{\omega}$ is countable [proposition 6.5.1]. As we have defined it, $\mathbf{Z}$ is a quotient of this set and hence countable as well. □

8.3 Rational numbers

Now we repeat the process, but this time we want to construct rational numbers of the form j/k, so we use ordered pairs (j,k) where $j \in \mathbf{Z}$ and $k \in \mathbf{Z} \setminus \{0_{\mathbf{Z}}\}$. We want

$$\frac{j}{k}=\frac{j'}{k'} \Leftrightarrow jk'=j'k,$$

so we define

$$(j,k) \sim (j',k') \Leftrightarrow jk'=j'k$$

to give us an equivalence relation $\sim$ on the set $\mathbf{Z} \times (\mathbf{Z} \setminus \{0_{\mathbf{Z}}\})$, and then let

$$\mathbf{Q} = (\mathbf{Z} \times (\mathbf{Z} \setminus \{0_{\mathbf{Z}}\}))/\sim.$$

Once more it is an easy matter to work out the right definitions. We want

$$\frac{j}{k} + \frac{j'}{k'} = \frac{jk' + j'k}{kk'},$$

so we define

$$[j, k] + [j', k'] = [jk' + j'k, kk'].$$

We want

$$\frac{j}{k}\frac{j'}{k'} = \frac{jj'}{kk'},$$

so we define

$$[j, k][j', k'] = [jj', kk'].$$

And we want

$$\frac{j}{k} \leqslant \frac{j'}{k'} \Leftrightarrow jk' \leqslant j'k$$

in the case when $k, k' \geqslant 0$, so we write

$$[j, k] \leqslant [j', k'] \Leftrightarrow jk' \leqslant j'k$$

in this case. Once more we have to check that all our definitions are consistent, but once more this is easy. Finally we wish to identify any integer k with the rational number $k/1$, so we define

$$k_{\mathbf{Q}} = [k, 1].$$

Many of the purely algebraic properties of the rational numbers thus defined are summarized by saying that they form an *ordered field*. The definition is as follows.

Definition. *An* ordered field *is a set F endowed with*

(O) a relation $\leqslant$,

(A) an operation $(x, y) \mapsto x + y$,

(M) an operation $(x, y) \mapsto xy$,

(I) a function $k \mapsto k_F$ from $\mathbf{Z}$ to F.

They are required to have the following properties:

(A1) $(\forall x, y, z \in F)\, x + (y + z) = (x + y) + z;$

(A2) $(\forall y \in F)\, y + 0_F = y;$

(A3) $(\forall y \in F)(\exists y' \in F)\, y + y' = 0_F;$

(A4) $(\forall x, y \in F)\, x + y = y + x;$

(M1) $(\forall x, y, z \in F)\, (xy)z = x(yz);$

(M2) $(\forall y \in F)\, y 1_F = y;$

(M3) $(\forall x, y \in F)\, xy = yx;$

(M4) $(\forall x \in F \setminus \{0_F\})(\exists x' \in F \setminus \{0_F\})\, xx' = 1_F;$

(AM) $(\forall x, y, z \in F)\, x(y + z) = xy + xz;$

(O) $\leqslant$ *is a total ordering on* F;

(OA) $(\forall x, y, z \in F)(x \leqslant y \Rightarrow x + z \leqslant x + z);$

(OM) $(\forall x, y, z \in F)(x \leqslant y \text{ and } z \geqslant 0 \Rightarrow xz \leqslant yz);$

(IA) $(\forall j, k \in \mathbf{Z})\, (j + k)_F = j_F + k_F;$

(IM) $(\forall j, k \in \mathbf{Z})\, (jk)_F = j_F k_F;$

(IO) $(\forall j, k \in \mathbf{Z})(j \leqslant k \Leftrightarrow j_F \leqslant k_F).$

We shall take for granted the development of the arithmetic of elements of an ordered field from this definition. Thus, for instance, we shall take it for granted that we can define the *absolute value* of an element x of an ordered field F by letting

$$|x| = \begin{cases} x & \text{if } x \geqslant 0 \\ 0 & \text{otherwise.} \end{cases}$$

Note that any ordered field is a line, since if $x < y$, then $x < \frac{1}{2}(x + y) < y$.

The definition we have given translates easily into a list of axioms in a first-order language containing as its non-logical symbols only addition, multiplication and the order relation. So the notion of an ordered field is finitely first-order axiomatizable. The development of the properties of ordered fields we are referring to is representable by means of derivations from these axioms in first-order logic.

(8.3.1) **Theorem.** $\mathbf{Q}$ *is an ordered field.*

(8.3.2) **Proposition.** $\mathbf{Q}$ *is countable.*

Proof. $\mathbf{Z}$ is countable [proposition 8.2.2], hence $\mathbf{Z} \times (\mathbf{Z} \setminus \{0\})$ is countable, hence $\mathbf{Z} \times (\mathbf{Z} \setminus \{0\})/\sim$ is countable. □

(8.3.3) **Corollary.** *The ordered set* $(\mathbf{Q}, \leqslant)$ *is isomorphic to* $(\mathbb{Q}, \leqslant)$.

Proof. $\mathbf{Q}$ is a line since it is an ordered field; and we have just shown that it is countable. The result now follows from Cantor's characterization of the rational line [theorem 7.1.2]. □

8.4 Real numbers

(8.4.1) **Theorem.** *There exists a complete ordered field.*

Proof. The real line $\mathbb{R}$ has a countable dense subset, which is isomorphic to the ordered set $\mathbf{Q}$. Now it is easy to check that the operation of addition on $\mathbf{Q}$ is normal in each variable. So we can use proposition 7.3.5 to extend it to an operation of addition on the completion $\mathbb{R}$. It is then a straightforward (if tedious) matter to check that all the properties of addition required by the definition of an ordered field (associativity, commutativity, etc.) are satisfied by the operation on $\mathbb{R}$ thus defined.

In much the same way it is easy to show that the operation of multiplication by a positive rational number is normal, so we can extend it so as to define multiplication by a positive real number. Then we can use this to define multiplication by a negative real number in the obvious way. Checking that the operation thus defined satisfies the requirements of the definition is even more tedious than in the case of addition. □

We have thus carried out (or, to be more accurate, sketched) the construction of a complete ordered field. The method of construction is not unique, of course: there are other quite different ways of proving theorem 8.4.1 (for instance, by using equivalence classes of Cauchy sequences of rational numbers). However, although the condition of being a complete ordered field does not determine a unique structure, all such structures are isomorphic (see corollary 8.7.6 below). So we are in a similar position to the one we found ourselves in when we defined the set of natural numbers: we need to pick one structure of a certain sort, but any we do pick has accidental properties we do not want. We shall not repeat the comments we made there about this dilemma.

Definition. *We choose some (pure) complete ordered field (of lowest possible birthday): we denote it* **R** *and call its members* real numbers.

We noted in the Introduction that a rigorous development of the calculus of functions of a real variable is available, based on the assumption that the real numbers form a complete ordered field. If we simply appended that rigorous development at this point, we would thus obtain a representation of a large part of mathematics in ZU.

But what would that show? At the very least, it would give us another relative consistency result: if ZU is consistent, so is the Weierstrassian theory of the continuum. But if the further step of reducing set theory to some sort of logic were possible, we could say much more, for we should then have achieved a species of logicism — a grounding of the calculus in logic. And if the logic in question were knowable independent of intuition, this would presumably constitute a final refutation of Kant's view. But, as we saw in part I, the project of reducing set theory to anything deserving the name of logic is fraught with difficulty. And in any case it is far from clear that the logic in question here could reasonably be regarded as analytic in Kant's sense. So the work presented here falls some way short of definitively refuting Kant.

Nonetheless, the grounding of analysis in set theory, even if it does not refute Kant outright, does something to weaken his case. For it seems much less persuasive in the case of set theory than it did in the theory of real numbers to argue that the intuitions we need to ground our knowledge of it are spatio-temporal in character. It is for this reason that the set-theoretic reduction of mathematics has given rise to a tradition, starting perhaps with Dedekind and then progressing through Gödel to Dummett, which conceives of reason as capable of constructing intuitions from the structure not of spatio-temporal experience but of thought itself.

8.5 The uncountability of the real numbers

From the work we have done we can deduce that $\mathbf{R}$ is uncountable, but our route to this conclusion has been rather indirect. In view of the historical and philosophical importance of the result, let us allow ourselves the luxury of a second, more direct proof. We start by defining a set much favoured by analysts as a tool in constructing counterexamples. Writing $[a, b]$ for the closed, bounded interval $\{x \in \mathbf{R} : a \leqslant x \leqslant b\}$ as usual, we let

$$\begin{aligned} K_0 &= [0, 1] \\ K_1 &= [0, \tfrac{1}{3}] \cup [\tfrac{2}{3}, 1] \\ K_2 &= [0, \tfrac{1}{9}] \cup [\tfrac{2}{9}, \tfrac{1}{3}] \cup [\tfrac{2}{3}, \tfrac{7}{9}] \cup [\tfrac{8}{9}, 1] \\ &\dots \end{aligned}$$

Then *Cantor's ternary set* is

$$K = \bigcap_{n \in \omega} K_n.$$

Otherwise put, K is the closed set we get from the unit interval if we remove its open middle third, remove the middle third of each of the two pieces remaining, remove the middle thirds of each of the four pieces remaining after that, and so on.

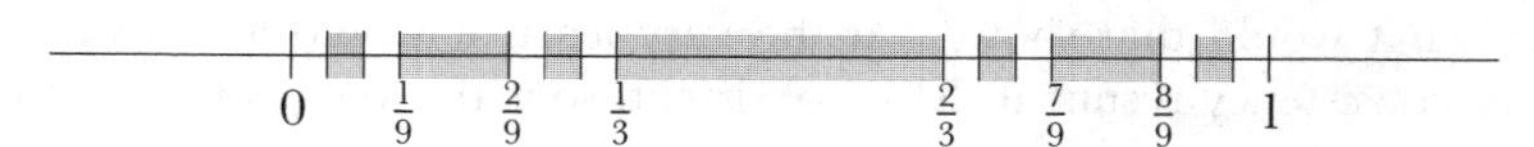

Yet another way of visualizing it is as the set of those real numbers of the form $\sum_{n=1}^{\infty} s_n/3^n$ with $s_n = 0$ or 2 for all n, i.e. numbers between 0 and 1 with ternary expansions (representations to the base 3) in which the digit 1 does not occur.

Suppose now that s is any sequence of real numbers. We define a sequence $(I_n(s))$ of closed intervals of Cantor's ternary set inductively as follows. We start by letting $I_0(s) = [0, 1]$. Then once $I_n(s) = [a_n, b_n]$ has been determined, we remove its middle third as above to leave the two closed intervals $[a_n, a_n + \frac{1}{3}(b_n - a_n)]$ and $[b_n - \frac{1}{3}(b_n - a_n), b_n]$; we then let $I_{n+1}(s)$ be the left-hand of these two remaining parts unless $s(n)$ belongs to that part, in which case we let it be the right-hand one. Now the sets

$$I_0(s), I_1(s), I_2(s), \ldots, I_n(s), \ldots$$

form a nested sequence of closed intervals of lengths

$$1, \tfrac{1}{3}, \tfrac{1}{9}, \ldots, \tfrac{1}{3^n}, \ldots.$$

So their left-hand endpoints form a bounded, increasing sequence. It therefore has a least upper bound by the completeness property, which is less than the right-hand endpoints of all the $I_n(s)$ and hence belongs to $\bigcap_{n\in\omega} I_n(s)$. If we call it $f(s)$, we have thus defined a function f from ${}^{\omega}\mathbf{R}$ into K. Now the point of defining f in this manner is that it follows easily from the construction that $f(s) \neq s(n)$ for all n. In other words, the function f generates, for each sequence of real numbers s, a real number $f(s)$ not in its range.

(8.5.1) **Proposition.** *K is uncountable.*

Proof. If K were countable, there would by definition be a sequence s such that $\text{im}[s] = K$. But $f(s) \in K \setminus \text{im}[s]$. Contradiction. □

(8.5.2) **Corollary.** *$\mathbf{R}$ is uncountable.*

Proof. Cantor's ternary set is a subset of $\mathbf{R}$. Since the former is uncountable, so is the latter. □

The function f defined above is called a *diagonal function*, and the use of it to prove the uncountability of $\mathbf{R}$ is an instance of a *diagonal argument*. This argument has been the focus of a great deal of criticism. One complaint that is certainly unfair is that the proof is inexplicit: it does not exhibit any *particular* real number that we cannot count, but then we would not expect it to; what it does is to generate from any sequence of real numbers a number not in the range of that sequence. A complaint that has more justice is that the proof is problematic not for want of explicitness but because it is *impredicative*. This is because the diagonalizing function f maps *down* the hierarchy, so that $f(s)$

is a set of lower birthday than s. (Recall that s is a *sequence* of real numbers and must therefore lie higher up the hierarchy than the members of the sequence.) This is the reason why the proof of uncountability is unacceptable to the constructivist: the objection is that although we have defined $f(s)$ explicitly, we have done so only in terms of s. In truth, the point might be better put by saying that the *definition* of the real numbers is impredicative, since the proof of uncountability does no more than exploit the definition. What it should remind us of, at any rate, is that the failure of the countable line to model the continuum arises only because we require *all* continuous functions to have the intermediate value property, including ones that are not definable in first-order terms.

8.6 Algebraic real numbers

The theory consisting of all the sentences in the *first-order* language of ordered fields that are true of the real numbers has non-isomorphic models, which are known as *real-closed fields*. The theory so described is obviously complete, but it also turns out, much less obviously, to be axiomatizable (Tarski 1948). Since the language of ordered fields is countable, model theory (the Löwenheim/Skolem theorem) tells us that the theory must have a countable model, and indeed that there must be a model of it that is a subfield of **R**. The minimal such example is the set of algebraic real numbers.

Definition. *A real number is said to be* algebraic *if it is a root of a polynomial equation with rational coefficients;* transcendental *if not. The set of all algebraic real numbers is denoted* **A**.

We shall not here make the detour into algebra necessary to demonstrate that **A** is indeed a real-closed field (or even that it is a field). However, we do not need to use either this last fact or the model-theoretic considerations just mentioned in order to show that the algebraic numbers are countable.

(8.6.1) **Proposition. A** *is countable.*

Proof. The set **Q** of rational numbers is countable [proposition 8.3.2]. So the set String(**Q**) of strings of rational numbers is countable [proposition 6.5.3]. So the set **Q**[x] of polynomials in the indeterminate x with rational coefficients is countable (since each of them can be represented by the string consisting of its coefficients). So **Q**[x] $\times\, \boldsymbol{\omega}$ is countable [proposition 6.5.1]. Now take any ordered pair (p, r) in this last set: define $f(p, r)$ to be the rth real root of the equation $p(x) = 0$, if such a root exists; and 0 otherwise. Then f maps the countable set **Q**[x] $\times\, \boldsymbol{\omega}$ onto **A** (since a polynomial of degree n has at most n real roots). Therefore **A** is countable too. □

(8.6.2) **Corollary.** *There exist transcendental real numbers.*

Proof. **A** is countable; **R** is not. □

Around the time that Cantor discovered this proof of their existence in (1874), other mathematicians were finding explicit examples of transcendental numbers: Liouville had shown (1844) that $\sum_{n=1}^{\infty} 1/k^{n!}$ is transcendental for any integer $k > 1$; then Hermite (1873) proved that e is transcendental, and Lindemann (1882) proved that π is. Various other examples were discovered subsequently: for instance, it was proved in 1930 that $2^{\sqrt{2}}$ is transcendental. So Cantor was not at this point proving something new. Moreover, although his proof is admittedly somewhat swifter than Liouville's (and of course reveals connections that the other does not), it has seemed clear to various authors (e.g. Kac and Ullam 1968; G. H. Moore 1982) that this brevity comes at the price that it proves the existence of a transcendental number without providing a means of finding an example. This is quite wrong, however: the difference between these proofs and Cantor's is *not* that they are any more explicit; the method we have given for counting the algebraic numbers could easily be made effective, and Cantor's diagonal argument would then give us an explicit construction of a transcendental real number.

The objection to Cantor's method of obtaining a transcendental number is not that it is not explicit, therefore, but only that it is not pretty. The dependence of the roots of a polynomial on its coefficients, although of course continuous, is computationally very unstable: Wilkinson (1959) notes, for instance, that if we perturb just one of the coefficients of the polynomial

$$(x+1)(x+2)(x+3)\dots(x+20)$$

by 2^{-23} we obtain a polynomial with only 10 real roots whose other 10 roots all have imaginary parts between 0.8 and 3. So although we could turn Cantor's procedure into an algorithm to calculate a transcendental number in decimal notation to any required degree of accuracy, some care is required to ensure that it is not too demanding computationally (see Davenport, Siret and Tournier 1993, §3.2.1). It turns out, in fact, that there *is* a reasonable algorithm of complexity $O(n^2 \log^2 n \log\log n)$ (Gray 1994), but plainly if what we want is a transcendental number expressed in decimal notation, Liouville's method is much simpler: putting $k = 10$ gives us, without doing any computational work, the explicit example

$$\sum_{n=1}^{\infty} \frac{1}{10^{n!}} = 0.110\,001\,000\dots$$

(known as Liouville's number). It is worth stressing, though, that the computational simplicity of a method is highly sensitive to the notation used: if we expressed Liouville's number to a base other than 10, for instance, its apparent simplicity would evaporate.

What distinguishes these different examples of transcendental numbers is thus not a matter of explicitness. They may, however, differ in regard to how natural they are, or how genuinely mathematical. These are distinct (though related) criteria for assessing examples, the first fairly precise, the second less so. An example is *natural* if it is independent of any arbitrary choice, such as a choice of representational scheme or of coding. Both Liouville's number and Cantor's transcendental numbers are on this criterion unnatural — indeed by varying the enumeration of the algebraic numbers that is diagonalized, Cantor's method can be made to produce any transcendental number whatever (Gray 1994) — whereas e, π and $2^{\sqrt{2}}$ are natural, since no arbitrary choices are involved in their definitions. An example counts as *genuinely mathematical*, on the other hand, if, roughly, its interest to mathematicians extends beyond the mere fact that it *is* an example. On this criterion e and π are plainly genuine by any lights, but what to say about $2^{\sqrt{2}}$ is somewhat less clear, and it takes a certain sort of number theorist to find Liouville's number genuinely mathematical.

What counts as genuine, in the sense in which we are now using the word, is evidently to some extent a psychological matter and cannot be given a precise definition. Indeed the scope of the term has no doubt varied as the interests of mathematicians have varied. Note that genuinely mathematical examples are very often natural, since the very arbitrariness of an unnatural example counts against its mathematical interest, but the linkage is not exceptionless, since a scheme of representation may itself become significant through long usage and utility. Most laymen (though few professional mathematicians) treat the decimal representation of numbers, for instance, as sufficiently central to their conception of them for examples dependent on that scheme to seem genuinely mathematical to them. (Hence, perhaps, the greater interest taken by laymen in such questions as whether the various digits occur with equal frequencies in the decimal expansion of π.)

8.7 Archimedean ordered fields

Suppose that F is an ordered field. The intersection of all the subfields of F (i.e. the subsets that are ordered fields with respect to the operations they inherit as subsets of F) is itself an ordered field, called the *prime* subfield of F and denoted F'. It is not hard to show that F' is always isomorphic to $\mathbf{Q}$. Moreover, the prime subfield of $\mathbf{Q}$ is just $\mathbf{Q}$ itself, so we may think of $\mathbf{Q}$ as being the *minimum* ordered field: every ordered field contains a copy of it. At the other extreme, however, there are in the corresponding sense no maximal ordered fields: ordered fields can be constructed of arbitrarily large cardinality. (This is a consequence of the fact that the notion of an ordered field can be given a first-order axiomatization.) In order to obtain maximality we need to add a further constraint known as *Archimedes' property*.

Definition. *An ordered field F is* archimedean *if its prime subfield is unbounded in F.*

Trivially $\mathbf{Q}$ is archimedean, since its prime subfield is itself. Moreover, the natural numbers are not bounded above in $\mathbf{Q}$, so an ordered field F is archimedean iff for each $x \in F$ we have

$$x < \underbrace{1_F + 1_F + \cdots + 1_F}_{n \text{ terms}}$$

for some natural number n. An element of F is said to be *infinitely large* if its absolute value is greater than every element of the prime subfield, and *infinitesimal* if its absolute value is smaller than every positive element of the prime subfield. So saying that F is archimedean is equivalent to saying that it contains no infinitely large elements, or (since x is infinitely large iff $1/x$ is infinitesimal) that it contains no non-zero infinitesimals.

As an example consider the set $\mathbf{Q}(\varepsilon)$ of rational functions in the indeterminate ε with rational coefficients, i.e. functions of the form

$$f(\varepsilon) = \frac{a_n\varepsilon^n + a_{n-1}\varepsilon^{n-1} + \cdots + a_0}{b_m\varepsilon^m + a_{m-1}\varepsilon^{m-1} + \cdots + b_0},$$

where all the coefficients $a_0, a_1, \ldots, a_n$ and $b_0, b_1, \ldots, b_m$ are rational numbers. These functions can be added and multiplied in the usual way, and we can order them by stipulating that the function is to count as positive if a_n and b_m have the same sign (both positive or both negative). It is easy to check that with these definitions $\mathbf{Q}(\varepsilon)$ becomes an ordered field. However, for any rational a we have $a - \varepsilon > 0$, i.e. $a > \varepsilon$, and so ε is an infinitesimal. Thus the field $\mathbf{Q}(\varepsilon)$ is non-archimedean.

(8.7.1) **Lemma.** *Any complete ordered field is archimedean.*

Proof. Suppose that F is a complete ordered field but F' is bounded in F, and let $a = \sup F'$. Now $a - 1 < a$, and so $a - 1$ is not an upper bound for F'; i.e. there exists $r \in F'$ such that $a - 1 < r$. But then $a < r + 1 \in F'$, contradicting the definition of a. □

It follows from lemma 8.7.1 that the prime subfield $\mathbf{R}'$ of $\mathbf{R}$ is isomorphic to $\mathbf{Q}$. The isomorphism is easy to give explicitly: a rational number m/n corresponds to the real number $m_{\mathbf{R}}n_{\mathbf{R}}^{-1}$. It is customary to identify $\mathbf{R}'$ with $\mathbf{Q}$, i.e. to make no distinction between a rational number and the corresponding real number.

(8.7.2) **Lemma.** *An ordered field F is archimedean if and only if its prime subfield F' is dense in F.*

Necessity. If F is archimedean and $x < y$ in F, then there exists $s \in F'$ such that $\frac{1}{y-x} < s$, i.e. $ys - xs > 1$. So there is an integer k such that $xs < k_F < ys$, i.e. $x < k_F s^{-1} < y$.

Sufficiency. Trivial. □

(8.7.3) **Proposition.** *The ordered set* $(\mathbf{R}, \leqslant)$ *is isomorphic to* $(\mathbb{R}, \leqslant)$.

Proof. The prime subfield of $\mathbf{R}$ is dense [lemma 8.7.2] and, since it is isomorphic to $\mathbf{Q}$, it is also countable [proposition 8.3.2]. Moreover, as an ordered set $(\mathbf{R}, \leqslant)$ is complete by definition. Hence it is a continuum and it follows that it is isomorphic to $(\mathbb{R}, \leqslant)$. □

Recall that $\mathbf{R}$ and $\mathbb{R}$ were both defined by arbitrary conventions: this proposition shows that we could have chosen them to be the same set.

(8.7.4) **Proposition.** *The ordered set* $(\mathbf{R} \setminus \mathbf{Q}, \leqslant)$ *is isomorphic to the Baire line* $({}^{\omega}\omega, \leqslant)$.

Proof. Immediate [corollary 7.5.2]. □

In fact, an isomorphism between $\mathbf{R} \setminus \mathbf{Q}$ and ${}^{\omega}\omega$ can be defined explicitly. The function which takes a sequence s in ${}^{\omega}\omega$ to the continued fraction

$$\cfrac{1}{1 + s(0) + \cfrac{1}{1 + s(1) + \cfrac{1}{1 + s(2) + \cfrac{1}{\vdots}}}}$$

is strictly increasing and its range consists of the irrational numbers between 0 and 1. To obtain an isomorphism with the whole of $\mathbf{R} \setminus \mathbf{Q}$, we have to compose this with a strictly increasing rational function which maps the unit interval $\{x : 0 < x < 1\}$ onto the whole real line, for instance

$$x \mapsto \frac{1 - 2x}{x(x-1)}.$$

The proof that this all works as advertised is in Truss 1997, ch. 10.

The isomorphism between the order structures of $\mathbf{R} \setminus \mathbf{Q}$ and ${}^{\omega}\omega$ is very far from unique, of course, but once we have chosen one, we can use it to encode any two-person game by means of a set of irrational numbers. By this means the study of such games is transformed into part of the theory of sets of real numbers.

(8.7.5) **Theorem.** *An ordered field is archimedean if and only if it is isomorphic to a subfield of* **R**.

Sufficiency. Clearly any subfield of an archimedean field is archimedean. **R** is archimedean [lemma 8.7.1], hence so is any field isomorphic to one of its subfields.

Necessity. Suppose that F is an archimedean ordered field. There is an isomorphism f between its prime subfield F' and the prime subfield of **R** (since they are both isomorphic to **Q**). This isomorphism is evidently normal. Moreover F' is dense in F since F is archimedean. So the isomorphism extends to a normal function $\overline{f}$ from F into **R** [proposition 7.3.5]. It is easy to check that $\overline{f}$ is an isomorphism of ordered fields. □

One consequence of this theorem is that Archimedes' principle cannot be expressed in the first-order language of ordered fields (since there is an upper bound to the cardinalities of its models). Another is that our construction in theorem 8.4.1 would not have worked if the ordered field we started from had not been archimedean: however we try to extend the operations from a non-archimedean ordered field to its order-completion, we shall inevitably find that some of the ordered field properties fail.

(8.7.6) **Corollary.** *Any complete ordered field is isomorphic to* **R**.

Proof. Let F be a complete ordered field. Then F is archimedean and therefore embeddable in **R**. The image of this embedding is complete and dense in **R** and must therefore be equal to **R**. □

Theorem 8.7.5 also provides us with a characterization of the ordered field of real numbers by a sort of maximality property: the real numbers are (up to isomorphism) the only ordered field in which every archimedean ordered field can be embedded.

In fact the whole of this theory can be generalized to the case of an ordered group, i.e. an ordered set on which only an *addition* function is defined satisfying (A1)–(A4), (O) and (OA) above. An ordered group is said to be *archimedean* if for all $x, y > 0$ there exists n such that

$$\underbrace{x + x + \cdots + x}_{n \text{ terms}} > y.$$

The only ordered group in which every archimedean ordered group can be embedded is (up to isomorphism) the additive group of the real numbers (see Warner 1968, §43). The significance of this is that it permits us to sidestep the concerns we had at the end of the last chapter about treating lines in space and time as continua. We can instead focus on convincing ourselves that points in space or time can be added together in such a way as to form an archimedean ordered group. If we succeed in convincing ourselves of this, the maximality condition can then be presented as a natural completion of the conception: a line in space, even if it is not itself complete, will be embeddable in the complete group of real numbers, bearing out Dedekind's idea of 'filling up its gaps in thought and thus making it continuous'.

8.8 Non-standard ordered fields

Suppose now that we expand our first-order language so as to include a constant for every real number and a relation symbol for every relation on **R** that is definable in set theory: call this the *extended* language. It is an easy consequence of the compactness theorem for first-order logic that the set of sentences of this language that are true about the real numbers has other non-isomorphic set-theoretic models. These other models are called *non-standard* ordered fields.

As A. Robinson (1961) was the first to observe, this gives us an elegant new method of proof: to prove that a sentence Φ in the extended language is true about **R**, move to the non-standard field $\mathbf{R}^*$ and prove that Φ is true there; it follows that Φ is true in **R** too. The method is non-constructive since its basis is model-theoretic: once we have a non-standard proof of Φ, we can be confident that a standard proof exists, but there is no general method of conversion, and the standard proof may make use of much higher levels in the set-theoretic hierarchy than the non-standard one (Henson and Keisler 1986).

Non-standard analysis therefore conforms to a pattern according to which an ideal theory (non-standard analysis) is superimposed on a real theory (the conventional theory of the Weierstrassian continuum). No requirement is imposed that the entities apparently referred to in the ideal theory (the infinitesimals) should be regarded as real objects. Their reliability is guaranteed rather by a proof of conservativeness. And their utility arises because they sometimes allow us to eliminate higher-order (hence abstract and therefore perhaps conceptually difficult) objects from proofs, with the result that the non-standard proofs may be easier for us humans to find and understand than the standard ones; or perhaps the proofs using infinitesimals are much shorter. (This is not a realization of Hilbert's programme because the proof of conservativeness is not only non-finitary but non-constructive: indeed it even makes essential use of the axiom of choice.)

A non-standard field must contain the real numbers as a proper subfield and hence is non-archimedean, but the converse need not hold: the non-archimedean field $\mathbf{Q}(\varepsilon)$, for instance, is not non-standard because there are first-order properties expressible in the extended language which distinguish it from **R**.

Proponents of non-standard analysis have amassed various quotations in support of the notion that it realizes the intentions of those earlier analysts who made use of the method of infinitesimals. Cauchy, for instance, thought that this method

> can and must be used as a means of discovery or of proof... but infinitesimal quantities should never, in my opinion, be admitted in the final equations, where their presence would be without purpose or utility. (1844, p. 13)

And Leibniz, the originator of the method, said (in a letter to Varignon of 1702) that

> if someone refuses to admit infinite and infinitesimal lines in a rigorous metaphysical sense and as real things, he can still use them with confidence as ideal concepts which shorten his reasoning, similar to what we call imaginary roots in ordinary algebra (for example $\sqrt{-2}$). ... Infinites and infinitesimals are grounded in such a way that everything in geometry, and even in nature, takes place as if they were perfect realities. (Leibniz 1996, pp. 252–4)

Whatever Leibniz's own views on the matter, however, it is clear that a further argument would be needed to convince us to place the real/ideal split at just this point. The analogy Leibniz drew with imaginary numbers is relevant: despite the terminology, modern mathematicians do not regard imaginary numbers as any less real than real numbers. Each of the extensions of the number system that we have discussed in this chapter could be presented as a means of adjoining ideal elements in order to solve a new class of equations: we move from the natural numbers to the integers so as to be able to solve $x + n = m$; we move to the rational numbers to solve $jx + k = 0$; we move to the real numbers to solve $f(x) = 0$ for any continuous f that changes sign; and we extend to the complex numbers to be able to solve $p(x) = 0$ for an arbitrary polynomial. None of these extensions seems to have much reason *prima facie* to be regarded as more ideal than any other.

But note that some of these extensions have a very particular sort of stability. Consider the first of them as an instance. We wish to be able to solve the equations $x + n = m$ for $m, n \in \boldsymbol{\omega}$, but within $\boldsymbol{\omega}$ we can do this only in the case when $m \geqslant n$, so we extend to $\mathbf{Z}$ and can then do it without restriction. But of course by extending our number system to $\mathbf{Z}$ we generate a whole new class of equations of the same form as before but now with m and n ranging over the whole of $\mathbf{Z}$. The stability we have just alluded to consists in the fact that we do not have to carry out a further extension to meet this point: $\mathbf{Z}$ already contains solutions to all these new equations as well as to the ones it was created to solve.

As we shall see, however, the extension from the standard to the non-standard real numbers does not possess the same sort of stability. The limited stability it does possess arises from a trade-off it attempts to exploit between first- and second-order formulations of analysis. The elementary theory of the real numbers, let us recall, is the theory of real-closed fields. The algebraic real numbers constitute a model of this theory. Within this model we can develop a coherent account of the calculus of polynomials, but the study of the phenomenon of differentiability in general is not available. For that to be possible we find ourselves adding a *second-order* principle (in our case the completeness axiom, although other, more limited principles might perhaps suffice). We now add to our language constants for all the objects of this second-order-specified domain and then restrict to the first-order fragment in order to apply the compactness theorem and obtain a non-standard model. It is evidently crucial to the success of this construction that Archimedes' principle should be irreducibly second-order, since it is this that the restriction to the first-order fragment exploits so as to obtain a non-archimedean extension of the real numbers that is nonetheless elementarily equivalent to it (i.e. shares all its first-order properties).

Now it is clearly quite implausible to ascribe to Cauchy or Leibniz even an inchoate understanding of a first-order transfer principle since, as we have already noted, the distinction between first- and second-order properties was made available only by the development of logic in the late 19th century, and even then it did not straightaway seem especially important. Even now very few mathematicians who are not logicians have any firm grip on which of the concepts they make use of are first- and which second-order.

But even when this point is granted and non-standard analysis is treated as no more than an allusive rational reconstruction of what the exponents of the method of infinitesimals intended, it is in danger of obscuring important insights. In particular, it is plain that many mathematicians who made use of infinitesimals did not regard them as mere ideal elements. This is by no means surprising: the pattern in mathematics has always been that elements which start out as mere posits become accepted and end up being treated as real. But if we take that course in this case and treat infinitesimals as real, we lose even the limited stability we obtained above and have no reason not to posit a new level of objects which are infinitesimal relative to the non-standard field. (To be more precise, we add to our language a constant standing for each non-standard number, and then apply the compactness theorem to obtain a proper extension which has all the same first-order properties as the non-standard field.)

We have evidently embarked now on a process which we can iterate in much the same manner as we iterated the construction of the set-theoretic hierarchy. If we do this, we arrive at a conception of the continuum which is very different from, and far richer than, the Weierstrassian one. But adopting this conception would have radical consequences too for the set-theoretic reduction we have been contemplating in this part of the book. For the proposal now under consideration is that we should conceive of the continuum as indefinitely *divisible* in much the same way as the hierarchy is indefinitely extensible, and it seems inevitable that if this idea is thought through it will eventually lead us to abandon the idea that the continuum is a *set* of points at all.

Notes

The set-theoretic construction of a complete ordered field which we have sketched here is essentially due to Dedekind (1872). Any reader who wants all the details will find them laid out in numerous textbooks: Landau's *Grundlagen der Analysis* (1930) is my own particular favourite for nostalgic reasons, but Henkin et al. 1962 is also commendably clear and detailed. Another construction with an equally long history uses the quotient of the set of Cauchy sequences of rational numbers by the relation which makes two sequences equivalent if the difference between their terms tends to zero.

The development of real analysis on some variant of the assumption that the real numbers form a complete ordered field can be found in numerous places: many generations of British university students learnt the material from Hardy 1910; among more recent texts my own favourite is Stromberg 1981.

We have mentioned only in passing the theory of real-closed fields, fields elementarily equivalent to the real numbers, the study of which goes back to Artin and Schreier (1927). For an exposition consult van der Waerden 1949, ch. 11. Tarski's proof that the theory is axiomatizable proceeds by the method of quantifier elimination and in fact yields a decision procedure for all sentences of the language. The fruitfulness of these ideas is well described by van den Dries (1988).

The theory of real-closed fields is closely related to that of Euclidean constructions, i.e. to the question of which geometrical constructions are possible if we restrict ourselves to the Euclidean tools of straightedge and compasses. The extraordinary hold this study has had on the imaginations of mathematicians focused for many centuries on three problems in particular — squaring the circle, duplicating the cube, and trisecting the angle. The first of these entered the language as a metaphor for any impossible task even before Lindemann's (1882) proof that π is transcendental had shown it to be insoluble.

The best introduction to non-standard analysis for those familiar with the standard version remains that of its originator, A. Robinson (1974), but Hurd and Loeb (1985) is also good. For a briefer account see Abian 1974. Tall (1982) describes a simple axiomatization of non-standard analysis. Keisler (1976) and Henle and Kleinberg (1980) give non-standard expositions of the calculus *ab initio* which do not rely on familiarity with the standard treatments.

Whether the standard Weierstrassian account correctly models the geometrical continuum of intuition is a question that has been discussed by a number of writers. There are some relevant remarks in Hobson 1921, vol. I, §§41–5 & 63.

Conclusion to Part II

In this part of the book we have shown how the classical theories of numbers and of functions of a real variable may be embedded in ZU. We can separate out two parts to this process: first we identify the axioms of the theory; then we construct a set-theoretic model of them. Any critique of the first part is specific to the discipline in question and, although we have paused briefly to discuss the issue in the case of the continuum, it is not strictly the concern of the set theorist. What, then, of the second part? What does it establish?

It is at any rate uncontroversial that we have established a series of non-negligible relative consistency results: if ZU is consistent, then so are Peano arithmetic and the theory of complete ordered fields. And we should be careful not to under-estimate now the genuine doubts some logicians in the 1920s and 1930s harboured as to the consistency of theories such as these which we do not now regard as being at all dubious.

On the other hand, even then no one thought set theory was *more* secure than number theory, so the interest of these relative consistency results is not that they increase our confidence in the weaker theories but rather that they provide us with a way of calibrating their relative strength. Moreover, this use of set theory to calibrate strength can be applied to individual theorems and their proofs. As we have seen, an explicit example of this is Dirichlet's theorem, whose original proof, if transcribed into our system, would make use of higher levels in the set-theoretic hierarchy than the elementary proof discovered much later.

Notice, though, that there is an important issue of stability to be addressed here. As we have mentioned several times already, the first stage of the modelling process, in which we specify the internal properties the set-theoretic model is to have, falls far short of determining uniquely the outcome of the second. Plainly any calibration of set-theoretic strength which turns out to be dependent on parochial features of the chosen embedding is of no special significance. What we want, therefore, is stability: perhaps it would be too much to expect a measure of strength that is always completely independent of the embedding, but we might at least hope for one that does not depend on it very much. Experience suggests, however, that even this is harder than it looks. The use our embedding makes of the height of the hierarchy is, for the overwhelming majority of mathematical applications, needlessly inefficient and could be much reduced by clever use of coding. In such cases it is not always as obvious as it should be what is the true measure of the abstractness of the mathematics involved.

Another theme has recurred throughout the last century: the fact that the theory of real numbers, and by extension most of the rest of mathematics, can be interpreted in set theory has been taken to show that they can be thought of in some significant sense as being *part* of set theory. The reasons that have led people to think this can be grouped into two sorts, one (in my view) distinctly more promising than the other.

The less promising argument takes as its ground a principle of ontological economy (Occam's razor): since the specification for the natural numbers which we drew up in the first stage of the process listed enough of their properties to characterize them up to isomorphism, and since we have shown that there is a *set* which has just these properties, it would multiply entities beyond necessity to suppose that the natural numbers are anything other than the members of the set-theoretic model. That this is a bad argument irrespective of the general merits of Occam's razor was lucidly exposed by Benacerraf (1965), who pointed out that the non-uniqueness of the set-theoretic model fatally flaws the claim that its members are really the natural numbers: if no one pure set has a privileged claim to that title, then none can have title at all.

I said, though, that there is a more promising argument. The historical origins of this more promising argument lie in Frege's failed logicism. Frege attempted to embed arithmetic in a theory not of sets but of classes, and he held that the properties of classes on which this embedding depended were *logical* truths which we could see to be analytic in a suitably extended sense of Kant's term. Frege's project, of course, was a failure, and hardly anyone now claims that the whole of mathematics could be embedded in logic, especially not on the rather narrow conception of logic lately in vogue. Nonetheless, the possibility remains live that set theory, if not part of logic, might have a privileged epistemological status not shared by the theory of real numbers, say. Set theory might — to put the point very loosely — be inherently more *foundational* than other branches of mathematics. If so, the embedding of the latter in the former would give us the outline of an argument that our knowledge of truths about number is in good standing just because our knowledge of set theory is.

Part III

Cardinals and Ordinals

Introduction to Part III

We shall begin this part of the book with a theory of the size of infinite sets, a theory developed almost in its entirety by Georg Cantor during the 1870s and 1880s. It is of course a precondition for the significance of this development that one should accept the consistency of the notion of an infinite set, and this acceptance was by no means universal in Cantor's time: as we have already noted, confusions between infinity and other sorts of limitlessness, which had encouraged the belief that the notion of an infinite set is inconsistent, were slow to disperse. But even when those concepts had been distinguished with sufficient clarity, other conceptual resources were still needed. One key step was the realization that the notion of equinumerosity provides us with a measure of size for infinite sets just as it does for finite ones, but more is required. Suppose we say that A is *more numerous* than B if A contains B but is not equinumerous with it. If A and B are finite, then A is more numerous than B just in case it properly contains B. For this reason we are inclined to apply a notion of size according to which 'is larger than' means 'is more numerous than' or 'properly contains' indiscriminately. Gregory of Rimini, a 14th century Augustinian monk, is sometimes credited with the realization that in dealing with the infinite case we must distinguish the two notions carefully, since otherwise we shall be perplexed by the fact that, for example, the set of natural numbers properly contains the set of even numbers and yet is equinumerous with it. The definition of equinumerosity gives rise to a coherent notion of size for infinite sets, therefore, but it is not yet clear that it is a fruitful one. For that to be so, one further thing needs to be true: there have to be sets of *different* infinite sizes. It is this that is quite unexpected, which is why Cantor's discovery of the existence of uncountable sets is pivotal.

Consider now the following problem. We are given a function f defined on a partially ordered set with the property that $f(x) \geqslant x$ for all x. For any a we wish to find $\overline{a} \geqslant a$ such that $\overline{a}$ is a fixed point of f, i.e. $f(\overline{a}) = \overline{a}$. The problem is simple enough to state, and in some cases at least it is simple to solve. The work we did in §5.1, for instance, deals with the case where $f(B) = B \cup r[B]$ for every subset B of the domain of some relation r, since if we let

$$\overline{A} = \bigcup_{n \in \omega} f^n(A),$$

then $\overline{A}$ is r-closed; i.e.

$$f(\overline{A}) = \overline{A}. \tag{1}$$

But the proof of the key identity (1) depended on special properties of the function f in the case in question; it is not hard to come up with examples where it fails. Suppose, for instance, that for any collection $\mathcal{A}$ of functions from $\mathbf{R}$ to $\mathbf{R}$ we let $f(\mathcal{A})$ be the set of functions which are pointwise limits of functions in $\mathcal{A}$. We saw earlier that the set $\mathcal{C}$ of continuous functions is not closed under the formation of pointwise limits, which is to say that $f(\mathcal{C}) \neq \mathcal{C}$. It is a natural problem, first studied by Baire (1898), to determine the smallest set of functions containing $\mathcal{C}$ that *is* closed under pointwise limits. If we mimic the construction we have just used and let

$$f^{\omega}(\mathcal{A}) = \bigcup_{n \in \omega} f^{n}(\mathcal{A}),$$

we might hope that the answer is $f^{\omega}(\mathcal{C})$, but in fact it is not: the process of forming pointwise limits leads to further functions not in this set.

What we have to do in order to solve our problem in this case, therefore, is to generalize the procedure: we need to keep on iterating our application of f until the process stabilizes. But how are we to express this idea? And how are we to show that the process does stabilize eventually? What we need now are symbols α to act in the exponent position of the notation $f^{\alpha}(A)$ as indices of the progress of the procedure even when it is iterated into the transfinite. The symbols in question are called *ordinal notations*, and Cantor invented them to deal with just such problems as this. But it quickly becomes clear that ordinals can be used to index many other processes that occur in mathematics. For example, we can use them to label the levels in the set-theoretic hierarchy, so that V_{α} is the level indexed by the symbol α.

Chapter 9

Cardinals

In this chapter we shall study the concept of equinumerosity which we introduced in §4.8 by means of the following definition.

Definition. *Two sets are said to be* equinumerous *if there is a one-to-one correspondence between them.*

As we noted in the introduction, one of the biggest steps is merely to see that this *is* a fruitful concept to study. This rather trite observation is well illustrated by the case of Bolzano, who was perhaps the first to note the characterization of the infinite which Dedekind took as his definition, and came tantalizingly close to developing a theory of cardinals in his book *Paradoxien des Unendlichen* (1851). He undoubtedly had an influence on Cantor, who mentioned the book in glowing terms, but he failed to make much progress because he failed to spot the key idea, which was left to Cantor to discover, that equinumerosity can be used as the basis of a theory of size.

9.1 Definition of cardinals

In order to develop this theory of size, it will be helpful to associate with each set an object called its *cardinality*, which we can look on as being (or at any rate representing) its size. What we need is that the cardinalities of two sets should be the same if and only if the sets are equinumerous. When the sets involved are finite, the natural numbers can be used to achieve precisely this, as we saw in §6.4. But what about the non-finite case? A naive conjecture might be that all non-finite sets are equinumerous and that we therefore need only one object ('infinity') to measure their size, but we have already seen that $\mathbf{R}$ is uncountable, and another way of expressing this is to say that $\boldsymbol{\omega}$ and $\mathbf{R}$ are of different cardinalities, so the theory we erect cannot be as straightforward as the naive conjecture suggests.

From a formal point of view what we are looking for is a term 'card(x)' with the following property (sometimes referred to in the literature, with dubious historical justification, as *Hume's principle*.

$$\text{card}(A) = \text{card}(B) \text{ iff } A \text{ and } B \text{ are equinumerous.}$$

There is, incidentally, an evident similarity between this and the abstraction principle for classes that we shall consider in appendix B. And just as in the case of classes, if all we do is to talk of cardinality as a way of expressing relations of equinumerosity, what we do is innocent enough.

This is indeed a reasonable account of what we find in Cantor's early work. His first paper on equinumerosity (1874) does not mention the notion of cardinality at all, and when it does appear in his 1878 paper, it is only in compound phrases, so that 'A has the same cardinality as B' is to be regarded simply as another way of saying 'A and B are equinumerous'. It seems not to be until his 1883 book that he treats cardinals as distinct objects to be manipulated in their own right.

If we wish to join Cantor in taking this step of regarding cardinals not merely as an eliminable *façon de parler* but as objects in their own right, it is at this point that a difference emerges with the case of classes, since there is no purely logical bar to supposing that every cardinal is an object, as we can show now by giving an explicit definition of sets that can serve as cardinals within our theory. To do this, we use the cutting-down trick we introduced in §4.4.

Definition. *The set* $\langle X : X$ *and* A *are equinumerous*$\rangle$ *is called the* cardinality *of* A *and denoted* card(A). *Anything that is the cardinality of some set is called a* cardinal number.

The point of this definition is solely that it delivers Hume's principle as a theorem.

(9.1.1) **Proposition.** card$(A) =$ card(B) *iff* A *and* B *are equinumerous.*

Necessity. Every set A is equinumerous with itself and so card$(A) \neq \emptyset$ [proposition 4.4.1]. Consequently, if card$(A) =$ card(B), there must exist a set X which belongs to both card(A) and card(B), and hence is equinumerous with both A and B: it follows that A and B must be equinumerous.

Sufficiency. If A and B are equinumerous, then the sets equinumerous with A are precisely those equinumerous with B, and so card$(A) =$ card(B). □

We could, of course, have chosen other definitions that would have delivered Hume's principle. Which definition we have chosen will make a difference at one point in the sequel (in the proof of proposition 9.2.5), where we need to make use of the fact that the cardinal of a set, as we have defined it, occurs no more than one level above the set in the hierarchy. Apart from this, however, all the properties of cardinals that we prove will be derived via Hume's principle and are therefore independent of the particular definition chosen.

It is worth emphasizing, though, that we are entitled to Hume's principle only because of the theoretical commitments we have already entered into: in particular, the justification we have given for it applies only if A and B are sets, i.e. *grounded* collections. A different justification would be needed if we wished to assert an analogue of proposition 9.1.1 for ungrounded collections (see appendix A).

From now on we shall reserve the lower-case Fraktur letters $\mathfrak{a}$, $\mathfrak{b}$, $\mathfrak{c}$, *etc. to denote cardinal numbers.*

9.2 The partial ordering

Definition. *Suppose that* $\mathfrak{a} = \text{card}(A)$ *and* $\mathfrak{b} = \text{card}(B)$. *We write* $\mathfrak{a} \leqslant \mathfrak{b}$ *if there exists a one-to-one function from* A *to* B, *and we write* $\mathfrak{a} < \mathfrak{b}$ *if* $\mathfrak{a} \leqslant \mathfrak{b}$ *and* $\mathfrak{a} \neq \mathfrak{b}$.

Note that this definition does not depend on the choice of the representative sets A and B. For if $\text{card}(A) = \text{card}(A')$ and $\text{card}(B) = \text{card}(B')$, then there are one-to-one correspondences f between A and A' and g between B and B' [proposition 9.1.1]. So if i is a one-to-one function from A to B, then $g \circ i \circ f^{-1}$ is a one-to-one function from A' to B'; and if i' is a one-to-one function from A' to B', then $g^{-1} \circ i' \circ f$ is a one-to-one function from A to B.

$$\begin{array}{ccc} A & \xrightarrow{i} & B \\ \big\downarrow{\scriptstyle f} & & \big\downarrow{\scriptstyle g} \\ A' & \xrightarrow{i'} & B' \end{array}$$

Having shown that the definition is correct, we next need to show that what it defines is a partial ordering.

(9.2.1) **Proposition.** *Suppose that* $\mathfrak{a}$, $\mathfrak{b}$, $\mathfrak{c}$ *are cardinals.*

(a) $\mathfrak{a} \leqslant \mathfrak{a}$;

(b) If $\mathfrak{a} \leqslant \mathfrak{b}$ *and* $\mathfrak{b} \leqslant \mathfrak{c}$ *then* $\mathfrak{a} \leqslant \mathfrak{c}$.

Proof. Trivial. □

Proving antisymmetry takes a little more work, however.

(9.2.2) **Lemma.** *If* $A \subseteq B \subseteq C$ *and* A *is equinumerous with* C, *then* B *is equinumerous with both* A *and* C.

Proof. Suppose that f is a one-to-one correspondence from C to A. It is an easy exercise to check that the function g from C to B given by

$$g(x) = \begin{cases} f(x) & \text{if } x \in \text{Cl}_f(C \setminus B) \\ x & \text{if } x \in C \setminus \text{Cl}_f(C \setminus B) \end{cases}$$

is a one-to-one correspondence too. □

(9.2.3) **Bernstein's equinumerosity theorem.** *If there exist one-to-one functions f from A to B and g from B to A, then there exists a one-to-one correspondence between A and B.*

Proof. $g[f[A]] \subseteq g[B] \subseteq A$ and $g \circ f$ is a one-to-one correspondence between A and $g[f[A]]$. So there exists a one-to-one correspondence h between A and $g[B]$ [lemma 9.2.2]. But g^{-1} is evidently a one-to-one correspondence between $g[B]$ and B. So $g^{-1} \circ h$ is a one-to-one correspondence between A and B as required. □

(9.2.4) **Corollary.** *If $\mathfrak{a}$ and $\mathfrak{b}$ are cardinals such that $\mathfrak{a} \leqslant \mathfrak{b}$ and $\mathfrak{b} \leqslant \mathfrak{a}$, then $\mathfrak{a} = \mathfrak{b}$.*

Proof. This is just a re-wording of Bernstein's equinumerosity theorem. □

Note that we have not claimed $\leqslant$ is a *total* ordering, i.e. any two cardinals are comparable: the reason for the omission is that this claim is not provable from the axioms now at our disposal; in fact we will show in §15.4 that it is equivalent to the axiom of choice.

(9.2.5) **Proposition.** *If Φ is any formula, the set $B = \{\mathfrak{a} : \Phi(\mathfrak{a})\}$ exists iff there is a cardinal $\mathfrak{c}$ such that $\mathfrak{a} \leqslant \mathfrak{c}$ whenever $\Phi(\mathfrak{a})$.*

Necessity. Suppose that B is a set and let $\mathfrak{c} = \text{card}(\text{V}(B))$. If $\mathfrak{a} \in B$, then for any $A \in \mathfrak{a}$ we have $A \subseteq \text{V}(B)$, so that $\mathfrak{a} = \text{card}(A) \leqslant \text{card}(\text{V}(B)) = \mathfrak{c}$ as required.

Sufficiency. Suppose that $\mathfrak{a} \leqslant \mathfrak{c}$ whenever $\Phi(\mathfrak{a})$, and let C be any set such that $\text{card}(C) = \mathfrak{c}$. Then

$$\begin{aligned}\Phi(\mathfrak{a}) &\Rightarrow \mathfrak{a} \leqslant \mathfrak{c}\\ &\Rightarrow \mathfrak{a} = \text{card}(X) \text{ for some } X \in \mathfrak{P}(C)\\ &\Rightarrow \mathfrak{a} \subseteq \text{V}(\mathfrak{P}(C)),\end{aligned}$$

and so B is a set. □

(9.2.6) **Theorem (Cantor 1892).** *If A is a set, there are one-to-one functions from A into $\mathfrak{P}(A)$, but there are no functions from A* onto *$\mathfrak{P}(A)$.*

Proof. The function from A to $\mathfrak{P}(A)$ given by $x \mapsto \{x\}$ is clearly one-to-one. For each function f from A to $\mathfrak{P}(A)$ let $B_f = \{x \in A : x \notin f(x)\}$. If $B_f = f(y)$, then

$$y \in B_f \Leftrightarrow y \notin f(y) \Leftrightarrow y \notin B_f,$$

which is absurd. So B_f is a subset of A that is not in the image of f. It follows that there cannot be a function from A *onto* $\mathfrak{P}(A)$. □

This proof of Cantor's theorem is evidently similar in structure to the proof that **R** is uncountable. In particular, it exhibits the same impredicativity: the set B_f is defined in terms of the function f, even though f lies several levels higher in the hierarchy than B_f.

(9.2.7) **Corollary.** *For every cardinal* $\mathfrak{a}$ *there is a cardinal* $\mathfrak{a}'$ *such that* $\mathfrak{a} < \mathfrak{a}'$.

Proof. It follows from Cantor's theorem that $\text{card}(A) < \text{card}(\mathfrak{P}(A))$ for any set A. □

We said above that $\leqslant$ partially orders *any* set of cardinals, not that it partially orders *the* set of cardinals. What is wrong with the latter way of speaking is simply that the cardinals do not form a set.

(9.2.8) **Proposition.** *The set of all cardinals does not exist.*

Proof. If there were a set of all cardinals, it would have a largest element [proposition 9.2.5], contradicting the corollary to Cantor's theorem we have just noted. □

If we were still in the grip of the idea that every property is collectivizing, this proposition would of course have just as much right as Russell's to be treated as a paradox, and it is of some historical significance in that role.

Exercises

1. Show that $\mathfrak{a} \leqslant \mathfrak{b} \Rightarrow \mathrm{V}(\mathfrak{a}) \subseteq \mathrm{V}(\mathfrak{b})$.
2. Write out the details of the proof of lemma 9.2.2.

9.3 Finite and infinite

We have already defined what it means for a set to be finite, infinite or countable. By a slight abuse of language we say that a cardinal $\mathfrak{a}$ is finite [resp. infinite, countable] if it is the cardinal of a finite [resp. infinite, countable] set.[1] In order to explain how these concepts fit into the theory of cardinals, we need to introduce a notation for the cardinal of the set of natural numbers.

Definition. $\aleph_0 = \text{card}(\boldsymbol{\omega})$.

(9.3.1) **Theorem.** *A set* A *is countable iff* $\text{card}(A) \leqslant \aleph_0$.

[1]The reason this is an abuse of language is that $\mathfrak{a}$ is itself a set and hence we have already defined what it means for it to be finite, etc. If there happen to be infinitely many individuals, finite cardinals will in fact be *infinite* sets.

Necessity. Suppose that A is countable. If A is empty, then trivially card$(A) \leqslant \aleph_0$. If it is not empty, then there is an onto function f from $\boldsymbol{\omega}$ to A. For each $x \in A$ the set of natural numbers mapped to x by f is therefore non-empty and has a least element $g(x)$. The function g from A to $\boldsymbol{\omega}$ thus defined is evidently one-to-one. Hence card$(A) \leqslant \aleph_0$.

Sufficiency. Suppose that card$(A) \leqslant \aleph_0$. If A is empty, then it is certainly countable. So suppose not, and choose an element $a \in A$. Now there is a one-to-one function g from A into $\boldsymbol{\omega}$. If $n \in g[A]$, let $f(n)$ be the unique $x \in A$ such that $g(x) = n$; and if $n \in \boldsymbol{\omega} \setminus g[A]$, let $f(n) = a$. This evidently defines a function f from $\boldsymbol{\omega}$ onto A, and so A is countable in this case too. □

(9.3.2) **Theorem.** *A set A is finite iff* card$(A) < \aleph_0$.

Necessity. Suppose that A is finite. So A has n elements for some $n \in \boldsymbol{\omega}$. Now $\boldsymbol{n} \subseteq \boldsymbol{\omega}$, so that obviously card$(A) \leqslant \aleph_0$. But if card$(A) = \aleph_0$, then $\boldsymbol{\omega}$ is finite and therefore has a greatest element, which is absurd. So card$(A) < \aleph_0$ as required.

Sufficiency. Suppose that card$(A) < \aleph_0$ but A is not finite. Note first that A is non-empty and countable [theorem 9.3.1] and so there exists a sequence $x_0, x_1, x_2, \ldots$ which enumerates all the elements of A. It is clear that if we can define a one-to-one function g from $\boldsymbol{\omega}$ to A we shall have the contradiction we require, since then card$(A) \geqslant \aleph_0$. We do this by recursion. If $g(m)$ has been defined for $m < n$, then $A \setminus g[\boldsymbol{n}]$ is not empty since A is not finite, and so we can let $g(n)$ be the element x_r of $A \setminus g[\boldsymbol{n}]$ of least possible index r. It is clear that the function g from $\boldsymbol{\omega}$ to A thus defined is one-to-one. □

(9.3.3) **Theorem.** *A set A is infinite iff* card$(A) \geqslant \aleph_0$.

Necessity. Suppose that A is infinite. So there is a one-to-one function f from A to A such that there exists $a \in A \setminus f[A]$. We shall show that the function g from $\boldsymbol{\omega}$ to A defined by $g(0) = a$ and $g(n+1) = f(g(n))$ is one-to-one. For if not, then there exist $m, n \in \boldsymbol{\omega}$ such that $m < n$ but $g(m) = g(n)$. Moreover, we can choose m minimal subject to this property. Now $n > 0$ and so $n = n_0 + 1$ for some $n_0 \in \boldsymbol{\omega}$. There are two cases to consider, depending on whether $m = 0$ or $m > 0$. If $m = 0$, then

$$a = g(0) = g(n) = g(n_0 + 1) = f(g(n_0)),$$

and so $a \in f[A]$. Contradiction. If $m > 0$, on the other hand, then $m = m_0 + 1$ for some $m_0 \in \boldsymbol{\omega}$. Hence

$$f(g(m_0)) = g(m_0 + 1) = g(m) = g(n) = g(n_0 + 1) = f(g(n)),$$

and so $g(m) = g(n)$ since g is one-to-one. Moreover, $m_0 < n_0$. This contradicts the minimality of m.

Sufficiency. Suppose that $\text{card}(A) \geqslant \aleph_0$. So there is a one-to-one function g from $\boldsymbol{\omega}$ to A. We can define a function f from A to itself by letting

$$f(x) = \begin{cases} g(g^{-1}(x) + 1) & \text{if } x \in g[\boldsymbol{\omega}] \\ x & \text{otherwise.} \end{cases}$$

This is clearly one-to-one, but it is not onto, since $g(0) \in A \setminus f[A]$. □

(9.3.4) **Corollary.** *These three assertions are equivalent:*

(i) $\text{card}(A) = \aleph_0$*;*

(ii) A is countably infinite;

(iii) A is countable and not finite.

Proof. Immediate. □

(9.3.5) **Corollary.** *A is uncountably infinite iff* $\text{card}(A) > \aleph_0$.

Proof. A is uncountably infinite iff $\text{card}(A) \nleqslant \aleph_0$ and $\text{card}(A) \geqslant \aleph_0$ [theorems 9.3.1 and 9.3.3] iff $\text{card}(A) > \aleph_0$. □

Exercise

Let f be a function from A to itself.

(a) If A is finite, show that f is one-to-one iff it is onto.

(b) If there exists an element $a \in f[A]$ such that $A = \text{Cl}_f(a)$, show that f is a one-to-one correspondence between A and itself. [First use the simple recursion principle to show that A is finite.]

9.4 The axiom of countable choice

We have already shown [corollary 6.4.2] that no set can be both finite and infinite. But is every set one or the other? The axioms we have stated so far are not sufficient to settle this question (Cohen 1966): what we need if we are to derive an affirmative answer is the following extra set-theoretic assumption.

Axiom of countable choice. *For every sequence* (A_n) *of non-empty sets there exists a sequence* (x_n) *such that* $x_n \in A_n$ *for all* $n \in \boldsymbol{\omega}$.

We call this an 'axiom' in deference to tradition, but we shall not treat it as such: we shall not, that is to say, add it to our default theory. Instead we shall state it explicitly as an assumption in any theorem that depends on it.

(9.4.1) **Theorem.** *It follows from the axiom of countable choice that every set is either finite or infinite.*

Proof. Suppose that A is a set which is not finite. Then for each $n \in \omega$ there are n-element subsets of A [proposition 6.4.4]. So by the axiom of countable choice there is a sequence $(A_n)_{n\in\omega}$ such that A_n is an n-element subset of A for each $n \in \omega$. Now the number of elements in A_{2^n} is 2^n, whereas the number in $\bigcup_{r<n} A_{2^r}$ is $\leqslant \sum_{r<n} 2^r = 2^n - 1 < 2^n$, and so it follows that the subset

$$B_n = A_{2^n} \setminus \bigcup_{r<n} A_{2^r}$$

of A is non-empty for each n. Hence (by the axiom of countable choice again) there is a function g from ω to A such that $g(n) \in B_n$ for each n. This function is one-to-one since the B_n are pairwise disjoint. Therefore A is infinite [theorem 9.3.3]. □

Otherwise put, the axiom of countable choice implies that every cardinal is comparable with $\aleph_0$.

(9.4.2) **Theorem.** *It follows from the axiom of countable choice that if (A_n) is a sequence of countable sets, then $\bigcup_{n\in\omega} A_n$ is also a countable set.*

Proof. Let us suppose for simplicity that each of the countable sets A_n is non-empty. So for each $n \in \omega$ the set of functions from ω onto A_n is non-empty. It follows from the axiom of countable choice that there is a sequence (f_n) such that f_n is a function from ω onto A_n for each n. So $(m, n) \mapsto f_n(m)$ is a function from $\omega \times \omega$ onto $\bigcup_{n\in\omega} A_n$. If we compose this with a one-to-one correspondence between ω and $\omega \times \omega$, we obtain a function from ω onto $\bigcup\{A_n : n \in \omega\}$ as required. □

The axiom of countable choice is used quite frequently to prove results in the theory of real numbers. For instance, it is one of the central notions of the general theory of integration that a set of real numbers is *null* if it can be covered by a sequence of intervals of arbitrarily small total length (where the *length* $l(I)$ of a bounded interval I is defined in the obvious manner as the difference in value between its endpoints). Let us try to prove that if (C_n) is a sequence of null sets, $\bigcup_{n\in\omega} C_n$ is also null. To do this, take any $\varepsilon > 0$ and cover each C_n by a sequence of intervals I_{nm} such that $\sum_{m=0}^{\infty} l(I_{nm}) < \varepsilon/2^{n+1}$. (This is possible since C_n is by hypothesis a null set.) Then the double sequence (I_{nm}) of intervals evidently covers $\bigcup_{n\in\omega} C_n$, and its total length is

$$\sum_{n=0}^{\infty}\sum_{m=0}^{\infty} l(I_{nm}) < \sum_{n=0}^{\infty} \frac{\varepsilon}{2^{n+1}} = \varepsilon.$$

This shows that we can cover $\bigcup_{n\in\omega} C_n$ by intervals of arbitrarily small total length, and hence that it is null. But embedded in the proof we have just given is an appeal to the axiom of countable choice: for each $n \in \omega$ we have to *choose* a sequence $(I_{nm})_{m\in\omega}$ of intervals covering C_n from among the many such sequences available, and yet we have not specified how the choice is to be made.

The sort of case that occurs most frequently in analysis may be expressed abstractly as follows. We have a theorem of the form

$$(\forall n \in \omega)(\exists x \in \mathbf{R})\Phi(n, x),$$

and wish to obtain a sequence (x_n) in $\mathbf{R}$ such that $(\forall n \in \omega)\Phi(n, x_n)$. This is trivially licensed by the axiom of countable choice: the sets $A_n = \{x \in \mathbf{R} : \Phi(n, x)\}$ are non-empty by hypothesis, and so the axiom ensures the existence of a sequence (x_n) in $\mathbf{R}$ such that for each n we have $x_n \in A_n$, i.e. $\Phi(n, x_n)$. But in textbooks on the calculus such uses of the axiom are rarely signalled explicitly. The one we have just described might be introduced simply by saying,

> For each natural number n let x_n be a real number x_n such that $\Phi(n, x_n)$.

Indeed this tendency to downplay uses of the axiom of countable choice mimics the historical situation. For an explicit example of this consider the following argument, which may well have been the first use of the axiom in mathematics when it was published by Heine (1872) — with an ascription to Cantor.

(9.4.3) **Proposition.** *The axiom of countable choice entails that any function from* $\mathbf{R}$ *to* $\mathbf{R}$ *which is sequentially continuous at a point is also continuous at it.*

Proof. Suppose that f is not continuous at a. So there is an interval J containing $f(a)$ such that for every interval I containing a we have $f[I] \not\subseteq J$. In particular, therefore, for each $n \in \omega$ we can (using the axiom of countable choice) choose x_n between $a - \frac{1}{n}$ and $a + \frac{1}{n}$ such that $f(x_n) \notin J$. Then we can conclude that the sequence (x_n) tends to a but $(f(x_n))$ does not converge to $f(a)$. □

Heine expresses the conclusion of the proposition unconditionally, and indeed the text makes no special comment whatever on the step in the proof which requires the axiom of countable choice for its justification. Between this publication and the end of the century the axiom was used implicitly on many occasions by Cantor, Dedekind, Borel, Baire and others. At first only Peano and his colleagues in Turin seem to have commented explicitly on its use: Peano (1890, p. 210) claimed that 'one cannot apply an infinite number of times an *arbitrary* law according to which a class a is made to correspond to an individual of that class'; and Bettazzi (1896, p. 512) criticized Dedekind's proof that every set is either finite or infinite on the ground that one must choose an object (correspondence) arbitrarily in each of the infinite sets, which does not seem rigorous; unless one wishes to accept as a postulate that such a choice can be carried out — something, however, which seems ill-advised to me.

Sometimes, though, it is possible to avoid using the axiom of countable choice by giving an explicit rule for defining the elements of the sequence in question. For example, if we have a theorem of the form $(\forall n \in \boldsymbol{\omega})(\exists x \in \mathbf{Q})\Phi(n, x)$, we can without using the axiom of countable choice define a sequence (x_n) in $\mathbf{Q}$ such that $\Phi(n, x_n)$ for all $n \in \boldsymbol{\omega}$. To do this, we note that $\mathbf{Q}$ is countable (unlike $\mathbf{R}$) and so there exists a sequence (a_r) whose image is $\mathbf{Q}$; if we let r_n be the least element of the set $\{r \in \boldsymbol{\omega} : \Phi(n, a_r)\}$ (which is non-empty by hypothesis) and let $x_n = a_{r_n}$ for each $n \in \boldsymbol{\omega}$, we obtain a sequence (x_n) with the property we require. A good example of this way of avoiding countable choice is the global version of the result about continuity which we proved a moment ago.

(9.4.4) **Proposition.** *A function from* $\mathbf{R}$ *to* $\mathbf{R}$ *which is sequentially continuous everywhere is also continuous everywhere.*

Proof. If f is sequentially continuous, then in order to prove that $f[I] \subseteq J$ it is enough to prove $f[I \cap \mathbf{Q}] \subseteq J$. So we can avoid the appeal to the axiom of countable choice in our earlier proof by choosing *rational* numbers at each stage. □

But it is not possible to avoid the axiom of countable choice in all cases, at least if the default theory is like the one we are using in this book: the axiom of countable choice cannot be proved in ZU (Fraenkel 1922*a*) or even in Z (Cohen 1963). The extent to which classical analysis depends on ineliminable uses of the axiom has been studied extensively and is now well understood: it has been shown, for example, that in the absence of the axiom not only may there be subsets of $\mathbf{R}$ which are neither finite nor infinite (Cohen 1966, p. 138), but there is even a model in which the continuum is a countable union of countable sets (Feferman and Levy 1963). It follows from this, moreover, that we cannot hope to eliminate the appeal to the axiom of countable choice from the proof we gave earlier that a countable union of null sets is null, since otherwise the model just mentioned would be one in which the whole real line is null, which is absurd.

A liberal constructivist might well think that the axiom of countable choice has a certain plausibility, since it appears to be only medically, rather than logically, impossible to make an infinite number of choices in a finite time by the device of performing each choice in half the time it took to perform the previous one (cf. Russell 1936, pp. 143–4). And, as we saw in part I, the constructivist already has to appeal to such supertasks to explain the existence of sets at infinite levels in the hierarchy: there seems to be no new reason to baulk at them now.

But to the platonist such thought experiments ought presumably to be completely beside the point: for him the issue concerns not what any being, however idealized, can do, but which sets exist. Moreover, nothing we have said so far has encouraged the idea that he should regard the distinction between countable and uncountable as an especially significant caesura. So it is at any rate hard to see what reason he might have to believe the axiom of countable choice that is not equally a reason to believe the unrestricted axiom of choice that we shall study in chapter 15.

Exercises

1. Assuming the axiom of countable choice, prove that for every cardinal $\mathfrak{a} > \aleph_0$ there exists exactly one $\mathfrak{b}$ such that $\mathfrak{a} = \aleph_0 + \mathfrak{b}$.

2. Prove without using the axiom of countable choice that **R** is not a countable union of finite sets.

Notes

The theory of cardinals really originates with Cantor, although the formal development we have given here deviates considerably from what is to be found in his *Beiträge* (1895; 1897), mainly because Cantor did not doubt the axiom of choice and therefore saw no reason to develop the theory in such a way as to isolate the points at which he appealed to it. The history of Cantor's development of his theory of cardinality is well described by Dauben (1990). A. W. Moore (1990) illuminates the earlier background. What is striking about this development is how few precursors Cantor had: only Bolzano (1851) came anywhere close to a coherent theory.

Cantor conjectured a special case of Bernstein's equinumerosity theorem in his 1883 book and the general result in an 1895 paper, but it was his pupil, Felix Bernstein, who proved it — in 1897 when he was barely 19. A slightly simplified version of Bernstein's proof was then published by Borel in an appendix to his *Leçons sur la théorie des fonctions* (1898). The proof we have given here, which uses Dedekind's theory of chains to avoid mention of the natural numbers, was found by Dedekind following a conversation with Bernstein[2] in 1897 and communicated to Cantor by letter in 1899. However, Dedekind never published this proof, and it was rediscovered by Zermelo in 1906 (see Poincaré 1906; Peano 1906; Zermelo 1908*b*). In Britain and Germany the result is usually called the 'Schröder/Bernstein theorem' because of an attempted proof by Schröder (1898), the fallacy in which was not exposed in print until Korselt (1911), although it had in fact been pointed out to Schröder

[2]A manuscript copy of a very similar proof, apparently dated 11 July 1887, was found among Dedekind's papers (Dedekind 1932, vol. IV, pp. 447–8).

and acknowledged by him in 1902. According to another tradition, common in France and Italy, the result is known as the 'Cantor/Bernstein theorem', perhaps in recognition of Cantor's statement of it in 1895 or possibly on account of an ambiguous footnote in Borel's *Leçons* (1898, p. 105n.). In any event it seems likely that Cantor never had a direct proof but could only deduce the theorem from a formulation of the cardinal comparability principle (discussed in §15.4), a result for which in turn he never obtained a convincing proof, as a letter he wrote in 1903 (Cantor 1991, p. 434) confirms.

Chapter 10

Basic cardinal arithmetic

The business of this chapter will be to study the elementary consequences of the following definitions.

Definition. *If* $\mathfrak{a} = \operatorname{card}(A)$ *and* $\mathfrak{b} = \operatorname{card}(B)$, *we let*

$$\begin{aligned} \mathfrak{a} + \mathfrak{b} &= \operatorname{card}(A \uplus B) \\ \mathfrak{a}\mathfrak{b} &= \operatorname{card}(A \times B) \\ \mathfrak{a}^{\mathfrak{b}} &= \operatorname{card}({}^{B}A). \end{aligned}$$

Strictly speaking, of course, we need to check that these definitions do not depend on our choice of the representative sets A and B, i.e. that if A is equinumerous with A' and B with B', then $A \uplus B$, $A \times B$ and ${}^{B}A$ are equinumerous with $A' \uplus B'$, $A' \times B'$ and ${}^{B'}A'$ respectively. However, the proofs of these facts are all straightforward, and we therefore omit them.

The definitions are formally consistent, then, but this does not explain why we have chosen them. A partial explanation will be supplied in the next section, where we shall show that for finite cardinals they simply reproduce the everyday operations of addition, multiplication and exponentiation with which we are already familiar. But this *is* only a partial explanation: other definitions are no doubt possible which coincide with these ones for finite cardinals but come apart from them in the infinite case. The fruitfulness of the definitions we have given is thus not something that can be judged straightaway but emerges only once the theory to which they give rise has been developed.

10.1 Finite cardinals

Definition. *For each* $n \in \boldsymbol{\omega}$ *we let* $|n| = \operatorname{card}(\boldsymbol{n})$.

With this notation we can say that a set A has n elements iff $\operatorname{card}(A) = |n|$.

(10.1.1) **Proposition.** $m \leqslant n \Leftrightarrow |m| \leqslant |n|$.

Proof. If $m \leqslant n$, then $\boldsymbol{m} \subseteq \boldsymbol{n}$ and so $|m| \leqslant |n|$. If, on the other hand, $|m| \leqslant |n|$ but $m > n$, then there exists a one-to-one function from $\boldsymbol{m}$ to $\boldsymbol{n}$: this function restricts to a one-to-one but not onto function from $\boldsymbol{n}$ to $\boldsymbol{n}$. Hence $\boldsymbol{n}$ is both finite and infinite, contradicting corollary 6.4.2. □

(10.1.2) **Corollary.** $|m| = |n| \Rightarrow m = n$.

Proof. Immediate [proposition 10.1.1]. □

(10.1.3) **Proposition.** *If* $m, n \in \boldsymbol{\omega}$, *then*

$$|m + n| = |m| + |n|$$
$$|mn| = |m||n|$$
$$|m^n| = |m|^{|n|}.$$

Proof. It is possible to show that the functions

$(r, i) \mapsto im + r$ from $\boldsymbol{m} \uplus \boldsymbol{n}$ to $\{r \in \boldsymbol{\omega} : r < m + n\}$

$(r, s) \mapsto rn + s$ from $\boldsymbol{m} \times \boldsymbol{n}$ to $\{r \in \boldsymbol{\omega} : r < mn\}$

$(n_r)_{r \in \boldsymbol{n}} \mapsto \sum_{r \in \boldsymbol{n}} n_r m^r$ from ${}^{\boldsymbol{n}}\boldsymbol{m}$ to $\{r \in \boldsymbol{\omega} : r < m^n\}$

are all one-to-one correspondences. The result follows. □

This shows that the function from $\boldsymbol{\omega}$ onto the set of finite cardinals given by $n \mapsto |n|$ preserves both the ordering [proposition 10.1.1] and the arithmetical structure [proposition 10.1.3] of the natural numbers. So it seems reasonable to suppose — especially since our choice of sets to call natural numbers was arbitrary in the first place — that little confusion will arise if we denote the cardinal $|n|$ by the symbol n and thus abandon the distinction between a natural number and the corresponding finite cardinal; we shall adopt this policy from now on whenever it is convenient (which will be almost always).

Exercise

Show that a cardinal $\mathfrak{a}$ is not finite iff $\mathfrak{a} > n$ for all $n \in \boldsymbol{\omega}$.

10.2 Cardinal arithmetic

Now that we are regarding the natural numbers as finite cardinals, it is natural to ask which of the familiar rules of the arithmetic of natural numbers generalize to the infinite case. An answer to this question is provided by the following proposition.

(10.2.1) **Proposition.** *Suppose that* $\mathfrak{a}$, $\mathfrak{b}$, $\mathfrak{c}$ *are cardinals.*

(*a*) $\mathfrak{a} + (\mathfrak{b} + \mathfrak{c}) = (\mathfrak{a} + \mathfrak{b}) + \mathfrak{c}$.

(*b*) $\mathfrak{a} + \mathfrak{b} = \mathfrak{b} + \mathfrak{a}$.

(*c*) $\mathfrak{a} + 0 = \mathfrak{a}$.

(*d*) $\mathfrak{a} \geqslant \mathfrak{b} \Leftrightarrow (\exists \mathfrak{d})(\mathfrak{a} = \mathfrak{b} + \mathfrak{d})$.

(*e*) *If* $\mathfrak{b} \leqslant \mathfrak{c}$, *then* $\mathfrak{a} + \mathfrak{b} \leqslant \mathfrak{a} + \mathfrak{c}$.

(*f*) $(\mathfrak{a}\mathfrak{b})\mathfrak{c} = \mathfrak{a}(\mathfrak{b}\mathfrak{c})$.

(*g*) $\mathfrak{a}\mathfrak{b} = \mathfrak{b}\mathfrak{a}$.

(*h*) $\mathfrak{a}0 = 0\mathfrak{a}, \ \mathfrak{a}1 = \mathfrak{a}, \ \mathfrak{a}2 = \mathfrak{a} + \mathfrak{a}$.

(*i*) $\mathfrak{a}(\mathfrak{b} + \mathfrak{c}) = \mathfrak{a}\mathfrak{b} + \mathfrak{a}\mathfrak{c}$.

(*j*) $\mathfrak{b} \leqslant \mathfrak{c} \Rightarrow \mathfrak{a}\mathfrak{b} \leqslant \mathfrak{a}\mathfrak{c}$.

(*k*) $(\mathfrak{a}^{\mathfrak{b}})^{\mathfrak{c}} = \mathfrak{a}^{\mathfrak{b}\mathfrak{c}}$.

(*l*) $(\mathfrak{a}\mathfrak{b})^{\mathfrak{c}} = \mathfrak{a}^{\mathfrak{c}}\mathfrak{b}^{\mathfrak{c}}$.

(*m*) $\mathfrak{a}^{\mathfrak{b}+\mathfrak{c}} = \mathfrak{a}^{\mathfrak{b}}\mathfrak{a}^{\mathfrak{c}}$.

(*n*) $\mathfrak{a}^0 = 1, \ \mathfrak{a}^1 = \mathfrak{a}, \ \mathfrak{a}^2 = \mathfrak{a}\mathfrak{a}$.

(*o*) *If* $\mathfrak{a} \leqslant \mathfrak{b}$ *and* $\mathfrak{c} \leqslant \mathfrak{d}$, *then* $\mathfrak{a}^{\mathfrak{c}} \leqslant \mathfrak{b}^{\mathfrak{d}}$.

Proof. The proofs are all straightforward. By way of illustration let us prove part (k). It will be enough to show that if A, B, C are sets, then ${}^{C}({}^{B}A)$ is equinumerous with ${}^{C\times B}A$: this is achieved by observing that if we take each function $f \in {}^{C}({}^{B}A)$ to the function in ${}^{C\times B}A$ given by $(c, b) \mapsto f(c)(b)$, then we obtain a one-to-one correspondence; its inverse takes each function $g \in {}^{C\times B}A$ to the function in ${}^{C}({}^{B}A)$ given by $c \mapsto (b \mapsto g(c, b))$. □

(10.2.2) **Lemma.** $\mathrm{card}(\mathfrak{P}(A)) = 2^{\mathrm{card}(A)}$.

Proof. We define the *characteristic function* c_B of a subset B of A by

$$c_B(x) = \begin{cases} 1 & \text{if } x \in B \\ 0 & \text{otherwise.} \end{cases}$$

It is evident that the function $B \mapsto c_B$ is a one-to-one correspondence between $\mathfrak{P}(A)$ and ${}^{A}\{0, 1\}$; the inverse function is given by $f \mapsto f^{-1}[1]$. Since the cardinality of ${}^{A}\{0, 1\}$ is $2^{\mathrm{card}(A)}$ by definition, the result follows. □

(10.2.3) **Proposition.** $\mathfrak{a} < 2^{\mathfrak{a}}$.

Proof. By lemma 10.2.2 this is no more than a translation into the language of cardinals of Cantor's theorem 9.2.6. □

10.3 Infinite cardinals

What the results of the last section show is that many of the familiar properties of the natural numbers generalize straightforwardly to the non-finite case. But the following proposition shows straightaway that there are some respects in which the arithmetic of infinite cardinals differs markedly from that of the natural numbers.

(10.3.1) **Proposition.** *$\mathfrak{a}$ is infinite iff* $\mathfrak{a} = \mathfrak{a} + 1$.

Necessity. Suppose that $\mathfrak{a} = \text{card}(A)$. If A is infinite, it is equinumerous with a proper subset A', so that $A \setminus A' \neq \emptyset$, i.e. $\text{card}(A \setminus A') \geqslant 1$; therefore

$$\mathfrak{a} = \text{card}(A) = \text{card}(A') + \text{card}(A \setminus A') \geqslant \mathfrak{a} + 1 \geqslant \mathfrak{a},$$

and hence $\mathfrak{a} = \mathfrak{a} + 1$.

Sufficiency. If $\mathfrak{a} = \mathfrak{a} + 1$, there are sets A, A' and an element x not in A' such that $\mathfrak{a} = \text{card}(A) = \text{card}(A')$ and $\text{card}(A) = \text{card}(A' \cup \{x\})$. So $A' \cup \{x\}$ is equinumerous with its proper subset A' and hence is infinite. But then A is infinite too. □

One immediate consequence of this is that we cannot define an operation of subtraction for infinite cardinals, as we might otherwise be tempted to do, by defining $\mathfrak{a} - \mathfrak{b}$ to be the cardinal of $A \setminus B$ when $\text{card}(A) = \mathfrak{a}$, $\text{card}(B) = \mathfrak{b}$, and $B \subseteq A$. What is wrong with this is that it fails the requirement for a good definition of being independent of the choice of the representative sets A and B. To see why, consider $\aleph_0 - \aleph_0$, i.e. the cardinal of $A \setminus B$ where A and B are both countably infinite. If we let $A = \omega$ and $B = \omega$, then of course $A \setminus B = \emptyset$ and we get $\aleph_0 - \aleph_0 = 0$. But if $B = \omega \setminus \{0\}$, then $A \setminus B = \{0\}$, so that $\aleph_0 - \aleph_0 = 1$; whereas if $B = \omega \setminus \{0, 1\}$, then $A \setminus B = \{0, 1\}$, so that $\aleph_0 - \aleph_0 = 2$; and so on. If $B = \{2n : n \in \omega\}$, on the other hand, then $A \setminus B = \{2n + 1 : n \in \omega\}$, so that $\aleph_0 - \aleph_0 = \aleph_0$. In other words, $\aleph_0 - \aleph_0$ could be *any* number between 0 and $\aleph_0$.

It does not follow from this that $\mathfrak{a} - \mathfrak{b}$ is *never* definable by the means proposed above: it is easy to show, for instance, that $2^{\aleph_0} - \aleph_0 = 2^{\aleph_0}$. Nevertheless, this example demonstrates vividly that we cannot expect the arithmetic of infinite cardinals to be just like that of finite ones. Another illustration of the difference is supplied by the following proposition.

(10.3.2) **Proposition.** $\aleph_0^2 = \aleph_0$.

Proof. Immediate [proposition 6.5.1]. □

There are a great many cardinal identities which follow more or less straightforwardly from the properties we have listed. Here are three examples.

(1) $\aleph_0 \leqslant 2\aleph_0 \leqslant \aleph_0\aleph_0 = \aleph_0^2 = \aleph_0$, and so $2\aleph_0 = \aleph_0$.

(2) The identity $\aleph_0^{\aleph_0} = 2^{\aleph_0}$ holds because

$$2^{\aleph_0} \leqslant \aleph_0^{\aleph_0} \leqslant (2^{\aleph_0})^{\aleph_0} = 2^{\aleph_0^2} = 2^{\aleph_0}.$$

(3) To show that $\aleph_0 2^{\aleph_0} = 2^{\aleph_0}$, we have only to note that

$$2^{\aleph_0} \leqslant \aleph_0 2^{\aleph_0} \leqslant (2^{\aleph_0})^2 = 2^{2\aleph_0} = 2^{\aleph_0}.$$

(10.3.3) **Proposition.** *$\mathfrak{a}$ is infinite iff $\mathfrak{a} + \aleph_0 = \mathfrak{a}$.*

Proof. If $\mathfrak{a}$ is infinite, then $\mathfrak{a} \geqslant \aleph_0$ [theorem 9.3.3] and so there exists $\mathfrak{b}$ such that $\mathfrak{a} = \mathfrak{b} + \aleph_0$ [proposition 10.2.1(d)], whence

$$\begin{aligned} \mathfrak{a} + \aleph_0 &= (\mathfrak{b} + \aleph_0) + \aleph_0 \\ &= \mathfrak{b} + (\aleph_0 + \aleph_0) \text{ [proposition 10.2.1(a)]} \\ &= \mathfrak{b} + 2\aleph_0 \text{ [proposition 10.2.1(h)]} \\ &= \mathfrak{b} + \aleph_0 \\ &= \mathfrak{a}. \end{aligned}$$

And if conversely $\mathfrak{a} + \aleph_0 = \mathfrak{a}$, then $\mathfrak{a} \geqslant \aleph_0$ [proposition 10.2.1(d)] so that $\mathfrak{a}$ is infinite [theorem 9.3.3]. □

In chapter 15 we shall study the simplifying effect that assuming the axiom of choice has on cardinal arithmetic. We have already come across one instance in §9.4, where we showed that every cardinal is either finite or infinite, but only on the assumption of the axiom of countable choice — a restricted version of the axiom of choice. Quite often, though, results proved using the axiom of choice have weaker variants that are provable without it but may in some circumstances be as useful. For example, the identity $\aleph_0 2^{\aleph_0} = 2^{\aleph_0}$, which we proved earlier, is a particular case of the general identity $\mathfrak{a}\mathfrak{b} = \max(\mathfrak{a}, \mathfrak{b})$, but this is provable only if we assume the axiom of choice. Another example — whose significance for the project of *Principia Mathematica* is nicely illuminated by Boolos (1994) — is the following weakening of the result that every cardinal is either finite or infinite.

(10.3.4) **Proposition (Whitehead and Russell 1910–13).** *If $\mathfrak{a}$ is any cardinal, either $\mathfrak{a}$ is finite or $2^{2^{\mathfrak{a}}}$ is infinite.*

Proof. If $\operatorname{card}(A) = \mathfrak{a}$ and A is not finite, the sets $\mathfrak{F}_n(A)$ are distinct, non-empty subsets of $\mathfrak{P}(A)$, i.e. $n \mapsto \mathfrak{F}_n(A)$ is a one-to-one function from $\boldsymbol{\omega}$ into $\mathfrak{P}(\mathfrak{P}(A))$, and so $2^{2^{\mathfrak{a}}}$ is infinite. □

The results on cardinal arithmetic which we have established also make it possible to calculate the cardinalities of a wide variety of sets. As an illustration let us determine the cardinality of the set $\mathcal{A}$ of all equivalence relations on ω. The technique we shall use is, for reasons which will quickly become apparent, often known as 'squeezing'. Observe first that every equivalence relation on ω is a subset of $\omega \times \omega$, and so

$$\operatorname{card}(\mathcal{A}) \leqslant \operatorname{card}(\mathfrak{P}(\omega \times \omega)) = 2^{\aleph_0^2} = 2^{\aleph_0}. \tag{1}$$

Now consider the function f from $\mathfrak{P}(\omega)$ to $\mathcal{A}$ which takes a set $B \subseteq \omega$ to the equivalence relation on ω whose equivalence classes are B and the singletons $\{n\}$ for $n \in \omega \smallsetminus B$: this function is not one-to-one since it does not distinguish between 1-element subsets of ω; but the restriction $f|\mathfrak{P}(\omega) \smallsetminus \mathfrak{F}_1(\omega)$ *is* one-to-one, and so

$$\operatorname{card}(\mathcal{A}) \geqslant \operatorname{card}(\mathfrak{P}(\omega) \smallsetminus \mathfrak{F}_1(\omega)).$$

Now $\operatorname{card}(\mathfrak{F}_1(\omega)) = \aleph_0$, $\operatorname{card}(\mathfrak{P}(\omega)) = 2^{\aleph_0}$, and $\operatorname{card}(\mathfrak{P}(\omega) \smallsetminus \mathfrak{F}_1(\omega)) \geqslant \aleph_0$. So $\operatorname{card}(\mathfrak{P}(\omega) \smallsetminus \mathfrak{F}_1(\omega)) = 2^{\aleph_0}$, and therefore

$$\operatorname{card}(\mathcal{A}) \geqslant 2^{\aleph_0} \tag{2}$$

It follows from (1) and (2) by Bernstein's equinumerosity theorem 9.2.3 that $\operatorname{card}(\mathcal{A}) = 2^{\aleph_0}$.

Exercises

1. Prove proposition 10.3.1.
2. Suppose that $\mathfrak{a}$, $\mathfrak{b}$ are cardinals.
   (a) $\mathfrak{a} \geqslant 2^{\aleph_0}$ iff $\mathfrak{a} + 2^{\aleph_0} = \mathfrak{a}$.
   (b) If $\mathfrak{a} \geqslant 2^{\aleph_0} \geqslant \mathfrak{b}$, then $\mathfrak{a} + \mathfrak{b} = \mathfrak{a}$.
   (c) $\mathfrak{a} = 2^{\aleph_0}$ iff $\mathfrak{a} \geqslant \aleph_0$ and $\mathfrak{a} + \aleph_0 = 2^{\aleph_0}$.
   (d) If $2 \leqslant \mathfrak{a} \leqslant \mathfrak{b} = \mathfrak{b}^2$, then $\mathfrak{a}^{\mathfrak{b}} = 2^{\mathfrak{b}}$.
3. Show that the set $B = \{\mathfrak{b} : \mathfrak{b}^2 = \mathfrak{b}\}$ does not exist. [*Hint.* $2^{\mathfrak{a}\aleph_0} \in B$ for all $\mathfrak{a}$.]
4. Prove that if $\mathfrak{a}$ is not finite, then $2^{2^{\mathfrak{a}}} \geqslant 2^{\aleph_0}$.
5. Find the cardinalities of the following sets:
   (a) The set of subsets of ω with more than one element;
   (b) The set of infinite subsets of ω;
   (c) The set of permutations of ω.

10.4 The power of the continuum

We have already shown (twice) that the continuum is uncountable. Now, however, we can be more precise.

(10.4.1) **Proposition.** $\operatorname{card}(\mathbf{R}) = 2^{\aleph_0}$.

Proof. Note first that the function f from $\mathbf{R}$ to $\mathfrak{P}(\mathbf{Q})$ given by $f(a) = \{x \in \mathbf{Q} : x < a\}$ is one-to-one since $\mathbf{Q}$ is dense in $\mathbf{R}$, and so

$$\operatorname{card}(\mathbf{R}) \leqslant \operatorname{card}(\mathfrak{P}(\mathbf{Q})) = 2^{\operatorname{card}(\mathbf{Q})} = 2^{\aleph_0}.$$

Recall now from §8.5 that Cantor's ternary set K is the subset of $\mathbf{R}$ consisting of those numbers between 0 and 1 expressible by means of a ternary expansion consisting only of 0s and 2s. So there is a one-to-one correspondence between ${}^{\omega}\{0, 2\}$ and K, taking any sequence (s_n) of 0s and 2s to the real number $\sum_{n\in\omega} s_n 3^{-n-1}$, and hence

$$\operatorname{card}(K) = \operatorname{card}({}^{\omega}\{0, 2\}) = 2^{\aleph_0}.$$

Therefore $\operatorname{card}(\mathbf{R}) \geqslant 2^{\aleph_0}$. It follows by Bernstein's equinumerosity theorem that $\operatorname{card}(\mathbf{R}) = 2^{\aleph_0}$. □

Because of proposition 10.4.1 the cardinal $2^{\aleph_0}$ is often called the *power of the continuum* ('power' here being an old synonym for 'cardinality'): some authors denote it $\mathfrak{c}$; others write $\beth_1$.

(10.4.2) **Corollary.** $\operatorname{card}(\mathbf{R}^n) = \operatorname{card}(\mathbf{R})$ *if* $n \geqslant 1$.

Proof. $\operatorname{card}(\mathbf{R}^n) = (2^{\aleph_0})^n = 2^{n\aleph_0} = 2^{\aleph_0} = \operatorname{card}(\mathbf{R})$. □

This result so astonished Cantor when he first discovered it that he exclaimed (in a letter to Dedekind of 1877), 'Je le vois, mais je ne le crois pas.'

Various other collections may be shown to have the power of the continuum. As an example let us consider the collection of open subsets of $\mathbf{R}$. Each such set is expressible as a countable union of disjoint open intervals [proposition 7.4.4] and hence may be coded by a pair of sequences of real numbers representing the endpoints of these intervals. So the number of open sets is no more than

$$\operatorname{card}({}^{\omega}\mathbf{R} \times {}^{\omega}\mathbf{R}) = ((2^{\aleph_0})^{\aleph_0})^2 = 2^{2\aleph_0^2} = 2^{\aleph_0}.$$

But there are plainly at least this many open subsets of the real line, and hence by squeezing there are exactly this many.

It follows at once from this that the set of closed subsets of $\mathbf{R}$ also has the power of the continuum, as does the set of perfect subsets. The set $\mathfrak{P}(\mathbf{R})$ of *all* subsets of $\mathbf{R}$, by contrast, has the strictly larger cardinality $2^{2^{\aleph_0}}$ ($= \mathfrak{c}^{\mathfrak{c}}$).

It is of some interest to extend proposition 10.4.1 to all the non-empty perfect sets. The idea of the proof is to generalize what we did before and show that any perfect set has a subset isomorphic to Cantor's ternary set.

(10.4.3) **Lemma.** *If P is a perfect subset of the real line and $\mathcal{P}$ is the set of closed bounded subsets of P with infinitely many points of P in their interior, there is a function which maps any member A of $\mathcal{P}$ to a pair consisting of two disjoint members A_0 and A_1 of P whose lengths are less than half the length of A.*

Proof. Suppose that $A \in \mathcal{P}$. There are infinitely many points of P in the interior of A. If we let a_0 and a_1 be any two of these points and set $\varepsilon = \frac{1}{3}|a_1 - a_0|$, then the sets $A_i = \{x \in P : a_i - \varepsilon \leqslant x \leqslant a_i + \varepsilon\}$ $(i = 0, 1)$ evidently have the required properties. However, nothing hinges on the exact endpoints selected for the A_i; we could easily arrange for the endpoints to be rational if we wish. This has the advantage that we can stipulate in advance an enumeration of the intervals with rational endpoints and require that A_0 and A_1 should be chosen as early as possible in this enumeration, thus avoiding appeal to any form of the axiom of choice in defining the function we want. □

(10.4.4) **Proposition.** *If B is any non-empty perfect subset of* $\mathbf{R}$, *then* $\operatorname{card}(B) = 2^{\aleph_0}$.

Proof. The idea is to use the lemma to construct an isomorphic copy of Cantor's ternary set. Start with any closed bounded subset A of P with infinitely many points of P in its interior. If s is a finite string of 0s and 1s of length n, let us write $s^\frown 0$ and $s^\frown 1$ for the strings of length $n + 1$ obtained by adding 0 and 1 respectively at the end of s. Then using the lemma we can recursively associate a closed set A_s with every s in such a way that $A_{s^\frown 0}$ and $A_{s^\frown 1}$ are both subsets of A_s of less than half the length. For every *infinite* sequence s in ${}^{\omega}\{0, 1\}$ let $f(s)$ be the unique element of P that belongs to $\bigcap_{n\in\omega} A_{s|n}$. It is clear that f is one-to-one, so its image is a subset of P with the power of the continuum. □

Notes

We are not finished with cardinal arithmetic yet: as we have already noted, chapter 15 will indicate how it can be simplified by assuming further set-theoretic axioms such as the axiom of choice or the generalized continuum hypothesis. But even without the axiom of choice there is much more to be said than there is space for here. Sierpinski 1965 remains the most comprehensive treatment, but Bachmann 1955 may also be consulted.

Chapter 11

Ordinals

The simple and general principles of induction are powerful tools for proving things about the natural numbers: we intend now to investigate ways in which they can be generalized to apply to a very much wider class of ordered sets than the subsets of $\boldsymbol{\omega}$. The basis of our study will be the observation that a version of induction can be applied to any ordered set with a property we shall call well-ordering. Our strategy will be to apply much the same techniques to the study of isomorphism between well-ordered sets as we used in chapter 9 to investigate equinumerosity between sets. Just as that work led to an arithmetic of cardinals, what we shall do in this chapter will lead to an arithmetic of ordinals.

11.1 Well-ordering

Definition. *A relation r on a set A is said to be* well-founded *if every non-empty subset of A has an r-minimal element.*

This definition is designed to be just what is required to prove a generalization of the general principle of induction.

(11.1.1) **Proposition.** *If Φ is any formula and r is a well-founded relation on a set A, then*

$$(\forall x \in A)((\forall y \; r \; x)\Phi(y) \Rightarrow \Phi(x)) \Rightarrow (\forall x \in A)\Phi(x).$$

Proof. Let $B = \{x \in A : \Phi(x)\}$ and suppose, if possible, that $B \neq A$. Then $A \smallsetminus B$ is a non-empty subset of A and hence by hypothesis has an r-minimal element x. But then there is no $y \in A \smallsetminus B$ such that $y \; r \; x$. So for all $y \in A$, if $y \; r \; x$, then $\Phi(y)$. Hence by hypothesis $\Phi(x)$ and so $x \in B$. Contradiction. □

(11.1.2) **Proposition.** *$r^{\mathbf{t}}$ is well-founded iff r is well-founded.*

Necessity. This is obvious since any $r^{\mathbf{t}}$-minimal element of a set is also r-minimal.

Sufficiency. Suppose that $r^{\mathbf{t}}$ is not well-founded. So there is a non-empty subset B of A with no $r^{\mathbf{t}}$-minimal element, i.e. $B \subseteq r^{\mathbf{t}}[B] = r[\mathrm{Cl}_r(B)]$. So

$$\begin{aligned}\mathrm{Cl}_r(B) &= B \cup \mathrm{Cl}_r(r[B]) \\ &= B \cup [\mathrm{Cl}_r(B)] \\ &= r[\mathrm{Cl}_r(B)].\end{aligned}$$

It follows that r is not well-founded. □

In this chapter we shall often use the notation

$$\mathrm{seg}_A(a) = \{x \in A : x < a\}.$$

We may also write simply $\mathrm{seg}(a)$ if there seems to be no danger of misunderstanding.

(11.1.3) **Lemma.** *If $(A, \leqslant)$ is a partially ordered set, then these three assertions are equivalent:*

(i) *The strict partial ordering $<$ is well-founded on A.*

(ii) *Every subset of A which has a strict upper bound in A has a least strict upper bound in A.*

(iii) *If $B \subseteq A$ and $(\forall a \in A)(\mathrm{seg}_A(a) \subseteq B \Rightarrow a \in B)$, then $B = A$.*

Proof. Exercise 1. □

Definition. *If $(A, \leqslant)$ is a [partially] ordered set which satisfies the equivalent conditions of lemma 11.1.3 above, then we say that $\leqslant$ is a* [partial] well-ordering *on A, that $<$ is a strict [partial] well-ordering on A, and that $(A, \leqslant)$ is a* [partially] well-ordered set.

A partially well-ordered set evidently cannot contain (the image of) a strictly decreasing sequence. (See §14.1 for a discussion of the converse.) Any partially ordered set in which every non-empty subset has a *least* (not just minimal) element is totally (and hence well) ordered.

Obviously every subset of a [partially] well-ordered set is [partially] well-ordered by the inherited ordering. In particular, the initial subset $\mathrm{seg}_A(a)$ is [partially] well-ordered if A is.

Suppose that $(A, \leqslant)$ is a well-ordered set. A has a least element $\perp$ iff A is non-empty. A need not have a greatest element: if $a \in A$, then either a is the greatest element of A or there exist elements of A greater than a, in which case the least of these is the unique successor of a, denoted a^+. An element of $A \setminus \{\perp\}$ need not be the successor of any other element of A: if it is not, it is called a *limit point* of A; this is the case iff $a = \sup \mathrm{seg}(a)$.

The motivating example of a well-ordered set is $(\omega, \leqslant)$: it has no limit points and no greatest element. Every finite [partially] ordered set is [partially] well-ordered [theorem 6.4.5]; indeed an ordered set is finite iff both it and its opposite are well-ordered.

(11.1.4) **Proposition.** *Every well-ordered subset of the real line is countable.*

Proof. Suppose that A is a well-ordered subset of $\mathbb{R}$ and let $(r_n)_{n\in\omega}$ be a sequence whose range is a dense subset of $\mathbb{R}$. For each x in A let x^+ be the least element of A that is greater than x (or, if there are none, let it be any real number greater than x); and let $g(x)$ be the element r_n of least index such that $x < r_n < x^+$. The function g from A to $\{r_n : n \in \omega\}$ thus defined is evidently one-to-one, and so A is countable. □

When we discussed the constructivist understanding of the process of set formation, we noted the difficulty that this conception would apparently limit us to finite sets. In order to liberate the constructivist from this limitation, we examined the possibility of appealing to supertasks — processes carried out repeatedly with increasing speed, so that an infinite number may be carried out within a finite period of time. The proposition we have just proved shows, however, that this method has a limit. This is because the tasks performed in a supertask are well-ordered in time. As a result, if we assume that the ordering of time is correctly modelled as a continuum, we can conclude that any supertask contains only countably many subtasks.

(11.1.5) **Theorem.** *If $(A, \leqslant)$ is a well-ordered set and f is a strictly increasing function from A to itself, then $x \leqslant f(x)$ for all $x \in A$.*

Proof. Suppose not. So there exist elements $x \in A$ such that $x > f(x)$; let x_0 be the least such x. Then

$$f(x_0) \leqslant f(f(x_0)) < f(x_0).$$

Contradiction. □

(11.1.6) **Corollary.** *If $(A, \leqslant)$ is a well-ordered set and $B \subseteq A$, then these three assertions are equivalent:*

(i) B is a proper initial subset of A.

(ii) There exists an element $a \in A$ such that $B = \operatorname{seg}_A(a)$.

(iii) B is an initial subset of A which is not isomorphic to A.

(i) ⇒ (ii). If B is a proper initial subset of A, then $A \setminus B$ is non-empty and therefore has a least element a; plainly $B = \operatorname{seg}_A(a)$.

(ii) $\Rightarrow$ *(iii)*. Suppose that $B = \text{seg}_A(a)$. If there is an isomorphism f between $(A, \leqslant)$ and $(B, \leqslant_B)$, then $f(a) \geqslant a$ [theorem 11.1.5] and so $f(a) \notin B$, which is absurd.

(iii) $\Rightarrow$ *(i)*. Obvious. □

(11.1.7) **Corollary.** *Let* $(A, \leqslant)$ *and* $(B, \leqslant)$ *be well-ordered sets. If there is an isomorphism from* $(A, \leqslant)$ *to* $(B, \leqslant)$*, then it is unique.*

Proof. Suppose that f and g are isomorphisms between $(A, \leqslant)$ and $(B, \leqslant)$. Then $g \circ f^{-1}$ is strictly increasing, so for all $x \in A$

$$f(x) \leqslant g(f^{-1}(f(x))) = g(x) \text{ [theorem 11.1.5]}$$

and similarly $g(x) \leqslant f(x)$. Hence $f = g$. □

(11.1.8) **Theorem.** *If* $(A, \leqslant)$ *and* $(B, \leqslant)$ *are well-ordered sets, then either* A *is isomorphic to an initial subset of* B *or* B *is isomorphic to an initial subset of* A *(or both).*

Proof. Let

$$f = \{(x, y) \in A \times B : \text{seg}_A(x) \text{ and } \text{seg}_B(y) \text{ are isomorphic}\}.$$

Note first that if $(x_1, y_1), (x_2, y_2) \in f$, then $x_1 < x_2 \Leftrightarrow y_1 < y_2$ since if, for example, $x_1 < x_2$ and $y_1 \geqslant y_2$, then $\text{seg}_B(y_1)$ is isomorphic to a proper initial subset of itself, contrary to corollary 11.1.6. It follows from this that f is strictly increasing and that $\text{dom}[f]$ and $\text{im}[f]$ are initial subsets of A and B respectively. Suppose now, if possible, that $\text{dom}[f] \neq A$ and $\text{im}[f] \neq B$. Then let a be the least element of $A \setminus \text{dom}[f]$ and b be the least element of $B \setminus \text{im}[f]$. Evidently, $\text{dom}[f] = \text{seg}_A(a)$ and $\text{im}[f] = \text{seg}_B(b)$, and so f is an isomorphism between $\text{seg}_A(a)$ and $\text{seg}_B(b)$. Hence $b = f(a)$. Contradiction. Thus f is the isomorphism we require. □

Exercises

1. Prove lemma 11.1.3.
2. Suppose that $(A, \leqslant)$ is a partially ordered set.

   (a) Show that $(A, \leqslant)$ is partially well-ordered iff $\text{seg}(a)$ is partially well-ordered for every $a \in A$.

   (b) Show that A is finite iff $\mathfrak{P}(A)$ is partially well-ordered by inclusion. [*Sufficiency.* Consider $\{B \in \mathfrak{P}(A) : A \setminus B \text{ is finite}\}$.]
3. If $(A, \leqslant)$ is a partially ordered set, show that these two assertions are equivalent:

   (i) $\text{seg}_A(a)$ is well-ordered for every $a \in A$;

   (ii) $(A, \leqslant)$ is partially well-ordered, and every directed subset of A is totally ordered by the inherited ordering.

$(A, \leqslant)$ is said to be a *tree* if it satisfies these conditions.

4. If A is a tree, show that B is a maximal totally ordered subset of A iff it is a totally ordered initial subset of A with no strict upper bound in A. B is said to be a *branch* of A if it satisfies these conditions.

5. A totally ordered set $(A, \leqslant)$ is said to be *perfectly ordered* if it satisfies:

   (1) A has a least element;

   (2) every element of A except the greatest (if there is one) has a unique successor;

   (3) every element of A can be obtained by applying the successor operation a finite number of times to either the least element of A or a limit point of A.

Show that every well-ordered set is perfectly ordered but that the converse does not hold.

11.2 Ordinals

We have now to introduce the notion of the order-type of a structure. The idea is that it should code whether structures are isomorphic. This is evidently analogous to the problem we faced in §9.1 of defining the cardinal of a set so as to encode only its size. So it is no surprise that the definition of order-types we adopt here uses the Scott/Tarski trick in just the same way.

Definition. *If (A, r) is a structure, then the set*

$$\{(B, s) : (B, s) \text{ is isomorphic to } (A, r)\}$$

is denoted ord(A, r) *and called the* order-type *of (A, r).*

(11.2.1) **Proposition.** *If (A, r) and (B, s) are structures, then (A, r) is isomorphic to (B, s) iff* $\text{ord}(A, r) = \text{ord}(B, s)$.

Proof. Straightforward. □

Definition. *The order-type of a well-ordered set is called an* ordinal.

From now on we shall generally use lower-case Greek letters α, β, γ, etc. for ordinals.

Definition. *Suppose that $\alpha = \text{ord}(A, \leqslant)$ and $\beta = \text{ord}(B, \leqslant)$. We shall write $\alpha \leqslant \beta$ to mean that there exists an isomorphism of $(A, \leqslant)$ onto an initial subset of $(B, \leqslant)$.*

It is an easy exercise to check that this definition does not depend on our choice of the representatives $(A, \leqslant)$ and $(B, \leqslant)$.

(11.2.2) **Proposition.**

(a) $(\alpha \leqslant \beta \text{ and } \beta \leqslant \gamma) \Rightarrow \alpha \leqslant \gamma$;

(b) $\alpha \leqslant \alpha$;

(c) $(\alpha \leqslant \beta \text{ and } \beta \leqslant \alpha) \Rightarrow \alpha = \beta$;

(d) $\alpha \leqslant \beta$ or $\beta \leqslant \alpha$.

Proof. Parts (a) and (b) are trivial; (c) follows from corollary 11.1.6, and (d) from theorem 11.1.8. □

(11.2.3) **Proposition.** *If the partially ordered set $(A, \leqslant)$ is the union of a chain $\mathcal{B}$ of initial subsets each of which is well-ordered by the inherited partial order, then A is also well-ordered and*

$$\operatorname{ord}(A, \leqslant) = \sup_{B \in \mathcal{B}} \operatorname{ord}(B, \leqslant_B).$$

Proof. Suppose that A has a non-empty subset C with no minimal element and choose an element $x \in C$. Now $x \in B$ for some $B \in \mathcal{B}$, so $B \cap C$ is non-empty and therefore has a minimal element y since B is partially well-ordered. Since C has no minimal element, there exists $z \in C$ such that $z < y$; moreover $z \in B$ since B is an initial subset of A. This contradicts the minimality of y in $B \cap C$. Hence $(A, \leqslant)$ is partially well-ordered. The remainder of the proof is left as an exercise. □

Definition. $\boldsymbol{\alpha} = \{\beta : \beta < \alpha\}$.

The set $\boldsymbol{\alpha}$ certainly exists for any ordinal α because its members all belong to $\mathfrak{P}(\mathrm{V}(\alpha))$.

(11.2.4) **Theorem.** *If α is an ordinal, then $(\boldsymbol{\alpha}, \leqslant)$ is a well-ordered set and* $\operatorname{ord}(\boldsymbol{\alpha}, \leqslant) = \alpha$.

Proof. Let $(A, \leqslant)$ be any well-ordered set such that $\operatorname{ord}(A, \leqslant) = \alpha$, and for each $x \in A$ let $f(x) = \operatorname{ord}(\operatorname{seg}_A(x), \leqslant)$. Then f is strictly increasing, since if $x < y$, then $\operatorname{seg}(x)$ is an initial subset of, but not isomorphic to, $\operatorname{seg}(y)$ [corollary 11.1.6] and so $f(x) < f(y)$. Moreover, the image of f is the whole of $\boldsymbol{\alpha}$, since if $\beta < \alpha$, then there exists $x \in A$ such that

$$\beta = \operatorname{ord}(\operatorname{seg}_A(x), \leqslant) = f(x) \text{ [corollary 11.1.6]}.$$

So $(A, \leqslant)$ is isomorphic to $(\boldsymbol{\alpha}, \leqslant)$. The result follows immediately. □

(11.2.5) **Corollary.** *Any non-empty set of ordinals has a least element.*

Proof. Let A be any non-empty set of ordinals and let $\alpha \in A$. If α is the least element of A, we are finished. If not, $A \cap \boldsymbol{\alpha}$ is non-empty and hence has a least element [theorem 11.2.4], which must also be the least element of A. □

(11.2.6) **Burali-Forti's paradox.** *$\{\alpha : \alpha$ is an ordinal$\}$ does not exist.*

Proof. Suppose on the contrary that $A = \{\alpha : \alpha \text{ is an ordinal}\}$ exists. It is well-ordered by the usual ordering [corollary 11.2.5]. So if $\alpha = \text{ord}(A, \leqslant)$, then $\text{ord}(\boldsymbol{\alpha}, \leqslant) = \alpha$ [theorem 11.2.4], and therefore $(\boldsymbol{\alpha}, \leqslant)$ is isomorphic to $(A, \leqslant)$ [proposition 11.2.1]. But $\boldsymbol{\alpha}$ is a proper initial subset of A (since $\alpha \in A$) and is therefore not isomorphic to it [corollary 11.1.6]. Contradiction. □

(11.2.7) **Proposition.** *If Φ is any formula, then the set $B = \{\beta : \Phi(\beta)\}$ exists iff there is an ordinal α such that $\Phi(\beta) \Rightarrow \beta < \alpha$.*

Necessity. Suppose that $B = \{\beta : \Phi(\beta)\}$. So there exists an ordinal α not in $\mathrm{V}(B)$, since if all the ordinals belonged to $\mathrm{V}(B)$, they would form a set, contradicting Burali-Forti's paradox. Now $\mathrm{V}(B) \in \mathrm{V}(\alpha)$, so

$$\begin{aligned} \Phi(\beta) &\Rightarrow \beta \in B \\ &\Rightarrow \mathrm{V}(\beta) \in \mathrm{V}(B) \\ &\Rightarrow \mathrm{V}(\beta) \in \mathrm{V}(\alpha) \\ &\Rightarrow \beta < \alpha. \end{aligned}$$

Sufficiency. If there is an ordinal α such that $\Phi(\beta) \Rightarrow \beta < \alpha$, then the ordinals satisfying Φ all belong to the set $\boldsymbol{\alpha}$. □

The least ordinal is denoted 0. There is evidently no greatest ordinal. For any ordinal α we let α^+ denote the least ordinal greater than α: it is called the *successor* of α. A non-zero ordinal which is not the successor of any other ordinal is called a *limit ordinal*.

(11.2.8) **Lemma.** *If $\alpha \neq 0$, then these three assertions are equivalent:*

- *(i)* *α is a limit ordinal;*
- *(ii)* $(\forall\beta)(\beta < \alpha \Rightarrow \beta^+ < \alpha)$;
- *(iii)* $\alpha = \sup_{\beta<\alpha} \beta$.

(i) ⇒ (ii). Suppose $\beta < \alpha$. Then certainly $\beta^+ \leqslant \alpha$. But if $\beta^+ = \alpha$, then α is not a limit. So $\beta^+ < \alpha$ as required.

(ii) ⇒ (iii). Let $\gamma = \sup_{\beta<\alpha} \beta$. Then $\gamma \leqslant \alpha$. But if $\gamma < \alpha$, then $\gamma^+ < \alpha$ by hypothesis and so $\gamma^+ < \gamma$, which is absurd. So $\gamma = \alpha$.

(iii) $\Rightarrow$ *(i)*. Suppose that $\sup_{\beta<\alpha}\beta = \alpha$ but α is not a limit. So there exists an ordinal γ such that $\alpha = \gamma^+$, and therefore

$$\gamma = \sup_{\beta<\alpha}\beta = \alpha = \gamma^+,$$

which is absurd. □

11.3 Transfinite induction and recursion

We now consider the generalizations to arbitrary ordinals of the simple and general principles of induction and recursion which we proved in chapter 9.

(11.3.1) **Proposition.** *If $A \subseteq \boldsymbol{\alpha}$ and if $\boldsymbol{\beta} \subseteq A \Rightarrow \beta \in A$, then $A = \boldsymbol{\alpha}$.*

Proof. Immediate [lemma 11.1.3 and theorem 11.2.4]. □

(11.3.2) **General transfinite induction scheme.** *If $\Phi(\alpha)$ is a formula and if $(\forall\beta)((\forall\gamma < \beta)\Phi(\gamma) \Rightarrow \Phi(\beta))$, then $(\forall\alpha)\Phi(\alpha)$.*

Proof. Let $A = \{\beta < \alpha : \Phi(\beta)\}$; then $A = \boldsymbol{\alpha}$ [proposition 11.3.1] and so $\Phi(\alpha)$ holds. □

(11.3.3) **Proposition.** *Suppose that A is a subset of $\boldsymbol{\alpha}$ with the following three properties:*

(1) $0 \in A$;

(2) $\beta \in A \Rightarrow \beta^+ \in A$ *for every* $\beta < \alpha$;

(3) $\boldsymbol{\lambda} \subseteq A \Rightarrow \lambda \in A$ *for every limit ordinal* $\lambda < \alpha$.

Then $A = \boldsymbol{\alpha}$.

Proof. Suppose, if possible, that $A \subset \boldsymbol{\alpha}$. So $\boldsymbol{\alpha} \setminus A$ is non-empty, and therefore has a least element [theorem 11.2.4]. The first property shows that this element is not zero, the second that it is not a successor, and the third that it is not a limit ordinal. Contradiction. □

(11.3.4) **Simple transfinite induction scheme.** *Suppose that $\Phi(\alpha)$ is a formula such that:*

(1) $\Phi(0)$;

(2) $(\forall\beta)(\Phi(\beta) \Rightarrow \Phi(\beta^+))$;

(3) If λ is a limit ordinal, then $(\forall\beta < \lambda)\Phi(\beta) \Rightarrow \Phi(\lambda)$.

Then $(\forall\alpha)\Phi(\alpha)$.

Proof. This follows from proposition 11.3.3 in the same way as the general principle of transfinite induction follows from proposition 11.3.1. □

(11.3.5) **General principle of transfinite recursion.** *If for each $\beta < \alpha$ there is a function s_β from $\mathfrak{P}(A)$ to A, then there exists exactly one function g from $\boldsymbol{\alpha}$ to A such that $g(\beta) = s_\beta(g[\boldsymbol{\beta}])$ for all $\beta < \alpha$.*

Uniqueness. This can be shown straightforwardly by using the general principle of transfinite induction.

Existence. We shall prove the existence of g by simple transfinite induction on α. This is trivial if $\alpha = 0$. Then if there is a function g from $\boldsymbol{\beta}$ to A with the required property, we can extend g to β^+ by letting $g(\beta) = s_\beta(g[\boldsymbol{\beta}])$. Finally, suppose that λ is a non-zero limit ordinal and that for each $\beta < \lambda$ there exists a (necessarily unique) function g_β from $\boldsymbol{\beta}$ to A such that $g_\beta(\gamma) = s_\gamma(g_\beta[\boldsymbol{\gamma}])$ for all $\gamma < \beta$. Any two of these functions g_β must agree on the intersection of their domains. So if $g = \bigcup_{\beta<\lambda} g_\beta$, then g is a function [proposition 4.8.1] and

$$\operatorname{dom}[g] = \bigcup_{\beta<\lambda} \operatorname{dom}[g_\beta] = \{\gamma : (\exists\beta < \lambda)(\gamma < \beta)\} = \boldsymbol{\lambda}.$$

It is easy to see that g is the function we require. This completes the proof by simple transfinite induction. □

(11.3.6) **Simple principle of transfinite recursion.** *If A is a set, f is a function from A to itself, a is a member of A, and s is a function from $\mathfrak{P}(A)$ to A, then for each ordinal α there exists a unique function g from $\boldsymbol{\alpha}$ to A such that*

(1) $g(0) = a$;

(2) $g(\beta^+) = f(g(\beta))$ if $\beta^+ < \alpha$;

(3) $g(\lambda) = s(g[\boldsymbol{\lambda}])$ for every limit ordinal $\lambda < \alpha$.

Proof. The proof is very similar to that of the general principle of transfinite recursion. □

As an illustration of the use of ordinals we shall prove a result about sets of real numbers that is central to descriptive set theory.

(11.3.7) **Theorem (Cantor 1883; Bendixson 1883).** *Every uncountable closed subset of* **R** *has a non-empty perfect subset.*

Proof. Let B be any closed subset of $\mathbf{R}$ and recursively define

$$B^{(0)} = B;$$
$$B^{(\alpha+1)} = (B^{(\alpha)})';$$
$$B^{(\lambda)} = \bigcap_{\alpha<\lambda} B^{(\alpha)} \text{ for } \lambda \text{ a limit ordinal.}$$

As B is closed, the $B^{(\alpha)}$ form a decreasing transfinite sequence of closed subsets of $\mathbf{R}$. If α_0 is the least ordinal α such that $B^{(\alpha+1)} = B^{(\alpha)}$, then $(B^{(\alpha_0)})' = B^{(\alpha_0+1)} = B^{(\alpha_0)}$, and so $B^{(\alpha_0)}$ is perfect. Let us show next that the set

$$B_0 = \bigcup_{\beta<\alpha_0} (B^{(\beta)} \smallsetminus B^{(\beta+1)})$$

is countable. We do this by mapping it into the countable set $\mathcal{U}$ of all the open intervals with rational endpoints. For each $\beta < \alpha_0$ and each x in $B^{(\beta)} \smallsetminus B^{(\beta+1)}$ we can select a member $U(x)$ of $\mathcal{U}$ which is disjoint from $B^{(\beta+1)}$ and whose intersection with $B^{(\beta)} \smallsetminus B^{(\beta+1)}$ consists only of x. Moreover, we can stipulate (so as to avoid using the axiom of countable choice) that $U(x)$ is chosen to occur as early as possible in some previously specified enumeration of $\mathcal{U}$. The function $x \mapsto U(x)$ thus defined from B_0 to $\mathcal{U}$ is evidently one-to-one, and hence B_0 is countable. But $B = B_0 \cup B^{(\alpha_0)}$, and so if B is uncountable, $B^{(\alpha_0)}$ is non-empty as required. □

(11.3.8) **Corollary.** *Every uncountable closed subset of* $\mathbf{R}$ *has the power of the continuum.*

Proof. We have already shown that a non-empty perfect set always has the power of the continuum [proposition 10.4.4]. □

The reason Cantor was interested in proving this result was that in 1878 he had made the conjecture which is now called the *continuum hypothesis*, namely that *every* uncountable subset of $\mathbf{R}$ has the power of the continuum. He therefore regarded the result we have just proved as a promising special case on the way to proving his conjecture. It is a very special case, though, as even a naive counting argument shows: the set of closed subsets of $\mathbf{R}$ has cardinal $\mathfrak{c} = 2^{\aleph_0}$, whereas the set of *all* subsets has cardinal $\mathfrak{c}^{\mathfrak{c}}$. We shall have more to say about this in chapter 15; it will emerge, in particular, that the difficulties in the way of extending the method of the Cantor/Bendixson theorem so as to yield a proof of the continuum hypothesis are at least as great as the naive argument suggests.

11.4 Cardinality

Definition. $|\alpha| = \text{card}(\boldsymbol{\alpha})$.

When we speak of the *cardinality* of an ordinal α, we shall mean not the cardinal number of that particular set (which is, after all, an accidental by-product of the definition of the ordinals that we are using) but the cardinal $|\alpha|$ just defined. We say that α is finite or infinite, countable or uncountable, according as $|\alpha|$ is. Evidently if $\text{ord}(A, \leqslant) = \alpha$, then $\text{card}(A) = |\alpha|$.

(11.4.1) **Proposition.** $\alpha \leqslant \beta \Rightarrow |\alpha| \leqslant |\beta|$.

Proof. Obvious. □

The converse fails in the infinite case, of course.

Definition. $\omega_0 = \text{ord}(\boldsymbol{\omega}, \leqslant)$.

So $|\omega_0| = \aleph_0$, and therefore ω_0 is the least infinite ordinal. If for each $n \in \boldsymbol{\omega}$ we let $n_0 = \text{ord}(\boldsymbol{n}, \leqslant)$, then the function from $\boldsymbol{\omega}$ to $\boldsymbol{\omega}_0$ given by $n \mapsto n_0$ is an isomorphism: from now on we shall identify the natural number n with the finite ordinal n_0 and write ω for ω_0.

A family $(x_\alpha)_{\alpha<\beta}$ indexed by $\boldsymbol{\beta}$ for some ordinal β is called a *transfinite sequence*. Since the ordinals $< \omega$ are (under the identification just introduced) simply the natural numbers, the special case where $\beta = \omega$ is that of a sequence in the sense of §5.3.

(11.4.2) **Theorem (Hartogs 1915).** *There is no cardinal* $\mathfrak{a}$ *such that* $|\alpha| \leqslant \mathfrak{a}$ *for every ordinal* α.

Proof. Suppose on the contrary that $\mathfrak{a}$ is an upper bound for all cardinals of the form $|\alpha|$. Then it is easy to show that every ordinal is contained in the fourth level after $V(\mathfrak{a})$. Hence $\{\alpha : \alpha \text{ is an ordinal}\}$ exists [axiom scheme of separation], contradicting Burali-Forti's paradox. □

This theorem is a useful tool for proving the existence of larger ordinals. It shows, for example, that not all ordinals are countable, since otherwise $|\alpha| \leqslant \aleph_0$ for all α, contradicting the theorem. The least uncountable ordinal is denoted ω_1, and we let $\aleph_1 = |\omega_1|$.

(11.4.3) **Proposition.** $\aleph_0 < \aleph_1 < 2^{2^{\aleph_0}}$.

Proof. Obviously $\aleph_0 \leqslant \aleph_1$; and indeed $\aleph_0 < \aleph_1$, since if $\boldsymbol{\omega}_1$ were countable, ω_1 would be a countable ordinal and we would have $\omega_1 < \omega_1$. If for each ordinal $\alpha < \omega_1$ we let $f(\alpha)$ be the set of all the well-orderings r on $\boldsymbol{\omega}$ such that $\text{ord}(\boldsymbol{\omega}, r) = \alpha$, then f is a one-to-one function from $\boldsymbol{\omega}_1$ into $\mathfrak{P}(\mathfrak{P}(\boldsymbol{\omega} \times \boldsymbol{\omega}))$. So

$$\aleph_1 = |\omega_1| \leqslant \text{card}(\mathfrak{P}(\mathfrak{P}(\boldsymbol{\omega} \times \boldsymbol{\omega}))) = 2^{2^{\aleph_0^2}} = 2^{2^{\aleph_0}}.$$ □

We have thus arrived at another example of an uncountable set, since one consequence of the proposition just proved is that the set $\boldsymbol{\omega}_1$ of countable ordinals is uncountable. Cantor called $\boldsymbol{\omega}$ the 'first number class' and $\boldsymbol{\omega}_1 \setminus \boldsymbol{\omega}$ the 'second number class'.

It is worth noting that the reasoning we have used to show that $\boldsymbol{\omega}_1$ is uncountable is impredicative, just as the reasoning for the corresponding result about real numbers was, although this time the impredicativity is hidden a little deeper: the proof of the Burali-Forti paradox depends on considering the ordinal of the set of all ordinals, which is evidently an impredicative specification, and our proof that $\boldsymbol{\omega}_1$ is uncountable inherits this impredicativity via Hartogs' theorem, since the proof of that result makes use of the Burali-Forti paradox. This impredicativity is significant because it is what takes the theory of ordinals beyond Cantor's original idea of 'ordinal notations'. For any system of notations is countable and will therefore fail to exhaust $\boldsymbol{\omega}_1$.

The method of labelling objects with ordinals is a powerful tool for proving results in set theory. In order to apply it, however, we need the objects we are studying to have a well-ordered structure. It therefore becomes a matter of interest to us whether there exists a well-ordering on a given set.

Definition. *A set A is said to be* well-orderable *if there exists a well-ordering on A.*

A set is well-orderable iff there exists a transfinite sequence whose range it is. If A and B are equinumerous, then evidently A is well-orderable iff B is. So whether or not a set is well-orderable depends only on its cardinality. Since $\boldsymbol{\omega}$ is well-orderable, it follows that every countable set is well-orderable too, but the converse is false: the well-orderable set $\boldsymbol{\omega}_1$ of all the countable ordinals is, as we have just seen, uncountable. We shall have more to say in §14.4 about which other uncountable sets are well-orderable.

Exercise

Show that the birthday of α is no more than four levels after the birthday of $|\alpha|$.

11.5 Rank

One of the most important applications of the system of ordinals is to measure how high in the hierarchy a set is located. The measure we shall use is called the *rank* of the set. To define this notion, we need to recall some facts from §3.6. A large part of what we established there may be expressed in the vocabulary we have introduced in this chapter by saying that the history of any level V (the set of all the levels belonging to V) is well-ordered: the lax ordering is by inclusion, and the associated strict ordering is by membership. We can now exploit this by making use of the ordinal of this history as an index of its position in the hierarchy.

Definition. *The* rank *of a set A is the ordinal*

$$\rho(A) = \mathrm{ord}(\{V : V \text{ is a level and } V \in \mathrm{V}(A)\}, \subseteq).$$

In words: the rank of A is the ordinal of the history of its birthday. Now

$$\rho(A) \leqslant \rho(B) \Leftrightarrow \mathrm{V}(A) \subseteq \mathrm{V}(B)$$

and so

$$\rho(A) < \rho(B) \Leftrightarrow \mathrm{V}(A) \in \mathrm{V}(B).$$

From this we can deduce that the rank of a set is an adequate surrogate for its birthday as a measure of its position in the hierarchy.

Definition. $\mathrm{V}_\alpha = \{x : x \text{ is an individual or } x \text{ is a set and } \rho(x) < \alpha\}$.

(11.5.1) **Proposition.**

$$\begin{aligned} \mathrm{V}_0 &= \{x : x \text{ is an individual}\}; \\ \mathrm{V}_{\alpha+1} &= \mathrm{V}_\alpha \cup \mathfrak{P}(\mathrm{V}_\alpha); \\ \mathrm{V}_\lambda &= \bigcup_{\alpha<\lambda} \mathrm{V}_\alpha \text{ if } \lambda \text{ is a limit ordinal.} \end{aligned}$$

Proof. Transfinite induction. □

(11.5.2) **Proposition.** *If* V_α *exists, then it is a level, and*

$$\alpha = \rho(\mathrm{V}_\alpha) = \mathrm{ord}(\{V : V \text{ is a level and } V \in \mathrm{V}_\alpha\}, \subseteq).$$

Proof. Transfinite induction again. □

No claim is being made at this stage that the level V_α exists, and indeed it is consistent with ZU that there should be ordinals α — even as small as the second limit ordinal — for which it does not. This issue will be central to our discussion in chapter 13.

We saw earlier that the hierarchy of levels contains within it a core consisting of the pure levels: these can be thought of as obtained by a somewhat similar process to the V_α but starting from Ø.

Definition. $\mathrm{U}_\alpha = \{a \in \mathrm{V}_\alpha : a \text{ is pure}\}$.

(11.5.3) **Proposition.**

$$\begin{aligned} U_0 &= \varnothing; \\ U_{\alpha+1} &= \mathfrak{P}(U_\alpha); \\ U_\lambda &= \bigcup_{\alpha<\lambda} U_\alpha \text{ if } \lambda \text{ is a limit ordinal.} \end{aligned}$$

Proof. As for proposition 11.5.1. □

It is easy to show that (in the notation of §5.4) $\text{card}(U_n) = 2_{n-1}$. Many authors extend this into the transfinite with a notation devised by Peirce according to which

$$\beth_\alpha = \text{card}(U_{\omega+\alpha})$$

for any ordinal α, so that for example $\beth_0 = \aleph_0$, $\beth_1 = 2^{\aleph_0}$, and $\beth_2 = 2^{2^{\aleph_0}}$. From this we can deduce that $\rho(\boldsymbol{\omega}) = \omega$ and $\rho(\mathbf{R}) = \omega + 1$ (but only, of course, because we stipulated that $\boldsymbol{\omega}$ and $\mathbf{R}$ should be as low in the pure hierarchy as possible).

(11.5.4) **Proposition.** *If r is a relation on a set A, then r is well-founded iff there exist an ordinal α and a function f from A into $\boldsymbol{\alpha}$ such that*

$$x \; r \; y \Rightarrow f(x) < f(y) \tag{1}$$

for all $x, y \in A$.

Necessity. The proof is by generalized recursion, defining $f(a)$ to be the least strict upper bound of $\{f(x) : x \; r \; a\}$.

Sufficiency. Straightforward. □

The least ordinal α for which a function f satisfying (1) exists is called the *rank* of the well-founded relation r. This notion of rank enables us to state a partial categoricity result for the second-order variant of our theory of sets.

Zermelo's categoricity theorem. *If T is a categorical theory, then the grounded parts of any two models of* Z2[T] *of the same rank are isomorphic.*

This is, of course, just what the discussion of the two principles of plenitude in part I encouraged us to expect. Once the individuals are given, the second-order separation scheme determines fully what sets there are at each level in the hierarchy — namely, all of them. The only other variable that needs to be fixed if we are to determine what sets there are is the ordinal measure of the height of the hierarchy that we have called its rank. For further reflections on this matter, see the conclusion to part IV.

Exercise

Complete the details of the proof of proposition 11.5.4.

Notes

Perhaps even more than for cardinals the theory of ordinals can be said to be the discovery of one man: Dauben (1990) describes the genesis of Cantor's theory in detail. Tait (2000) describes how in *Grundlagen* Cantor went beyond his earlier conception of ordinals as a system of notations to serve as indices for recursive definitions.

The history of Burali-Forti's paradox is somewhat convoluted. To explain it, we shall (temporarily) adopt the name 'quasi-ordinal' for the order-type of a perfectly ordered set (see exercise 5 of §11.1). With this terminology, what Burali-Forti (1897*b*) proved was that the set of all quasi-ordinals is not totally ordered in its natural partial order. He also asserted that every quasi-ordinal is an ordinal, thereby contradicting Cantor's (1897) theorem that the ordinals are totally ordered. But there was a mistake here: Burali-Forti had misunderstood Cantor's definition of a well-ordering, and it is in fact the converse implication — every ordinal is a quasi-ordinal — which holds. So the contradiction dissolves: there is nothing absurd about a non-totally ordered set having a totally ordered subset. Burali-Forti corrected his error in his 1897*a* and added the rather enigmatic remark that 'the reader can check which propositions in (1897*b*) are verified also by the well-ordered classes'. He seems not to have taken his own advice, however: it was left to Russell (1903) to do the checking and hence resurrect the paradox by noticing that Burali-Forti's argument shows without any essential change that the ordinals are not totally ordered. This last result *is* genuinely paradoxical, and because of Russell's attribution it has become universally known as 'Burali-Forti's paradox', although Burali-Forti himself denied as late as 1906 (in a letter to Couturat) that any contradiction was involved. (The letter as Couturat published it is hopelessly confused: it would be charitable to Burali-Forti to suppose that it has been mistranslated.) Cantor had in any event discovered the paradox independently some time before 1899.

Transfinite induction as a means of proof and transfinite recursion as a means of definition are now a commonplace in the toolkit of any pure mathematician: textbooks of classical analysis such as Hobson 1921 contain frequent illustrations.

Of special interest to logicians has been the discovery that transfinite induction can be used in harness with cut elimination in proof theory to demonstrate the consistency of various formal systems. This idea goes back to Hilbert (1925), who hoped to achieve this aim by *finite* induction on the natural numbers. Gödel (1931) showed that this was too ambitious, at least for the formal systems Hilbert was interested in, but interest in the project was revived when Gentzen (1936) used transfinite induction as far as the countable ordinal ε_0 (defined in §12.5 below) to show that PA is consistent. Subsequent work has extended Gentzen's result to various more or less predicative systems of analysis, but the obvious goal of achieving an informative consistency proof for classical analysis, or even for a system of set theory such as Z, remains elusive.

Chapter 12

Ordinal arithmetic

In this chapter we shall define operations of addition, multiplication and exponentiation for ordinals. We shall take as our model the recursive definitions of the corresponding operations for natural numbers (§5.4): the form the extended definitions should take at successor ordinals is clear; the form of the additional clause defining the behaviour of the operations at limit ordinals is also clear if we require that the operations must all be normal in their second variable.

12.1 Normal functions

(12.1.1) **Proposition.** *If $(A, \leqslant)$ is well-ordered and $(B, \leqslant)$ is partially ordered, then a function f from A to B is normal [resp. strictly normal] iff it satisfies:*

If $a \in A$ is a limit point of A, then $f(a) = \sup_{x<a} f(x)$;

$f(a^+) \geqslant f(a)$ [resp. $f(a^+) > f(a)$] for all a in A apart from the greatest (if any).

Necessity. Obvious.

Sufficiency. Let us prove first that f is increasing. For if it is not, then there exists $a \in A$ such that for some $b < a$ we have $f(b) \nleqslant f(a)$: choose the element a as small as possible. There are two cases to consider:

(1) a is a limit point of A. In this case

$$f(a) = \sup_{x<a} f(x) \nleqslant f(a).$$

Contradiction.

(2) $a = c^+$ for some $c \in A$. In this case for all $x < a$ we have $x \leqslant c$ and hence $f(x) \leqslant f(c)$ (by the minimality of a), so that $c = b$ and hence $f(c) \nleqslant f(a)$. But $f(a) = f(c^+) \geqslant f(c)$. Contradiction.

Therefore f is indeed increasing. [The proof that f is strictly increasing under the stronger version of the second hypothesis is similar.]

Now let us show that f is normal. Suppose that C is a non-empty bounded subset of A and let $c = \sup C$. Now if $x \in C$, then $x \leqslant c$ and so $f(x) \leqslant f(c)$; hence $f(c)$ is an upper bound for $f[C]$. Now consider two possibilities:

(1) $c \in C$. In this case $f(c) \in f[C]$ and so $f(c)$ is the supremum of $f[C]$.

(2) $c \notin C$. In this case c is a limit point of A. Now if $x < c$, then there exists $y \in C$ such that $x \leqslant y$, and so if z is an upper bound for $f[C]$ in B, then $f(x) \leqslant f(y) \leqslant z$. Hence $f(c) = \sup_{x<c} f(x) \leqslant z$. It follows that $f(c)$ is the supremum of $f[C]$ in this case too.

Thus f is [strictly] normal. □

(12.1.2) **Proposition.** *If $(A, \leqslant)$ is a well-ordered set and f is a strictly normal function from A to itself, then for every $a \geqslant f(\bot)$ there exists a greatest element $b \in A$ such that $f(b) \leqslant a$.*

Proof. If a is the greatest element of A, then it is clearly the element we are looking for. If not, then $f(a^+) > f(a) \geqslant a$ and so there exist elements $c \in A$ such that $f(c) > a$: choose the least such element c. Now $c > \bot$ since $f(c) > f(\bot)$. And if c were a limit point of A, then we would have

$$f(c) = \sup_{x<c} f(x) \leqslant a < f(c),$$

which is absurd. So there exists $b \in A$ such that $c = b^+$. Clearly $f(b) \leqslant a$; and if $x > b$ then $x \geqslant c$, so that $f(x) \geqslant f(c) > a$. Thus b is the element we want. □

12.2 Ordinal addition

Consider first the case of ordinal addition. We want to define an operation with the following properties:

$$\begin{aligned} \alpha + 0 &= \alpha; \\ \alpha + \beta^+ &= (\alpha + \beta)^+; \\ \alpha + \lambda &= \sup_{\beta<\lambda}(\alpha + \beta) \text{ if } \lambda \text{ is a limit ordinal.} \end{aligned}$$

When we gave the definition of addition for the natural numbers, we justified it directly from Dedekind's theorem on the validity of definition by recursion. So we might hope to be able to justify addition of ordinals by appeal to the corresponding theorem on the validity of definition by transfinite recursion.

But unfortunately there is a difficulty with this. In the third clause of the definition, dealing with the case of a limit ordinal λ, we need an independent argument to show that the ordinals $\alpha + \beta$ for $\beta < \lambda$ are bounded above, since if they are not, then $\sup_{\beta<\lambda}(\alpha + \beta)$ will not exist.

As it turns out, the easiest way to meet this difficulty is to sidestep it. We do not use transfinite recursion in the definition at all, but instead define ordinal addition explicitly by synthetic means. Once we have done this, it will be a straightforward matter to check that the operation we have defined does indeed satisfy the recursion equations just given.

Definition. *If* $(A, \leqslant)$ *and* $(B, \leqslant)$ *are ordered sets, then we define the* ordered sum $A + B$ *of* A *with* B *to be the disjoint union* $A \uplus B = (A \times \{0\}) \cup (B \times \{1\})$ *with the ordering defined so that* $(x, i) \leqslant (y, j)$ *iff*

$i = 0$ *and* $j = 1$

or $i = j = 0$ *and* $x \leqslant y$

or $i = j = 1$ *and* $x \leqslant y$.

This definition amounts to placing a copy of B after a copy of A. It is clear that the ordered sum of two ordered sets is ordered. For it to be useful to us here, however, it must also be well-ordered if its components are. Fortunately, this is easily seen to be the case.

(12.2.1) **Lemma.** *The ordered sum of two well-ordered sets is well-ordered.*

Proof. Suppose that C is a non-empty subset of $A \uplus B$. So

$$C = (D \times \{0\}) \cup (E \times \{1\})$$

for some $D \subseteq A$ and $E \subseteq B$. If D is non-empty, then it has a least element a in A (since A is well-ordered), in which case $(a, 0)$ is the least element of C; whereas if D is empty, E is non-empty and has a least element b in B (since B is well-ordered), in which case $(b, 1)$ is the least element of C. □

This lemma provides the justification for the following definition.

Definition. *If* $\alpha = \text{ord}(A, \leqslant)$ *and* $\beta = \text{ord}(B, \leqslant)$, *then we define* $\alpha + \beta$ *to be the ordinal of the ordered sum of* $(A, \leqslant)$ *and* $(B, \leqslant)$.

So $\omega + 2$, for example, is the ordinal of a well-ordered set consisting of a copy of $\boldsymbol{\omega}$ followed by a copy of $\mathbf{2}$.

$$\begin{array}{cccccc} \bullet & \bullet & \bullet & \cdots & \bullet & \bullet \\ 0 & 1 & 2 & & 0' & 1' \end{array}$$

(12.2.2) **Proposition.** $|\alpha + \beta| = |\alpha| + |\beta|$.

Proof. Trivial. □

(12.2.3) **Proposition.** *Ordinal addition is characterized by the following recursion equations:*

(a) $\alpha + 0 = \alpha$;

(b) $\alpha + \beta^+ = (\alpha + \beta)^+$;

(c) $\alpha + \lambda = \sup_{\beta<\lambda}(\alpha + \beta)$ *if* λ *is a limit ordinal.*

Proof. Straightforward. □

It is clear from proposition 12.2.3 that for finite ordinals (i.e. natural numbers) this definition of addition coincides with the one given in §5.4.

Since $\alpha + 1 = \alpha + 0^+ = (\alpha + 0)^+ = \alpha^+$, proposition 12.2.3(b) can be rewritten, rather more familiarly, as

$$\alpha + (\beta + 1) = (\alpha + \beta) + 1.$$

(12.2.4) **Corollary.** *For each* α *the function given by* $\beta \mapsto \alpha + \beta$ *is strictly normal on any set of ordinals.*

Proof. Immediate [propositions 12.1.1 and 12.2.3]. □

(12.2.5) **Subtraction algorithm.** *If* $\beta \leqslant \alpha$*, then there exists a unique ordinal* ρ *such that* $\alpha = \beta + \rho$.

Existence. There exists a greatest ordinal ρ such that $\beta + \rho \leqslant \alpha$ [corollary 12.2.4 and proposition 12.1.2]. But if $\beta + \rho < \alpha$, then

$$\beta + (\rho + 1) = (\beta + \rho) + 1 \leqslant \alpha,$$

contradicting our choice of ρ. Hence $\beta + \rho = \alpha$ as required.

Uniqueness. This follows from the fact that the function $\rho \mapsto \beta + \rho$ is one-to-one [corollary 12.2.4]. □

If $\beta \leqslant \alpha$, the unique ordinal ρ such that $\alpha = \beta + \rho$ is sometimes written $\alpha - \beta$.

(12.2.6) **Proposition.** *Suppose that* α, β, γ *are ordinals.*

(a) $\beta < \gamma \Rightarrow \alpha + \beta < \alpha + \gamma$.

(b) $\alpha + \beta = \alpha + \gamma \Rightarrow \beta = \gamma$.

(c) $\alpha + \sup_{\beta \in B} \beta = \sup_{\beta \in B}(\alpha + \beta)$ *for every non-empty set of ordinals* B.

(d) $\alpha + (\beta + \gamma) = (\alpha + \beta) + \gamma$.

(e) $\alpha \leqslant \beta \Rightarrow \alpha + \gamma \leqslant \beta + \gamma$.

(f) $\alpha \leqslant \beta \Leftrightarrow (\exists\delta)\beta = \alpha + \delta$.

(g) $\alpha < \beta \Leftrightarrow (\exists\delta > 0)\beta = \alpha + \delta$.

Proof of (a), (b) and (c). Immediate [corollary 12.2.4].

Proof of (d). We use transfinite induction on γ. The case $\gamma = 0$ is trivial. If $\alpha + (\beta + \gamma) = (\alpha + \beta) + \gamma$, then

$$\begin{aligned}\alpha + (\beta + (\gamma + 1)) &= \alpha + ((\beta + \gamma) + 1)\\ &= (\alpha + (\beta + \gamma)) + 1\\ &= ((\alpha + \beta) + \gamma) + 1\\ &= (\alpha + \beta) + (\gamma + 1).\end{aligned}$$

And if λ is a limit ordinal such that $\alpha + (\beta + \gamma) = (\alpha = \beta) + \gamma$ for all $\gamma < \lambda$, then

$$\begin{aligned}\alpha + (\beta + \lambda) &= \sup_{\gamma<\lambda}(\alpha + (\beta + \gamma))\\ &= \sup_{\gamma<\lambda}((\alpha + \beta) + \gamma)\\ &= (\alpha + \beta) + \lambda.\end{aligned}$$

This completes the proof by transfinite induction.

Proof of (e). This is also proved by a straightforward transfinite induction on γ.

Proof of (f) and (g). Immediate [corollary 12.2.4 and subtraction algorithm]. □

Note that addition of ordinals (unlike addition of cardinals) is *not* commutative since, for example,

$$1 + \omega = \sup_{n<\omega}(1 + n) = \omega < \omega + 1.$$

This non-commutativity runs quite deep: the analogues for right-addition of α of parts (a), (b), (c) and (g) of proposition 12.2.6 are not generally valid.

Exercises

1. Show that $\alpha \geqslant \omega$ iff $\alpha = 1 + \alpha$.

2. Is it true in general that $\alpha = (\alpha - \beta) + \beta$?

3. Find necessary and sufficient conditions on the ordinals α, β for $\alpha + \beta$ to be a limit ordinal.

4. The function $\alpha \mapsto \alpha + \beta$ is increasing for all β [proposition 12.2.6(e)]. Show, however, that it is strictly increasing iff β is finite, and normal iff $\beta = 0$.

12.3 Ordinal multiplication

We turn now to multiplication. Again the recursion equations we are aiming for are obtained by extension from the finite case:

$$\alpha 0 = 0;$$
$$\alpha(\beta + 1) = \alpha\beta + \alpha;$$
$$\alpha\lambda = \sup_{\beta<\lambda} \alpha\beta \text{ if } \lambda \text{ is a limit ordinal.}$$

But again we approach them indirectly by defining ordinal multiplication synthetically.

Definition. *If* $(A, \leqslant)$ *and* $(B, \leqslant)$ *are two ordered sets, we define their* ordered product *to be the cartesian product* $A \times B$ *with the ordering defined so that* $(x_1, y_1) \leqslant (x_2, y_2)$ *iff* ***either*** $y_1 < y_2$ **or** $y_1 = y_2$ *and* $x_1 \leqslant x_2$.

This ordering is frequently called the *reverse lexicographic* ordering since it corresponds to the order in which words appear in a Persian dictionary. Another way of thinking of it is as what results if we take a copy of B and replace each member of it with a copy of A in order.

(12.3.1) **Lemma.** *The ordered product of two well-ordered sets is well-ordered.*

Proof. If C is a non-empty subset of $A \times B$, then from among those elements of C whose B-coordinate is the least possible choose the one whose A-coordinate is the least possible: this is evidently the least element of C. □

Definition. *If* $\alpha = \text{ord}(A, \leqslant)$ *and* $\beta = \text{ord}(B, \leqslant)$, $\alpha\beta$ *is defined to be the ordinal of the ordered product of* $(A, \leqslant)$ *with* $(B, \leqslant)$.

So $\omega 2$, for example, is the ordinal of a well-ordered set consisting of copies of $\boldsymbol{\omega}$ corresponding to the members of **2**, arranged in the same order as those members.

$$\begin{array}{cccc} \bullet & \bullet & \bullet & \cdots \\ 0 & 1 & 2 & \\ \bullet & \bullet & \bullet & \cdots \\ 0' & 1' & 2' & \end{array}$$

The meaning of the notation is therefore perhaps best represented by reading $\alpha\beta$ as 'α, β times', and so it might seem more natural to write it as $\beta\alpha$: indeed Cantor did just this in his first paper on the subject (1883). He was induced to reverse the notation by the observation that the formal properties are thus made much neater; for example, proposition 12.4.6(e) below would in Cantor's original notation take on what he called the 'repulsive' (1887, p. 86) form

$$\alpha^{(\beta+\gamma)} = \alpha^{(\gamma)}\alpha^{(\beta)}.$$

(12.3.2) **Proposition.** $|\alpha\beta| = |\alpha||\beta|$.

Proof. Trivial. □

(12.3.3) **Proposition.** *Ordinal multiplication is characterized by the following recursion equations:*

(a) $\alpha 0 = 0$;

(b) $\alpha(\beta + 1) = \alpha\beta + \alpha$;

(c) $\alpha\lambda = \sup_{\beta<\lambda} \alpha\beta$ *if* λ *is a limit ordinal.*

Proof. Straightforward. □

It follows at once from proposition 12.3.3 that for finite ordinals (i.e. natural numbers) this definition coincides with the one given in §5.4. As with addition, however, commutativity fails in the infinite case: for instance,

$$2\omega = \sup_{n<\omega} 2n = \omega < \omega + \omega = \omega 2.$$

(12.3.4) **Corollary.** *The function* $\beta \mapsto \alpha\beta$ *is normal (strictly normal if* $\alpha > 0$*) on any set of ordinals.*

Proof. Immediate [propositions 12.1.1 and 12.3.3]. □

(12.3.5) **Division algorithm.** *If* $\beta \neq 0$, *then there exist unique ordinals* δ *and* $\rho < \beta$ *such that* $\alpha = \beta\delta + \rho$.

Existence. If we choose the greatest ordinal δ such that $\beta\delta \leqslant \alpha$ [corollary 12.3.4 and proposition 12.1.2], then there exists ρ such that $\alpha = \beta\delta + \rho$ [proposition 12.2.6(f)]. Hence

$$\beta\delta + \beta = \beta(\delta + 1) > \alpha = \beta\delta + \rho,$$

and so $\rho < \beta$ by cancellation.

Uniqueness. Suppose that $\alpha = \beta\delta + \rho$ as in the theorem, but that δ is not the greatest ordinal such that $\beta\delta \leqslant \alpha$. Then

$$\alpha \geqslant \beta(\delta + 1) = \beta\delta + \beta > \beta\delta + \rho = \alpha,$$

which is absurd. This proves the uniqueness of δ; the uniqueness of ρ follows from the subtraction algorithm 12.2.5. □

The division algorithm is particularly useful in the case when $\beta = \omega$: any ordinal can be written uniquely in the form $\omega\delta + n$ with $n < \omega$.

(12.3.6) **Corollary.** α *is a limit ordinal iff* $\alpha = \omega\beta$ *for some* $\beta > 0$.

Necessity. $\alpha = \omega\beta + n$ for some β and $n < \omega$ [division algorithm]. If $n \neq 0$, then $n = m + 1$ for some $m < \omega$ and so

$$\alpha = \omega\beta + (m + 1) = (\omega\beta + m) + 1,$$

so that α is a successor ordinal, contrary to assumption. Thus $n = 0$ and therefore $\alpha = \omega\beta$ as required.

Sufficiency. Suppose that $\omega\beta$ is not zero or a limit. So $\omega\beta = \gamma + 1$ for some γ. Now $\gamma = \omega\sigma + n$ for some $\sigma \leqslant \alpha$ and $n < \omega$. Then

$$\omega\beta = (\omega\sigma + n) + 1 = \omega\sigma + (n + 1).$$

By the uniqueness of the decomposition in the division algorithm, $n + 1 = 0$. Contradiction. □

(12.3.7) **Proposition.** *Suppose that α, β, γ are ordinals.*

(a) *If $\alpha \neq 0$, then $\beta < \gamma \Rightarrow \alpha\beta < \alpha\gamma$.*

(b) *If $\alpha \neq 0$, then $\alpha\beta = \alpha\gamma \Rightarrow \beta = \gamma$.*

(c) *$\alpha \sup_{\beta \in B} \beta = \sup_{\beta \in B}(\alpha\beta)$ for any non-empty set of ordinals B.*

(d) *$\alpha(\beta + \gamma) = \alpha\beta + \alpha\gamma$.*

(e) *$\alpha(\beta\gamma) = (\alpha\beta)\gamma$.*

(f) *$\alpha \leqslant \beta \Rightarrow \alpha\gamma \leqslant \beta\gamma$.*

Proof of (a), (b) and (c). Immediate [corollary 12.3.4].

Proof of (d). We shall prove this by simple transfinite induction on γ. The case $\gamma = 0$ is trivial. If $\alpha(\beta + \gamma) = \alpha\beta + \alpha\gamma$, then

$$\begin{aligned}\alpha(\beta + (\gamma + 1)) &= \alpha((\beta + \gamma) + 1)\\ &= \alpha(\beta + \gamma) + \alpha\\ &= (\alpha\beta + \alpha\gamma) + \alpha\\ &= \alpha\beta + (\alpha\gamma + \alpha)\\ &= \alpha\beta + \alpha(\gamma + 1).\end{aligned}$$

And if λ is a limit ordinal such that $\alpha(\beta + \gamma) = \alpha\beta + \alpha\gamma$ for all $\gamma < \lambda$, then

$$\alpha(\beta + \lambda) = \sup_{\gamma<\lambda}(\alpha(\gamma + \lambda)) = \sup_{\gamma<\lambda}(\alpha\beta + \alpha\gamma) = \alpha\beta + \alpha\lambda.$$

This completes the proof by induction.

Proof of (e) and (f). Simple transfinite induction on γ again. □

Note that the analogues for multiplication on the right by α of parts (a) to (d) of proposition 12.3.7 are not valid in general. For example,

$$(2+3)\omega = 5\omega = \omega < \omega 2 = 2\omega + 3\omega.$$

And $(\omega+1)n = \omega n + 1$, so that

$$(\omega+1)\omega = \sup_{n<\omega}((\omega+1)n) = \sup_{n<\omega}(\omega n + 1) = \omega\omega < \omega\omega + \omega.$$

Exercises

1. Using the notation of the division algorithm 12.3.5, show that if $\delta < \tau$, then $\alpha < \beta\tau$.

2. Show that $\beta\omega$ is a limit ordinal for all $\beta \neq 0$ but that not all limit ordinals are of this form.

3. Show that $\alpha + 1 + \alpha = 1 + \alpha 2$.

4. If $\alpha \neq 0$ and β is a limit ordinal, show that $(\alpha+1)\beta = \alpha\beta$.

5. The function given by $\alpha \mapsto \alpha\beta$ is increasing for all β; show that it is strictly increasing iff β is a successor ordinal.

6. Show that if $\beta \neq 0$ and $2 < n < \omega$, then β is a limit ordinal iff $n\beta = \beta$.

7. Find necessary and sufficient conditions on the ordinals α, β for $\alpha\beta$ to be a limit ordinal.

12.4 Ordinal exponentiation

Definition. *If $(A, \leqslant)$ and $(B, \leqslant)$ are ordered sets and B has a least element $\bot$, then the* ordered exponential *of $(A, \leqslant)$ to $(B, \leqslant)$ is defined to be the set ${}^{(A)}B$ of all the functions f from A to B such that $f(x) = \bot$ for all but finitely many $x \in A$, with the ordering defined so that $f < g$ iff $f \neq g$ and $f(x_0) < g(x_0)$ where x_0 is the greatest element of the finite set $\{x \in A : f(x) \neq g(x)\}$.*

The choice of this ordering is determined purely by our desire to obtain a definition of ordinal exponentiation which obeys the appropriate recursive conditions (proposition 12.4.3 below), and it is much harder to picture than either the ordered sum or the ordered product of A and B.

(12.4.1) **Lemma.** *If $(A, \leqslant)$ and $(B, \leqslant)$ are well-ordered, then the ordered exponential $({}^{(A)}B, \leqslant)$ of A with B is also well-ordered.*

Proof. Suppose that $\mathcal{F}$ is a non-empty subset of B. So it has an element f. Let $a_0, a_1, \ldots, a_{n-1}$ be the elements $a \in A$ such that $f(a) \neq \bot$, arranged so that $a_0 > a_1 > \cdots > a_{n-1}$. Then recursively define elements $b_0, b_1, \ldots, b_{n-1}$ of B so that

$$b_r = \min\{g(a_r) : g \in \mathcal{F} \text{ and } g(a_p) = b_p \text{ for } p < r\}.$$

Now let f_0 be given by $f_0(a_r) = b_r$ for $r < n$ and $f_0(x) = \bot$ for $x \in A \smallsetminus \{a_0, a_1, \ldots, a_{n-1}\}$. It is easy to check that f_0 is the least element of $\mathcal{F}$. □

Definition. *If* $\alpha = \operatorname{ord}(A, \leqslant)$ *and* $\beta = \operatorname{ord}(B, \leqslant)$, *then we let* $\beta^{(\alpha)}$ *denote the ordinal of the ordered exponential* $({}^{(A)}B, \leqslant)$.

(12.4.2) **Proposition (Schönflies 1913).** *If* α *and* β *are both infinite, then*

$$|\beta^{(\alpha)}| = \max(|\alpha|, |\beta|).$$

Proof. This follows from the definition and a result on well-orderable cardinals which we shall prove later [proposition 15.3.3].[1] □

(12.4.3) **Proposition.** *Suppose that* β *is an ordinal.*

(a) $\beta^{(0)} = 1$.

(b) $\beta^{(\alpha+1)} = \beta^{(\alpha)}\beta$.

(c) $\beta^{(\lambda)} = \sup_{\alpha<\lambda} \beta^{(\alpha)}$.

Proof. Straightforward. □

Thus the definition coincides with the familiar one when the ordinals in question are finite.

Notice in particular that

$$2^{(\omega)} = \sup_{n<\omega} 2^n = \omega.$$

One consequence of this is that

$$|2^{(\omega)}| = |\omega| = \aleph_0 < 2^{\aleph_0} = |2|^{|\omega|}.$$

So the exponentiation of ordinals does not mesh neatly with that of cardinals in the way addition and multiplication do (cf. propositions 12.2.2 and 12.3.2). This is the reason I have departed from the standard notation and written $\beta^{(\alpha)}$ for ordinal exponentiation, rather than the more usual β^α: I wanted to highlight how different ordinal exponentiation is from cardinal exponentiation. Having made this point, however, I shall occasionally write β^α when it is more

[1] The practice of using results which have not yet been proved is not generally to be recommended, but it is easy to check that on this occasion no logical circle is involved.

convenient; no confusion with cardinal exponentiation should arise because no ordinal is also a cardinal.

Notice, though, that the difference just mentioned between ordinal and cardinal exponentiation is inevitable so long as we insist on the (surely desirable) requirement that $\alpha \mapsto 2^{(\alpha)}$ be a normal function. In fact, even if we dropped this requirement, we still could not define ordinal exponentiation in a way that matched it up with cardinal exponentiation. For if we could, we would in particular have defined an ordinal 2^{ω} such that $|2^{\omega}| = |2|^{|\omega|} = 2^{\aleph_0}$, and hence we would have defined a well-ordering on $\mathfrak{P}(\boldsymbol{\omega})$. But this is known to be impossible to do in our theory, even if we include the axiom of choice (Feferman 1965).

(12.4.4) **Corollary.** *The function $\alpha \mapsto \beta^{(\alpha)}$ is normal if $\beta > 0$ (strictly normal if $\beta > 1$).*

Proof. Immediate [propositions 12.1.1 and 12.4.3]. □

(12.4.5) **Logarithmic algorithm.** *If $\alpha > 0$ and $\beta > 1$, then there exist unique ordinals γ, δ, ρ such that $\alpha = \beta^{(\gamma)}\delta + \rho$ with $0 < \delta < \beta$ and $\rho < \beta^{(\gamma)}$.*

Existence. Choose the greatest ordinal γ such that $\beta^{(\gamma)} \leqslant \alpha$ [corollary 12.4.4 and proposition 12.1.2] and then use the division algorithm 12.3.5 to obtain ordinals δ and ρ such that $\rho < \beta^{(\gamma)}$ and $\beta^{(\gamma)}\delta + \rho = \alpha$. If we had $\delta = 0$, we would have $\rho = \alpha \geqslant \beta^{(\gamma)} > \rho$, which is absurd. And if we had $\delta \geqslant \beta$, we would have

$$\alpha < \beta^{(\gamma+1)} = \beta^{(\gamma)}\beta \leqslant \beta^{(\gamma)}\delta \leqslant \beta^{(\gamma)}\delta + \rho = \alpha,$$

which is also absurd: hence $0 < \delta < \beta$ as required.

Uniqueness. Suppose that $\alpha = \beta^{(\gamma)}\delta + \rho$ as in the proposition, but that γ is not the greatest ordinal such that $\beta^{(\gamma)} \leqslant \alpha$. Then

$$\alpha \geqslant \beta^{(\gamma+1)} = \beta^{(\gamma)}\beta \geqslant \beta^{(\gamma)}(\delta + 1) = \beta^{(\gamma)}\delta + \beta^{(\gamma)} > \beta^{(\gamma)}\delta + \rho = \alpha,$$

which is absurd. This proves the uniqueness of γ. The uniqueness of δ and ρ now follows from the division algorithm 12.3.5. □

(12.4.6) **Proposition.** *Suppose that α, β and γ are ordinals.*

(a) $\beta < \gamma \Leftrightarrow \alpha^{(\beta)} < \alpha^{(\gamma)}$ provided that $\alpha > 1$.

(b) $\alpha^{(\beta)} = \alpha^{(\gamma)} \Rightarrow \beta = \gamma$ provided that $\alpha > 1$.

(c) $\alpha^{(\sup B)} = \sup_{\beta \in B} \alpha^{(\beta)}$ for any non-empty set of ordinals B.

(d) If $\alpha > 1$, then $\beta \leqslant \alpha^{(\beta)}$.

(e) $\alpha^{(\beta+\gamma)} = \alpha^{(\beta)}\alpha^{(\gamma)}$.

(f) $(\alpha^{(\beta)})^{(\gamma)} = \alpha^{(\beta\gamma)}$.

g) $\alpha \leqslant \beta \Rightarrow \alpha^{(\gamma)} \leqslant \beta^{(\gamma)}$.

Proof. (a), (b) and (c) follow from corollary 12.4.4; (d) follows from (a) and theorem 11.1.5; (e), (f) and (g) may all be proved by simple transfinite induction on γ. □

Exercises

1. If $\gamma < \tau$ in the decomposition of 12.4.5, show that $\alpha < \beta^{(\tau)}$.

2. Show how to define well-orderings on the set $\boldsymbol{\omega}$ whose ordinals are $\omega + 2$, $\omega 2$, ω^2 and $\omega^{(\omega)}$.

3. Find conditions on the ordinals α, β which are necessary and sufficient to ensure that $\alpha^{(\beta)}$ is a limit ordinal.

4. Show that the function $\beta \mapsto \beta^{(\alpha)}$ is strictly increasing iff α is a successor ordinal.

5. Find an example where $(\alpha\beta)^{(\omega)} \neq \alpha^{(\omega)}\beta^{(\omega)}$.

12.5 Normal form

(12.5.1) **Theorem.** *If α is any ordinal and $\beta > 1$, then there exist unique finite sequences $(\delta_r)_{r<n}$ and $(\alpha_r)_{r<n}$ of ordinals such that $\alpha_0 > \alpha_1 > \cdots > \alpha_{n-1}$, $0 < \delta_r < \beta$ $(r < n)$, and*

$$\alpha = \beta^{(\alpha_0)}\delta_0 + \beta^{(\alpha_1)}\delta_1 + \cdots + \beta^{(\alpha_{n-1})}\delta_{n-1}.$$

Existence. Apply the logarithmic algorithm 12.4.5 repeatedly: the process must stop after a finite number of steps since otherwise we would obtain a strictly decreasing sequence of ordinals, contradicting the fact that the ordinals are well-ordered.

Uniqueness. This follows from the uniqueness of the logarithmic algorithm 12.4.5. □

The expression for α given in theorem 12.5.1 is called the *normal form* of α to the base β. Two particular cases are noteworthy: when $\beta = 2$, we obtain the *dyadic normal form*

$$\alpha = 2^{(\alpha_0)} + 2^{(\alpha_1)} + \cdots + 2^{(\alpha_{n-1})};$$

and when $\beta = \omega$, we obtain the *Cantorian normal form*

$$\alpha = \omega^{(\alpha_0)}m_0 + \omega^{(\alpha_1)}m_1 + \cdots + \omega^{(\alpha_{n-1})}m_{n-1}.$$

For example,

$$\begin{aligned}(\omega 2 + 1)(\omega + 1)3 &= ((\omega 2 + 1)\omega + \omega 2 + 1)3 \\ &= (\omega^{(2)} + \omega 2 + 1)3 \\ &= \omega^{(2)} + \omega 2 + 1 + \omega^{(2)} + \omega 2 + 1 + \omega^{(2)} + \omega 2 + 1 \\ &= \omega^{(2)} + \omega^{(2)} + \omega^{(2)} + \omega 2 + 1 \\ &= \omega^{(2)}3 + \omega 2 + 1,\end{aligned}$$

which is in Cantorian normal form.

When we are considering the expression of ordinals in normal form to the base β, one ordinal that is important is the least $\gamma > \beta$ such that $\beta^{(\gamma)} = \gamma$. We call this the *salient ordinal* for β. The reason it is important is that it is the first ordinal for which the normal form does not supply a *reduction*: if γ_0 is the salient ordinal for β, its normal form to the base β is just $\gamma_0 = \beta^{(\gamma_0)}$, whereas for any $\alpha < \gamma_0$ we obtain

$$\alpha = \beta^{(\alpha_0)}\delta_0 + \beta^{(\alpha_1)}\delta_1 + \cdots + \beta^{(\alpha_{n-1})}\delta_{n-1},$$

where $0 < \delta_r < \beta$ and $\alpha_{n-1} < \alpha_{n-2} < \cdots < \alpha_0 < \alpha$. Suppose now that we express each of $\alpha_0, \ldots, \alpha_{n-1}$ in normal form to the base β; then express all the exponents in *these* expressions in normal form; and so on. Since the exponents are dropping at each step, in a finite number of steps we must arrive at exponents which are all less than β. This expression is called the *complete normal form* of α to the base β.

Two cases of this are worth highlighting. If the base being used is a finite ordinal $N > 1$, it is easy to see that the salient ordinal is ω, so that a reduction to complete normal form is possible only for finite ordinals (i.e. natural numbers). Thus, for example, we can calculate the complete normal form of 2350 to the base 3 by writing

$$2350 = 3^7 + 3^4 . 2 + 1 = 3^{3.2+1} + 3^{3+1} . 2 + 1;$$

and similarly the complete normal form of 8192 to the base 2 is

$$8192 = 2^{13} = 2^{2^3+2^2+1} = 2^{2^{2+1}+2^2+1}.$$

If the base is ω, on the other hand, the salient ordinal, i.e. the least solution of the equation $\omega^{(\alpha)} = \alpha$, is

$$\varepsilon_0 = \omega^{\omega^{\omega^{\cdot^{\cdot^{\cdot}}}}},$$

which is standardly denoted ε_0. Every ordinal $< \varepsilon_0$ can be put in complete normal form to the base ω; i.e. it can be expressed in finite form using only finite numerals and ω together with the expressions for addition, multiplication and exponentiation.

The ordinal ε_0 is also of considerable significance for understanding the first-order theory PA (a fact we shall touch on in the next chapter). We can get a sense of the size of this ordinal if we consider its place in the transfinite sequence of ordinals

$$0, 1, 2, 3, \ldots, \omega, \omega + 1, \omega + 2, \ldots, \omega 2, \omega 2 + 1, \ldots, \omega 3, \ldots,$$
$$\omega^2, \omega^2 + 1, \ldots, \omega^2 + \omega, \ldots, \omega^3, \ldots, \omega^\omega, \ldots, \omega^{\omega^\omega}, \ldots, \varepsilon_0, \ldots, \omega_1, \ldots$$

So from an ordinal perspective ε_0 is evidently very big indeed, since so many limiting processes are involved in getting to it. But notice, too, that $|\varepsilon_0| = \aleph_0$, i.e. ε_0 is *countable*, so from a cardinal perspective it is not very big at all.

Exercises

1. Find the Cantorian normal form of $\omega(\omega + 1)(\omega^{(2)} + 1)$.
2. Find the Cantorian normal form of $\omega^{(n+1)} - (1 + \omega + \omega^{(2)} + \cdots + \omega^{(n)})$.
3. Find the Cantorian normal form of $2^{(\omega\alpha+n)}$, where $n < \omega$.
4. Show that $\alpha = \omega\alpha$ iff $\alpha = \omega^{(\omega)}\beta$ for some β. [Consider the Cantorian normal form of α.]
5. An ordinal α is said to be *indecomposable* if $\beta + \alpha = \alpha$ for all $\beta < \alpha$.
   (a) Show that α is indecomposable iff $\alpha = \beta + \gamma \Rightarrow (\alpha = \beta$ or $\alpha = \gamma)$.
   (b) Show that if $\beta \neq 0$ then α is indecomposable iff $\beta\alpha$ is.
   (c) If α is indecomposable and $0 < \beta < \alpha$, show that $\alpha = \beta\gamma$ for some indecomposable γ.
   (d) Show that for any $\alpha \neq 0$ the least indecomposable ordinal $> \alpha$ is $\alpha\omega$.
   (e) Show that an ordinal is indecomposable iff it is 0 or of the form $\omega^{(\beta)}$ for some β.
6. An ordinal α is said to be *critical* if $\beta\alpha = \alpha$ whenever $0 < \beta < \alpha$.
   (a) Show that α is critical iff $\beta\gamma = \alpha \Rightarrow (\beta = \alpha$ or $\gamma = \alpha)$.
   (b) Show that if $\beta > 1$, then the least critical ordinal $> \beta$ is $\beta^{(\omega)}$.
   (c) Deduce that the critical ordinals are 0, 1, 2 and those of the form $\omega^{(\omega^{(\beta)})}$.

Notes

We have already mentioned how Cantor's conception of ordinals developed beyond that of mere notations when he conceived of the second number class, but it was when he defined arithmetical operations on them that it was possible to regard them as *numbers*. The basic theory outlined here is clearly described in Cantor's *Beiträge* (1895; 1897). A programme of research between then and the 1930s by Hessenberg and others showed that a surprising amount of number theory generalizes to the transfinite case. For the details see Sierpinski 1965.

Conclusion to Part III

That a pure mathematical theory such as we have developed in this part of the book might generate any philosophical perplexities is the sort of suggestion that reduces many mathematicians to a mixture of impatience and despair. A field of *applied* mathematics can have its own distinctive philosophical problems, right enough, because there may be a substantial question about what it amounts to for it to model reality or how we can know that it does. But in the case of *pure* mathematics, once we are satisfied at a mathematical level that the theorems are correct, how can our theory raise any issues that are not merely instances of the quite general ones that philosophy seeks to address concerning mathematics as a whole?

What this retort misses, however, is the extent to which a theory such as the one we have developed here is indeed part of applied mathematics. We have given precise definitions to words — cardinality, larger, smaller — with whose ordinary meanings we are already familiar. In application to finite sets we take the theory to formalize the practice of counting which we have practised since childhood. And — the crucial step — in application to infinite sets we take the theory to be a natural conceptual extension of the finite case.

One measure of the success of this extension is that the material we have presented in this part of the book is nowadays a commonplace in the toolkit of most pure mathematicians.[1] Another measure of success is that hardly any mathematician now thinks the existence of infinite sets might be logically inconsistent, or even incoherent. If some are finitists (and very few are), it is because they are unconvinced by the positive arguments for the existence of infinite sets, not because they think there is a negative argument which shows that there are none.

It is hard to overstate what a radical shift this is in the mathematicians' way of conceptualizing what they do. Even if they do not really believe that every real number is obtainable from $\emptyset$ by transfinite iterations of the power-set operation (as their deference to ZF as a framework would, if taken at face

[1] If the theory of ordinals is less well known than that of cardinals, it is because of the existence of a device for avoiding them in many of their mathematical applications, popularized especially by Bourbaki: see §14.5.

value, entail), they certainly believe — most of the time, at least — that the real numbers form a set. Two hundred years ago no one thought that.

Yet this reconceptualization, while it vastly extends the scope of mathematics and provides us with new tools even in relation to very old problems, does not of itself lead to revisions in classical mathematics, which seems to have a content that is neutral as between the differing interpretations. 'Water is wet' has some content which I, who know that it is (mostly) H_2O, can grasp in common with those who lived before Lavoisier was born; and in much the same way there is a mathematical content which can be transmitted to me from an 18th century mathematics book even if I do not enter wholly into 18th century modes of thought when I read it.

The reason for belabouring this point here is that it indicates how differences in the interpretation of a piece of mathematics can arise when there is no *mathematical* dispute about its correctness. And this problem of interpretation is one which anyone studying the theory of cardinals and ordinals must face.[2] We have proved that $2^{\aleph_0} > \aleph_0$. But is this true in the *same* sense that $4 > 2$? From a narrowly formal perspective, of course, we can give a positive answer to this question, since both are instances of Cantor's theorem. But that only answers the question to the extent that the formalism in which the common proof of the two inequalities is formulated has a sense. What we really wanted was to be taught how to think about infinite sets, and the answer which the theory urges upon us is that we should think of them, as far as possible, as being just like finite sets. But how far is that? Analogy is one of the most important tools in mathematical thought, but the analogy we are dealing with here is one that we can be sure will at some point fail us.

[2]The consideration of large cardinal axioms, to which we shall turn in the last part of the book, only makes this question even more pressing.

Part IV

Further Axioms

Introduction to Part IV

The last two parts of the book have been almost entirely positive in character: we have shown how to develop within our default theory **ZU** not only traditional mathematics — arithmetic, calculus and (by extension) geometry — but also Cantor's theories of infinite numbers. But now we must revert to the more uncertain tone of the first part in order to introduce some possible additions to the default theory which we left undiscussed then.

The mere existence of axioms not yet included in the default theory will not surprise anyone who adheres to the iterative conception of set. The axioms we stated in part I guarantee the existence of levels

$$V_0, V_1, V_2, \ldots, V_\omega, V_{\omega+1}, V_{\omega+2}, \ldots$$

but say nothing as to the existence or non-existence of a level $V_{\omega+\omega}$. And yet the second principle of plenitude, which we used in chapter 4 to justify the lower levels, seems to entail the existence of $V_{\omega+\omega}$ as well. Our first task, then, will be to study propositions (known as *higher axioms of infinity*) asserting the existence of this or higher levels in the hierarchy.

Higher axioms of infinity are responsible for many of the perplexities that the iterative conception throws up, since it seems to force on us not merely one such axiom but an endlessly growing hierarchy of them. We want to know how many levels there are in the hierarchy, but any answer we give can immediately be seen to be defective since there could be (and hence, according to the second principle of plenitude, are) further levels beyond that. Each case we meet is an instance of the by now familiar phenomenon of indefinite extensibility writ very large.

What these axioms exhibit, then, is that there is a sort of incompleteness implicit in the iterative conception itself. However, we have already encountered examples of incompleteness that do not seem to be of quite this sort. Souslin's hypothesis, that a complete line in which every pairwise disjoint collection of open intervals is countable must be a continuum, is independent of the axioms of first-order set theory. But if there is a counterexample to the hypothesis, it has cardinality at most $2^{\aleph_0}$ and so is isomorphic to a counterexample whose domain is of rank $\omega + 1$. Souslin's hypothesis is therefore *decided* by the second-order theory **Z2**. This is in stark contrast to the higher axioms of infinity just mentioned, which represent incompletenesses in the first- and second-order theories indifferently.

However, the incompleteness exemplified by Souslin's hypothesis marks no mere inattention on our part in formulating the axioms: it is an inevitable consequence of the decision to restrict our attention to theories which are capable of being fully formalized. The reason for this, of course, is Gödel's incompleteness theorems, which show that in the language of any fully formal set theory there is a sentence (known as a Gödel sentence) which is true but not provable, assuming that the axioms themselves are true. If what is in question is a first-order theory U such as we considered in part I, the formalization $\mathrm{Con}(U)$ of the claim that U is consistent will be a Gödel sentence.

So, for realists at least, it is inevitable that a first-order theory will not exhaust the claims about sets that we are, on reflection, prepared to accept as true. (The response of formalists to incompleteness will evidently be somewhat different, unless they can give an account of a notion of truth independent of what follows from the axioms.)

The central difference between first- and second-order theories of sets lies, as we noted in part I, in the fact that the first-order axiom scheme of separation falls far short of expressing all the instances of the second-order axiom that the platonist would be willing to accept. The first-order axiomatization therefore does not characterize the operation which takes us from one level in the hierarchy to the next. Or, since the next level after V is $V \cup \mathfrak{P}(V)$, it amounts to much the same to say that the axiomatization fails to capture the operation that takes a set to its power set.

One of the simplest ways in which this failure manifests itself is in relation to the question of the cardinality of the power set. If $\mathrm{card}(A) = \mathfrak{a}$, then $\mathrm{card}(\mathfrak{P}(A)) = 2^{\mathfrak{a}}$, and we know from Cantor's theorem that $2^{\mathfrak{a}}$ is bigger than $\mathfrak{a}$. But how much bigger? More precisely, are there any cardinals lying in between? The answer to this question turns out to be independent of our default theory. Even the most basic case with $\mathfrak{a} = \aleph_0$, which is known as the *continuum problem* because $2^{\aleph_0}$ is the cardinal of the continuum, is not settled by any of the first-order theories we have been discussing, and hence is an instance of the same sort of incompleteness as Souslin's hypothesis.

But by far the most important of the set-theoretic propositions not decided by the axioms of our default theory is the axiom of choice, which asserts the existence at each level in the hierarchy of a supply of sets with certain convenient properties. The mathematical importance of this axiom lies principally in the large number of important propositions that have been shown to be underivable without it, and the large number of elegant propositions in disparate areas of pure mathematics that have been shown to be equivalent to it. It has also been important philosophically, because it has been the historical focus for one of the central debates between platonists and constructivists over the nature of mathematical existence.

Chapter 13

Orders of infinity

The theory of levels that we developed in chapter 3 was wholly neutral as to how many levels there are in the hierarchy, since it was based solely on the axiom scheme of separation. Only in chapter 4 did we add assumptions — the axioms of creation and infinity — which can be thought of as saying that the hierarchy extends to a certain height. But there are evidently other ways of extending the theory so as to assert that the hierarchy goes to other heights. We shall call such assertions *axioms of infinity*. Typically they can be put in the form of a claim that the level V_α exists for certain ordinals α: the axiom of infinity we stated in §4.9 asserts that V_ω exists and hence counts as an axiom of infinity in this new sense. We shall not attempt to give a precise characterization of what constitutes an axiom of infinity, and indeed we shall see shortly that no formal characterization of the scope of the term is possible. Notice, however, that on our intended meaning axioms of infinity can be ordered according to strength in a natural way. Let us say that a formal theory U' which extends the theory of levels is *strictly stronger* than another extension U if we can prove in U' the existence of a level V such that $(V, \in)$ is a model of U. It is not hard to persuade oneself that there is no limit to the strength, in this sense, of the axioms of infinity that can be devised. What will interest us in this chapter is rather whether there is a limit to the strength of axioms of infinity that are *true* and, if so, what it is.

But before we go into that question, we should consider how it relates to the at first sight much less abstract questions of quotidian interest to mathematicians. The connection arises because of the fact, familiar from recursive function theory, that if T is a formal theory, there is a sentence $\mathrm{Con}(T)$ in the language of arithmetic which can be read via some coding as expressing the consistency of T, but which can also, like any arithmetical sentence, be interpreted set-theoretically as saying something about the set $\boldsymbol{\omega}$. And if U' is strictly stronger than U, we can expect to be able to prove in U' that $\mathrm{Con}(U)$ is true about $\boldsymbol{\omega}$; but, by Gödel's incompleteness theorem, we cannot prove this in U.

It is hardly surprising, of course, that there is a set-theoretic claim provable in U' but not in the weaker theory U: what is significant is that it is an elementary arithmetical one; i.e. it is obtained from a sentence in the first-order language of arithmetic simply by interpreting the quantifiers in that sentence as ranging over $\boldsymbol{\omega}$. In fact, we can say more: the Gödel sentence $\mathrm{Con}(U)$ is Π_1, i.e. expressible in the form $(\forall x) f(x) = 0$ for some primitive recursive function f. (Gödel himself sometimes described such sentences as being 'of Goldbach type' because the celebrated problem of number theory known as Goldbach's conjecture is expressible in this form.)

The pattern, then, is that whenever we add a stronger axiom of infinity to our set theory (i.e. stronger in the sense just enunciated), we also extend the range of Π_1 sentences of arithmetic that we can prove to be true in $\boldsymbol{\omega}$. This is a graphic illustration of the fact, already noted, that the set of natural numbers is defined within our system by means of a first-order approximation to a full second-order characterization. It is defined, that is to say, by requiring that no subset of $\boldsymbol{\omega}$ should be a counterexample to the induction property; but because of the limitations of the first-order language in which we have presented the theory, this can only mean that no subset *definable in the language of the theory* is a counterexample. If we strengthen the axiom of infinity, we enlarge the range of sets that are definable, constraining further the definition of $\boldsymbol{\omega}$ and hence increasing the range of arithmetical sentences that are true in all interpretations.

There is nothing special about the natural numbers in this regard, however. We focus on them as the simplest (and hence perhaps the most troubling) case, but we can expect the phenomenon to occur for any other infinite structure (the real line, for example) which we present by means of a categorical characterization, since any such characterization is second-order, and its first-order approximation will therefore have its interpretation constrained in the manner just mentioned by any strengthening of the axiom of infinity.

13.1 Goodstein's theorem

The work we did in chapter 5 gives us a proof in **ZU** that **PA** has a model. This can be formalized to generate a proof in **ZU** that the Gödel sentence $\mathrm{Con}(\mathsf{PA})$ is true in $\boldsymbol{\omega}$, although not provable in **PA**. But what *is* $\mathrm{Con}(\mathsf{PA})$? Gödel's proof is constructive, so if we want an answer to this question, all we have to do is to choose an explicit numerical coding of the language of arithmetic and then follow through the details to obtain the Gödel sentence explicitly. We know that it will have the overall form $(\forall x) f(x) = 0$, but beyond that there is no reason to expect it to be intrinsically interesting.

Our proof in **ZU** that the Gödel sentence is true, i.e. that $(\forall n \in \boldsymbol{\omega}) f(n) = 0$, depends on knowing the coding, since it is via the coding that the sentence can be read as expressing something we have the means to prove in **ZU**, namely that **PA** is consistent. Different codings will give different Gödel sentences, and if I presented you with the sentence without telling you the coding I had used to obtain it, you would have no privileged route to finding a proof of it. Indeed there is no reason to expect the sentence to be simple enough to be remotely memorable, so it is unlikely that I could even persuade you to be interested in finding the proof. In short, the Gödel sentence is neither natural nor genuinely mathematical.

There is therefore, at least from a practical psychological perspective, some interest in the question whether we can find an explicit example of an incompleteness in PA that is not only natural (i.e. independent of any coding) but also, if possible, genuinely mathematical. It turns out that the work on ordinal arithmetic which we carried out in the last chapter puts us in a position to produce such an example.

Recall our demonstration at the end of the last chapter that every natural number r has an expression, called its *complete normal form* to the base n, in which no number $> n$ appears. If we replace n in this expression with $n + 1$, evaluate the natural number so described, and then subtract one from the answer, we obtain a natural number which we shall denote $F_n(r)$. For example,

$$2350 = 3^7 + 3^4 . 2 + 1 = 3^{3.2+1} + 3^{3+1} . 2 + 1,$$

which is a complete normal form to the base 3, and so

$$F_3(2350) = 4^{4.2+1} + 4^{4+1} . 2.$$

Similarly,

$$8192 = 2^{13} = 2^{2^3+2^2+1} = 2^{2^{2+1}+2^2+1},$$

which is a complete normal form to the base 2, and so

$$F_2(8192) = 3^{3^{3+1}+3^3+1} - 1.$$

In general the function F_n thus defined is primitive recursive. We now use these functions F_n to obtain, for any natural number m as starting point, what we shall call the *Goodstein sequence* of m.

Definition. *The* Goodstein sequence $(g(m, n))_{n \geqslant 1}$ *starting from a natural number m is defined recursively as follows:*

$$g(m, 1) = m$$
$$g(m, n + 1) = F_{n+1}(g(m, n)).$$

The Goodstein sequence starting from m is thus

$$m, F_2(m), F_3(F_2(m)), F_4(F_3(F_2(m))), \ldots$$

Expressed less compactly but perhaps more comprehensibly: the first term of the Goodstein sequence is m; and once the nth term has been calculated, the $(n + 1)$th is obtained by expressing the nth in complete normal form to the base $n + 1$, changing all the occurrences of $n + 1$ in this expression to $n + 2$, and subtracting 1 from the result.

What we are interested in is the behaviour of Goodstein sequences for different starting points. The Goodstein sequence starting at 2 is trivial, terminating at the 4th step.

$$\begin{aligned} g(2,1) &= 2; \\ g(2,2) &= 3 - 1 = 2; \\ g(2,3) &= 1; \\ g(2,4) &= 0. \end{aligned}$$

And the sequence starting at 3 terminates at the 6th step.

$$\begin{aligned} g(3,1) &= 3 = 2 + 1 \\ g(3,2) &= 3 \\ g(3,3) &= 4 - 1 = 3 \\ g(3,4) &= 2 \\ g(3,5) &= 1 \\ g(3,6) &= 0. \end{aligned}$$

But the very next case is somewhat different. Although the sequence starting at 4 does not increase very rapidly, after a while it does become quite large.

$$\begin{aligned} g(4,1) &= 4 = 2^2 \\ g(4,2) &= 3^3 - 1 = 26 = 2.3^2 + 2.3 + 2 \\ g(4,3) &= 2.4^2 + 2.4 + 1 = 41 \\ g(4,4) &= 2.5^2 + 2.5 = 60 \\ g(4,5) &= 2.6^2 + 2.5 = 83 \\ g(4,6) &= 2.7^2 + 7 + 4 = 109 \\ g(4,7) &= 2.8^2 + 8 + 3 = 139 \\ &\cdots \\ g(4,96) &= 11,327 \\ &\cdots \end{aligned}$$

And for only slightly larger starting points the numbers become more than astronomical in astonishingly few steps. The sequence for 51, for instance, starts as follows.

$$
\begin{aligned}
g(51,1) &= 51 = 2^{2^2+1} + 2^{2^2} + 2 + 1 \\
g(51,2) &= 3^{3^3+1} + 3^{3^3} + 3 \sim 10^{13} \\
g(51,3) &= 4^{4^4+1} + 4^{4^4} + 3 \sim 10^{155} \\
g(51,4) &= 5^{5^5+1} + 5^{5^5} + 2 \sim 10^{2185}; \\
g(51,5) &= 6^{6^6+1} + 6^{6^6} + 1 \sim 10^{36,306} \\
g(51,6) &= 7^{7^7+1} + 7^{7^7} \sim 10^{695,975} \\
g(51,7) &= 8^{8^8+1} + 8^{8^8} - 1 \sim 10^{15,151,337} \\
&\dots
\end{aligned}
$$

The question this raises is whether a Goodstein sequence with arbitrary starting point eventually terminates at 0. As we have just seen, the sequence starting at 3 terminates in 6 steps. But direct calculation is of no help whatever in any larger case than this: whether the Goodstein sequence starting at 4 eventually terminates cannot be determined by actually calculating all the terms, since even after 10^{30} terms, which is far more than we can feasibly calculate, all that is apparent from the numbers alone is that the sequence is continuing to increase.

Nevertheless, it *does* eventually terminate, and so does every other Goodstein sequence. It is a very striking illustration of the power of the technique of ordinal notations that we can use it to give a quick proof of this. The trick is to form from any Goodstein sequence a parallel sequence of ordinals, which we shall call the *Goodstein ordinal sequence*, obtained by changing all the occurrences of $n + 1$ into ω rather than $n + 2$.

In order to express this formally, let us first write $\delta_n(r)$ for the ordinal obtained by expressing natural number r in complete normal form to the base n and then replacing each n in this expression with ω. It is easy to see that $\delta_n(r) < \varepsilon_0$ and that

$$\text{if } r < s \text{ then } \delta_n(r) < \delta_n(s). \tag{1}$$

Moreover,

$$\delta_{n+1}(F_n(r) + 1) = \delta_n(r). \tag{2}$$

The Goodstein ordinal sequence of a natural number m is then defined to be the sequence of ordinals

$$\gamma(m,1), \gamma(m,2), \gamma(m,3), \dots$$

where

$$\gamma(m,n) = \begin{cases} \delta_{n+1}(g(m,n)) & \text{if } g(m,n) \neq 0 \\ 0 & \text{if } g(m,n) = 0. \end{cases}$$

Thus, for example, the Goodstein ordinal sequence of 3 looks like this:

$$\begin{aligned}
\gamma(3, 1) &= \omega + 1; \\
\gamma(3, 2) &= \omega; \\
\gamma(3, 3) &= 3; \\
\gamma(3, 4) &= 2; \\
\gamma(3, 5) &= 1; \\
\gamma(3, 6) &= 0.
\end{aligned}$$

And the Goodstein ordinal sequence of 51 begins as follows:

$$\begin{aligned}
\gamma(51, 1) &= \omega^{\omega^{\omega}+1} + \omega^{\omega^{\omega}} + \omega + 1; \\
\gamma(51, 2) &= \omega^{\omega^{\omega}+1} + \omega^{\omega^{\omega}} + \omega; \\
\gamma(51, 3) &= \omega^{\omega^{\omega}+1} + \omega^{\omega^{\omega}} + 3; \\
\gamma(51, 4) &= \omega^{\omega^{\omega}+1} + \omega^{\omega^{\omega}} + 2; \\
\gamma(51, 5) &= \omega^{\omega^{\omega}+1} + \omega^{\omega^{\omega}} + 1; \\
\gamma(51, 6) &= \omega^{\omega^{\omega}+1} + \omega^{\omega^{\omega}}; \\
&\dots
\end{aligned}$$

In these two instances we observe that the Goodstein ordinal sequences are strictly decreasing from the very start. And in fact that is always the case.

(13.1.1) **Lemma.** *If* $g(m, n-1) \neq 0$, *then* $\gamma(m, n) < \gamma(m, n-1)$.

Proof.

$$\begin{aligned}
\gamma(m, n) &= \delta_{n+1}(g(m, n)) \\
&= \delta_{n+1}(F_n(g(m, n-1))) \text{ if } g(m, n-1) \neq 0 \\
&< \delta_{n+1}(F_n(g(m, n-1)) + 1) \text{ by ((1))} \\
&= \delta_n(g(m, n-1)) \text{ by ((2))} \\
&= \gamma(m, n-1).
\end{aligned}$$

□

From this lemma the result we want follows at once.

(13.1.2) **Theorem (Goodstein 1944).** $(\forall m \in \boldsymbol{\omega})(\exists n \in \boldsymbol{\omega})(g(m, n) = 0)$.

Proof. Suppose on the contrary that there is a natural number m such that $g(m, n) \neq 0$ for all $n \in \boldsymbol{\omega}$. Then

$$\gamma(m, 1), \gamma(m, 2), \gamma(m, 3), \dots$$

is an infinite strictly decreasing sequence of ordinals less than ε_0, contradicting the fact that the set of ordinals less than ε_0 is well-ordered. □

Another way of expressing this result is via the following definition.

Definition. *The* Goodstein function G *is defined by letting* $G(m)$ *be the least natural number* n *such that* $g(m, n) = 0$.

Goodstein's function is obviously partial recursive, which is to say that there is in principle a mechanical method for calculating its value whenever it is defined: one merely calculates the relevant Goodstein sequence until it terminates, and then counts the number of steps. What is not obvious, and is supplied by Goodstein's theorem above, is the information that the function G *is* everywhere defined, i.e. that it is total recursive. But it increases with quite awesome rapidity:

$$G(1) = 2, \;\; G(2) = 4, \;\; G(3) = 6, \;\; G(4) \sim 10^{121,210,694}.$$

And it is this rapidity of growth that makes the function relevant to our purposes here. An analysis of the form of proofs in PA can be used to show that there is a limit to the rapidity of growth of any recursive function which can be proved to be total in PA; moreover, the Goodstein function G exceeds this rate of growth. So although G *is* total recursive, the fact that it is is not provable in PA (see Kirby and Paris 1982). In other words, the sentence $(\forall x)(\exists y)g(x, y) = 0$ in the first-order language of arithmetic is an example of an incompleteness in PA, a sentence which is true in $\boldsymbol{\omega}$ but not provable in PA. It is undoubtedly natural, in contrast to the Gödel sentence, since it does not depend on any arbitrary choice of coding. But further, it is plainly possible to argue that it is genuinely mathematical: it does not seem implausible that one might become interested in whether it is true independently of its role as an instance of incompleteness.

It is worth noting, incidentally, that although the Goodstein sentence is not provable in PA, any instance of it is. This is because an instance is Σ_1, and *any* true Σ_1 sentence is provable in PA. (If the sentence $(\exists x)f(x) = 0$ is true in $\boldsymbol{\omega}$, then there exists a natural number n such that $f(n) = 0$: the calculation which shows this, followed by an application of existential generalization, constitutes a proof in PA of the original sentence.)

In order to obtain a mathematical incompleteness in PA, we have had to pay a price in logical complexity: the Gödel sentences mentioned earlier were Π_1, i.e. of the form $(\forall x)f(x) = 0$, whereas the Goodstein sentence is Π_2, of the form $(\forall x)(\exists y)f(x, y) = 0$. The importance of this difference emerges when we consider them in the light of Hilbert's programme. Suppose that I am committed to the truth of PA: in that case a proof in PA of an arithmetical sentence certainly gives me a straightforward reason to believe that it is true. But suppose now that I am given a set theory U extending PA and believe not that it is true but only that it is formally consistent. Does a proof in U of the arithmetical sentence give me any reason to believe it?

If the sentence Φ which we have proved in U is Π_1, then it does. For if Φ were false, its negation would be a Σ_1 truth about $\boldsymbol{\omega}$: it would therefore (as we noted earlier) be provable in PA and hence in U, contradicting the consistency of U. (Even if we had two *incompatible* — i.e. mutually inconsistent — but individually consistent extensions U_1 and U_2 of PA, the Π_1 arithmetical consequences of both theories would all be true.)

If, on the other hand, the sentence proved in U is Σ_1 (or *a fortiori* if it is Π_2, as in the case of Goodstein's theorem), we cannot argue in this fashion. For if Φ is any true Π_1 sentence not provable in PA (e.g. a Gödel sentence), its negation is a false Σ_1 sentence which is consistent with PA and hence provable in some consistent set theory extending PA.

13.2 The axiom of ordinals

For which ordinals α does the αth level V_α exist? The level V_ω exists by the axiom of infinity. Hence, by the axiom of creation, so do the levels $V_{\omega+1}$, $V_{\omega+2}$, etc. But the axioms of ZU do not entail that there is a second limit level $V_{\omega+\omega}$. Ought we then to add another axiom to our system to ensure that this level exists? Let us denote by ZfU and Zf the extensions of ZU and Z respectively obtained by adding the following axiom.

Axiom of ordinals. *For each ordinal α there is a corresponding level* V_α.

The effect of this axiom is to strengthen the axiom of infinity by ensuring the existence of many further levels V_α with $\alpha \geqslant \omega + \omega$, one for each ordinal α whose existence we could prove in ZU. However, the effect of the axiom on the height of the hierarchy is very much amplified beyond this because the ordinals are not an independent measure of calibration but an integral part of the theory: the existence of the new levels just mentioned ensures the existence of further ordinals α that were not previously available, and the axiom of ordinals applied to *them* guarantees the existence of corresponding levels V_α. And so on. The hierarchy described by the new theory is therefore colossally higher than anything we could have countenanced before.

One fruit of the extra headroom which ZfU affords is that it permits an elegant characterization of the structure of the membership relation on the pure transitive sets. This characterization, which is known as Mostowski's collapsing lemma, is a very useful tool for studying the iterative hierarchy, and for that reason set theorists will regard ZfU as a more attractive theory than ZU. (Of course, the pure hierarchy described by Zf is from this point of view more attractive still, because it does not have any non-set-theoretic objects to complicate its properties.) We should not under-estimate the extent to which axiom selection is influenced by set theorists themselves, who want an elegant structure to study. We saw earlier how much of the initial motivation for the axiom of foundation was of this sort, and similar considerations evidently count in favour of systems such as Zf.

Let us also note here another consequence of the axiom of ordinals (although one that will be more important when we discuss the standard axiomatizations of set theory in appendix A than in the current context): the axiom of ordinals permits the theory of ordinals to be developed using a quite different idea from the one we employed in chapter 11. Let us give the name *von Neumann ordinals* to the sets α' defined by transfinite recursion as follows:

$$0' = \emptyset$$
$$(\alpha + 1)' = \alpha \cup \{\alpha'\}$$
$$\lambda' = \bigcup_{\alpha<\lambda} \alpha' \text{ for any limit ordinal } \lambda.$$

The idea we want to consider is that we could have used the von Neumann ordinals instead of ordinals in the formal treatment. The key point is that if the theory we are working in is ZU, this idea is stillborn because $\rho(\alpha') = \alpha$: in ZU we cannot prove the existence of a level $V_{\omega+\omega}$ and hence cannot prove the existence of the corresponding von Neumann ordinal $(\omega+\omega)'$. So there are not enough von Neumann ordinals to give us a workable theory. In ZfU, on the other hand, things are different: we can prove that the von Neumann ordinal α' exists for every ordinal α, and hence can, if we wish, use von Neumann ordinals as proxies for the ordinals.

There would obviously be no practical advantage to be gained from doing this unless we could develop the theory of von Neumann ordinals independently by defining them *directly* without using recursion on the standard ordinals, but this can in fact be done quite easily. The autonomous definition of a von Neumann ordinal is as follows.

Definition. A von Neumann ordinal *is a transitive set* A *such that* $\in_A$ *is a transitive relation on* A.

When we decouple von Neumann ordinals from the definition by recursion in this manner, we discover that this autonomous theory of ordinals works very smoothly, mainly because of the elegant fact that

$$\alpha < \beta \Leftrightarrow \alpha' \in \beta'.$$

We shall not go through the details, which may be found in almost any set theory textbook: suffice it to say that this definition does correctly characterize the sets that we want. In summary, then, we could if we wished develop the theory of ordinals using the von Neumann definition, as long as we assumed the axiom of ordinals.

What reason do we have, though, to believe our new axiom? If von Neumann's theory of ordinals were the only one available, that would give us a sort of regressive reason to believe the axiom. And this is certainly of some historical importance, since the possibility of the alternative theory of ordinals presented in part III was not discovered until the 1950s and did not really become widely known until some time later. So in the 1920s the evident usefulness of ordinals in proving mathematical theorems might have seemed to give the axiom of ordinals strong support.

In the context of ZU, however, this hardly provides much support for the axiom once we see that another treatment, which does not require the axiom, is available. There is a general moral here. Regressive arguments for any set-theoretic axiom depend on a prior belief in the *mathematical* truth of some consequences of the axiom, but the fact that they are consequences of it depends in turn on an embedding of part of mathematics in set theory: a different embedding may not require the same axiom, and so the regressive justification is relative to the embedding. It is of interest, therefore, to enquire into whether the axiom of ordinals has any mathematical consequences that are less dependent on our choice of embedding.

We showed in part II that two of the structures of central interest to mathematicians, the natural numbers and the real numbers, can be modelled by structures of low infinite rank in the hierarchy: to be precise, there is a set of rank ω that will serve for the set of natural numbers, and one of rank $\omega + 1$ that will do for the real numbers. Of course, we need not just these sets but the familiar relations and functions defined on them, which will be of slightly higher rank. Then, in order to do ordinary mathematics, we shall want to define other functions which may be of slightly higher rank still.

Some parts of mathematics are often said to be more abstract than others. Functional analysis, for instance, is more abstract than the calculus of functions of one real variable. This use of the word 'abstract', which is quite familiar to most mathematicians, seems to be represented quite well by the *ranks* of the objects referred to in this set-theoretic modelling of the parts of mathematics in question: functional analysis is more abstract than the calculus because the objects it deals with are modelled by sets of higher rank.

One of the trends we can trace in the development of mathematics, especially during the 20th century, is a move towards greater abstractness in the sense just defined. Nevertheless, the overwhelming majority of 20th century mathematics is straightforwardly representable by sets of fairly low infinite ranks, certainly less than $\omega + 20$. So although the Gödelian considerations alluded to earlier tell us that there are sentences in the language of arithmetic, such as Con(ZU), that are provable with the axiom of ordinals but not without, there is once again a further question whether there are any examples that might count as genuinely mathematical in the sense in which we have been using that term, i.e. as being of interest to mathematicians who are not set theorists.

In fact there are. We proved in §7.5 that the game on every open, closed, countable or co-countable subset of the Baire line is determined. It is very natural, having proved these results, to wonder what larger class of games is determined. One obvious class to consider in this connection is the *Borel sets*, i.e. the sets obtainable from the closed and open subsets of the line by taking countable unions and complements as often as we wish, and this is the case which turns out to supply us with an example of a mathematical incompleteness in ZU: it can be proved in ZfU (Martin 1975), but not in ZU (Friedman 1971), that the game on every Borel set is determined.

On its own, of course, this does not give us a regressive argument for the *truth* of the axiom of ordinals: for that we would have to have an *independent* reason to believe that the game on every Borel set is determined. What it does supply, however, is an extrinsic reason for mathematicians to be interested in the axiom, since it demonstrates that the axiom of ordinals has consequences in the theory of games on Borel sets, which no doubt deserves to be regarded as a genuine branch of mathematics, rather than merely within set theory itself. Moreover, the result is especially striking, since it seems to be the sort of claim that one might formulate quite independently of any search for sentences whose proofs make essential use of higher axioms of infinity.

On the other hand, a claim about Borel sets is not the ideal example to impress mathematicians generally: the branch of mathematics which studies the properties of such sets — descriptive set theory — is, as its name might suggest, practised largely by set theorists themselves. Friedman has more recently turned his attention to number theory and devised methods for generating various sentences whose proofs involve differing degrees of abstractness. Although these are certainly natural, i.e. independent of coding, one might question whether those that have been published so far are genuinely mathematical. There therefore remains room for scepticism about the relevance of the axiom of ordinals to mathematical practice. The turning point, of course, would be a genuine conjecture of number theory — one made by a *number theorist* — whose proof turned out to require higher-order methods, but to date no such example has been found.

13.3 Reflection

If the relevance of the axiom of ordinals to the practice of mathematics is still very limited, and if the regressive support it gains from the few applications that have been discovered is negligible, the principal argument for it must be intuitive. The most obvious such argument is that there is no good reason why the hierarchy should not extend beyond $V_{\omega+\omega}$, and so by some version of the second principle of plenitude we conclude that it does so extend. In this section I want to investigate whether this intuitive argument, or an extension of it, can be made to justify an even stronger class of axioms of infinity known collectively as the axiom scheme of reflection. They are intended to capture the idea that every property of the whole universe is reflected in some sub-universe.

Axiom scheme of reflection. *For each formula Φ with free variables $x_1, x_2, \ldots, x_n$ this is an axiom:*

$$(\forall x_1, \ldots, x_n)(\exists V)(\Phi \Rightarrow \Phi^{(V)}).$$

This axiom is to be read, of course, according to the convention of §3.4, which has fallen dormant but is hereby revived for the remainder of this section, that the letters V, V', etc. range only over levels. So the axiom asserts the existence of a *level* V such that $\Phi \Rightarrow \Phi^{(V)}$; this level is said to *reflect* the formula Φ. The system whose only axioms are the instances of the schemes of separation and reflection is denoted ZFU; its pure variant is denoted ZF. We can illustrate how the axiom scheme of reflection is employed in practice by proving the axioms of infinity, creation and ordinals from it.

(13.3.1) **Proposition** (ZFU). *There exists a level.*

Proof. Reflection on any sentence whatever shows that there is at least one level. □

(13.3.2) **Proposition** (ZFU). *For every level there is another level after it.*

Proof. If V is any level, use reflection on the fact that $(\exists a)(a = V)$ to obtain $(\exists V')(\exists a \in V')(a = V)$, i.e. $(\exists V')(V \in V')$. □

(13.3.3) **Proposition** (ZFU). *There exists a limit level.*

Proof. From proposition 13.3.2 we obtain

$$(\forall a)(\exists V)(a \in V).$$

Reflection on this gives us a level V' such that

$$(\forall a \in V')(\exists V \in V')(a \in V).$$ □

(13.3.4) **Proposition** (ZFU). *For every ordinal α the level V_α exists.*

Proof. By induction on α. The case $\alpha = 0$ is already taken care of by proposition 13.3.1, and the case of a successor ordinal follows from proposition 13.3.2. So suppose now that λ is a limit ordinal and V_α exists for all $\alpha < \lambda$. So there exists a level V which reflects this, i.e. $\mathrm{V}_\alpha \in V$ for all $\alpha < \lambda$. It follows that $\mathrm{V}_\lambda = \bigcup_{\alpha<\lambda} \mathrm{V}_\alpha$ exists as required. □

In summary, then, we have shown that ZFU is at least as strong as ZfU. In fact, it is stronger: the following theorem (or, to be strict, theorem scheme) is provable in ZFU but not in ZfU.

(13.3.5) **Proposition** (ZFU). *If τ is a term, then*

$$(\forall a)((\forall x \in a)(\tau(x) \text{ is a set}) \Rightarrow \{\tau(x) : x \in a\} \text{ is a set}).$$

Proof. Suppose that a is a set. Then by reflection on

$$(\forall x \in a)(\exists y)(y = \tau(x))$$

we obtain a level $V \ni \mathrm{V}(a)$ such that

$$(\forall x \in a)(\exists y \in V)(y = \tau(x)),$$

from which it follows that $\{\tau(x) : x \in a\}$ is a subset of V. □

Indeed, ZFU is not merely stronger than ZfU, but much stronger: no consistent extension of ZU by a finite number of axioms can give us a theory of the strength of ZFU (Montague 1961). If reflection is so strong, then, why should we believe it? As we have noted, regressive arguments for the axiom of ordinals are already very weak. When we get to reflection, they become weaker still. The likelihood that we would need something stronger than the axiom of ordinals in order to prove a result we already had other reasons for believing true seem slim indeed. The best we could hope for in this direction might be a weaker extrinsic argument, perhaps to the effect that the theory with full reflection is more convenient or more elegant than the theory without it.

We have seen that the axiom of ordinals can easily be put in the form of a particular instance of the reflection scheme, so there is some prospect that any intuitive argument we have for the axiom of ordinals might apply more generally to all the other instances. The most prominent example of this strategy is an argument of Gödel, who is reported by Wang (1974, p. 536) as observing that it

> does not have the same kind of *immediate* evidence (previous to any closer analysis of the iterative concept of set) which the other axioms have. This is seen from the fact that it was not included in Zermelo's original system of axioms. [Gödel] suggests that, heuristically, the best way of arriving at it from this standpoint is the following. From the very idea of the iterative concept of set it follows that if an ordinal α has been obtained, the operation of power set $\mathfrak{P}$ iterated α times leads to a set $\mathfrak{P}^{\alpha}(\emptyset)$. But, for the same reason, it would seem to follow that if, instead of $\mathfrak{P}$, one takes some larger jump in the hierarchy of types, ... $\mathfrak{Q}^{\alpha}(\emptyset)$ likewise is a set. Now, to assume this for any conceivable jump operation (even for those that are defined by reference to the universe of all sets or by use of the choice operation) is equivalent to the axiom.

What is most interesting about this argument is that it attempts to justify an axiom scheme by *generalizing* from an argument for a particular instance. In other words, it does not conform to the pattern we met in part I, where axiom schemes such as separation were seen as gaining their justification from our prior belief in the second-order axioms to which they were (poor) approximations.

But the form of the axiom scheme of reflection suggests that this difference is inevitable, since the reflection of a formula by a level is an inherently syntactic notion that is highly sensitive to the form of expression used. Indeed, even the consistency of the reflection scheme (never mind its truth) is sensitive to the forms of expression available in the language. The platonist presumably believes that the universe of sets has some height or other, whether or not we can express it. But if we *could* express in absolute terms what the height is, and apply reflection to that expression, we would obtain a level *within* the universe with the very same height, contrary to our intentions. In the context of our formal theory, of course, this serves to remind us that the height of the hierarchy is, in a certain sense, inexpressible from within. But it also suggests that a careless generalization of the reflection scheme might turn out to be inconsistent. And this danger has been confirmed: third-order reflection in its apparently most natural formulation is inconsistent (Tait 1998).

It follows, therefore, that *any* argument for the reflection scheme will have to be sensitive to distinctions of level concerning what can be expressed. Something of this may be traced in Cantor's idea that the universe of sets in some way represents the Absolute, and hence that it would be a sort of blasphemy to suppose that a finite being can express it. This sort of thought has been popular in theology: according to St Gregory, for instance, 'no matter how far our mind may have progressed in the contemplation of God, it does not attain to what He is but only to what is beneath Him'. But work would evidently be required to establish that the class of all sets is in the relevant sense a representation of Him.

A non-theological argument for reflection has been offered by Fraenkel et al., who suggest that

> when we try to reconcile the image of the ever-growing universe with our desire to talk about the truth or falsity of statements that refer to *all* sets we are led to assume that some temporary universes are as close an 'approximation' to the ultimate unreached universe as we wish. In other words, there is no property *expressible in the language of set theory* which distinguishes the universe from some 'temporary universes'. (1958, p. 118, latter emphasis mine)

In other words, the fact that our attempts at justifying reflection have been syntactic generalizations from particular instances is not accidental but is an inevitable consequence of the nature of the reflection scheme. But in that case it seems likely that any such justification will have a distinctly constructivist spin. A direct argument in support of reflection based on a platonistic understanding of the iterative notion of set would indeed be 'a coup' (Aken 1986, p. 1001), but for the reason just stated the prospects for such an argument are poor.

13.4 Replacement

Let us consider now an alternative strategy for justifying the reflection scheme which stems from a technical result concerning the following scheme.

Axiom scheme of replacement. *If $\tau(x)$ is any term, this is an axiom:*

$$(\forall x \in a)(\tau(x) \text{ is a set}) \Rightarrow \{\tau(x) : x \in a\} \text{ is a set.}$$

We have already shown in proposition 13.3.4 that every instance of this scheme is a theorem of ZFU. But the technical result I want to consider is the converse of this: if we let ZFU^r be the theory whose axioms are those of ZU together with the axiom scheme of replacement, then we can *prove* reflection in ZFU^r.

(13.4.1) **Theorem** (ZFU^r). *If Φ is any formula, then*

$$(\exists V)(\Phi \Rightarrow \Phi^{(V)}).$$

Proof. Every first-order sentence is logically equivalent to a *prenex* sentence, i.e. one of the form

$$\mathsf{Q}_1 x_1 \mathsf{Q}_2 x_2 \ldots \mathsf{Q}_n x_n \Psi(x_1, \ldots, x_n),$$

where each Q_r is either $\exists$ or $\forall$ and Ψ is quantifier-free. So we may assume without loss of generality that Φ is already of this form. Write Ψ_r as an abbreviation for

$$\mathsf{Q}_{r+1} x_{r+1} \mathsf{Q}_{r+2} x_{r+2} \ldots \mathsf{Q}_n x_n \Psi(x_1, \ldots, x_n),$$

so that Ψ_n is Ψ and Ψ_0 is Φ itself. For $1 \leqslant r \leqslant n$, if Q_r is an existential quantifier, let $V_r(x_1, \ldots, x_{r-1})$ be the lowest level V such that

$$(\exists x_r)\Psi_r(x_1, \ldots, x_r) \Rightarrow (\exists x_r \in V)\Psi_r(x_1, \ldots, x_r);$$

and if Q_r is a universal quantifier, let $V_r(x_1, \ldots, x_{r-1})$ be the lowest V such that

$$(\exists x_r)\text{not}\Psi_r(x_1, \ldots, x_r) \Rightarrow (\exists x_r \in V)\text{not}\Psi_r(x_1, \ldots, x_r).$$

For any level V, let $f_r(V)$ be the earliest level containing $V_r(x_1, \ldots, x_{r-1})$ for all $x_1, \ldots, x_{r-1} \in V$ (which exists by replacement), and let

$$f(V) = f_1(V) \cup f_2(V) \cup \cdots \cup f_n(V).$$

Now let

$$\begin{aligned} V^0 &= \mathrm{V}_0 \\ V^{p+1} &= f(V^p) \\ V^\omega &= \bigcup_{p \in \omega} V^p. \end{aligned}$$

Then V^{ω} is a limit level. Suppose now that $x_1, \ldots, x_n \in V^{\omega}$. Then $x_1, \ldots, x_n \in V^{p}$ for some $p \in \boldsymbol{\omega}$, and so if $1 \leqslant r \leqslant n$, then

$$V_r(x_1, \ldots, x_{r-1}) \in V^{p+1} \in V^{\omega}.$$

It follows by the definition of the V_r that

$$(\mathsf{Q}_r x_r)\Psi_r(x_1, \ldots, x_r) \Leftrightarrow (\mathsf{Q}_r x_r \in V^{\omega})\Psi_r(x_1, \ldots, x_r).$$

Now Ψ_n has no quantifiers, so trivially

$$\Psi_n(x_1, \ldots, x_n) \Leftrightarrow \Psi_n^{(V^{\omega})}(x_1, \ldots, x_n).$$

Hence successively

$$\Psi_r(x_1, \ldots, x_r) \Leftrightarrow \Psi_r^{(V^{\omega})}(x_1, \ldots, x_r)$$

for all $x_1, \ldots, x_r \in V^{\omega}$. In the case where $r = 0$ we obtain

$$\Phi \Leftrightarrow \Phi^{(V^{\omega})},$$

whence the result. □

This technical result thus offers us the prospect of a wholly different route to an intuitive justification for ZFU: instead of trying to justify reflection directly, we could concentrate on trying to justify replacement and then use the result just mentioned to derive reflection as a theorem. In this connection it is worth noting a feature of the proof of reflection that marks it out as unusual. The reflection principle is, of course, strictly speaking not a single theorem but a theorem scheme. This is not the first such scheme we have come across, but in all our previous encounters with the genre the schematic formula has occurred throughout the proof unanalysed: just as the scheme $\ldots \Phi \ldots$ we are trying to prove consists of instances of a second-order formula $\ldots X \ldots$, so the *proof* of the scheme has consisted of instances of a second-order proof involving the variable X. The proof we have just given of reflection, however, does not fall into this pattern at all: the proof strategy for each instance of the scheme depends on the logical complexity of the instance of Φ occurring in it, and so it cannot be represented as an instance of a second-order proof. We saw earlier that reflection is unusual because we cannot justify it just by viewing it as a restriction of a single second-order axiom: the feature we have just noted is powerful confirmation of this.

In this respect reflection differs from replacement, which *can* be viewed as a special case of a single second-order axiom.

Axiom of replacement. $(\forall F)(\forall a)(a \textit{ is a set} \Rightarrow \{F(x) : x \in a\} \textit{ is a set.}$[1]

[1]This is second-order because the first quantifier is intended to range over all functions in the logical sense.

So if we can find an argument for this second-order principle, we can then assert the first-order scheme as an approximation to it, on the same pattern as our justification of separation in chapter 3. The platonist, in particular, might regard this as a more plausible strategy, since it avoids the difficulty we faced at the end of the last section: that a justification for reflection has to be sensitive to the limits of what can be described, and hence seems inevitably to have a constructivist slant.

13.5 Limitation of size

But even if our new strategy is more plausible for the platonist, it is still very problematic. The argument that has most often been advanced to support the second-order replacement principle, and hence derivatively the first-order replacement scheme, is of a wholly different sort from the motivation for the iterative conception which we discussed in chapter 3. This new sort of motivation is known as *limitation of size*. We group together under this heading those principles which classify properties as collectivizing or not according to how *many* objects there are with the property. The origins of such principles are quite old: Cantor outlined a theory based on them during the 1890s, for example, although he did not publish it and communicated it only in letters (to Hilbert in 1897 and Dedekind in 1899). And Russell (independently, one assumes) sketched another such theory, under the description 'limitation of size', in his 1906*a* discussion of possible solutions to the paradoxes. The theory Russell eventually settled on does not owe anything to limitation of size, however, and little seems to have been made of the idea for some years thereafter; but from the late 1920s it gained popularity, perhaps because of the advocacy of von Neumann. The result of this was that it entered the consciousness of mathematicians, from which it has yet to be wholly dislodged. It lives on, for example, in the practice of calling a mathematical entity (such as a category) 'small' if it is a collection.

In discussing limitation of size it will be helpful to follow Boolos (1989) in distinguishing two variants, weak and strong.

Weak limitation-of-size principle. *If there are no more Fs than Gs and the Gs form a collection, then the Fs form a collection.*

Strong limitation-of-size principle. *A property F fails to be collectivizing iff there as as many Fs as there are objects.*

What these two principles have in common is the idea that there is a limit to the size of a collection. The strong principle evidently implies the weak but not conversely, since the weak principle allows the possibility that the limit on size might fall short of the size of the universe. When we come to examine the details of the argument for limitation of size, there is, as one might expect, a substantial difference between the platonist and constructivist accounts. We shall consider what the constructivist has to say first, as it seems rather more straightforward.

The constructivist's motivation for affirming the limitation-of-size principle, weak or strong, is that there are limits on how many objects can form a collection — limits which result from constraints on our ability to comprehend the objects to be collected. Let us follow Cantor in calling a cardinality *absolutely infinite* if it is too large for that number of objects to be collected together. It is easy to see how the very attempt to collect so many things together could be presented as an instance of that recurrent theme of human folly — hubris in the face of the incomprehensible. Such a view is very old — it is a commonplace among the Greeks — and Cantor espoused it with enthusiasm in the argument he gave for the weak limitation-of-size principle. The difference was that ancient authors had identified the Absolute with the infinite, whereas Cantor now proposed to split them apart so that on his account there would be cardinalities which are infinite but not absolutely infinite (hence are comprehensible by human thought).

Any theological argument for a mathematical principle will strike many modern eyes as something close to a category mistake, and we are entitled to wonder whether a different argument is available. What we need from the constructivist is more detail on what is involved in comprehending some objects as a collection. It cannot be merely a matter of understanding the property they share, since we presumably understand the property of self-identity even though the view requires us to deny that we can form a collection containing everything which has this property. Perhaps, then, what is required is that we should run through the objects in our thought. If so, it is easy to see why pre-Cantorians identified the absolutely infinite with the simply infinite, since it is by no means clear how a finite being — even an idealized one — is expected to be able to run through an infinity of things. To make any sense of this, we have to be willing to accept the coherence of the notion of a supertask.

Even supertasks have their limits, however: as we noted earlier (§11.1), if a supertask is performed in time, and the structure of time is that of the real line, then every supertask is countable. So this account would make all uncountable cardinalities absolutely infinite. Some constructivists might be untroubled by this, of course, but it is certainly inadequate to the intentions of Cantor himself, whose theory of uncountable cardinals was one of his greatest achievements. The strong suspicion must be that Cantor's talk of sets being 'collected together in thought' is not to be taken at its constructivist face value.

But if instead we adopt the platonist conception, the difficulty we face is that it is very hard to see what argument remains for the weak limitation-of-size principle. If it is quite independent of our thought whether some things form a collection, why should it matter that there are a great many of them?

And even if we succeed in motivating the weak limitation-of-size principle, there remains the difficulty of saying how big the absolutely infinite cardinalities are. The platonist is committed, on pain of contradiction, to saying that there is no collection of everything, and hence that the number of everything is absolutely infinite. Many from von Neumann on have claimed that this is the *only* absolutely infinite cardinality, but it is not clear on what ground. If all that interested us were paradox-barring, it might *suffice* to rule out just this one case, but why should we think that this is the only case that *has* been ruled out?

Let us turn now to the question of which of the axioms of ZFU^r are justified by limitation of size. The easiest cases are the two axiom schemes, separation and replacement, which are obviously justified by weak limitation of size. But these two schemes, so powerful in the context of the other axioms, do no work at all on their own, as is easily seen by observing that they are both trivially satisfied in a universe consisting only of the empty set. (The situation here is thus analogous to the one faced by the iterative conception, which did not justify the existence of any levels in the hierarchy until it was afforced by the second principle of plenitude.) This problem is not solved if we move to the strong version of the limitation-of-size principle. Consider, for instance, a universe with a countable infinity of individuals, and let the only sets there are be hereditarily finite — finite sets of individuals, finite sets of such sets, etc. The total number of objects in such a universe would then still be countably infinite, and the strong limitation-of-size principle would be satisfied, but there would be no set of all individuals, and so even the temporary axiom of §4.2 would not be satisfied.

Rather similar considerations apply to the axiom of creation. For this to follow from limitation of size, it would have to be the case that if $\mathfrak{a}$ is small, i.e. less than the total number of objects, then $2^{\mathfrak{a}}$ is too. But why should we think that? There can be found in the literature somewhat half-hearted attempts to assert that $2^{\mathfrak{a}}$ is 'not much bigger' than $\mathfrak{a}$, but even if that were so (which is open to doubt), a further argument would be required to conclude that the total number of things cannot be $2^{\mathfrak{a}}$.

The difficulty is well illustrated by the fate of recent neo-logicist attempts to base a theory of sets on the so-called *New V*, a second-order axiom encapsulating the strong limitation-of-size principle.

$$\{x : Fx\} = \{x : Gx\} \Leftrightarrow ((\forall x)(Fx \Leftrightarrow Gx) \text{ or there are as many } F\text{s as } G\text{s}).$$

It is not, as some commentators (e.g. Ketland 2002) have thought, 'surprising' that New V turns out to be incapable of generating a workable theory unless we add a strong axiom as to how many sets there are, but merely an indication of the weakness of limitation of size taken on its own. But it is very hard to see how an axiom about the size of the universe is to be motivated. There are no doubt arguments to be had about how many non-sets there are or must be

— see, for example, the continuing debate about whether there could have been nothing — but that is of no use to the limitation-of-size theorist since, as we have just seen, it is quite consistent with limitation of size for there to be infinitely many individuals without there being infinite sets. What we need, in other words, is an argument as to the size of the *set-theoretic* part of the universe. But that is simply to ask how many sets there are, which is just the question that limitation of size cannot answer.

We have focused so far on what limitation of size has to say about sets, but we need also to consider whether it provides any motivation for this focus: does it, in other words, give us any reason to rule out ungrounded collections? It seems clear that it does not. A non-well-founded set theory such as Aczel's (1988) allows there to be a collection a specified by the stipulation that $a = \{a\}$. There is certainly no logical inconsistency in this: Aczel's theory can be shown to be consistent if **ZF** is. But if the requirements of bare consistency do not rule a out, the limitation-of-size theorist is committed to its existence, for according to the weak limitation-of-size principle, if any singleton at all exists, then a does.

13.6 Back to dependency?

The conclusions we have reached, then, are that if we come to believe the limitation-of-size account, there is little prospect that it will provide support for the axiom of infinity, and none whatever for the axiom of foundation; and, worse, that it is very hard to see what ground there might be for thinking that the account is true. If we want an intuitive argument to support our theory of sets, therefore, the dependency account seems to hold out a better prospect of success. But the replacement axiom has been thought to be something of an embarrassment for this approach, since it is unclear how the dependency account is supposed to justify it. Limiting case platonism might justify it, after a fashion, but we have rejected that as unsatisfactory. It has therefore been a commonplace of recent philosophical discussions (e.g. Boolos 1989) to say that neither the iterative conception *nor* limitation of size justifies the whole theory, which comes to be seen as an incoherent amalgam of two distinct conceptions.

This conclusion need not worry us unduly here, of course, since the replacement axiom — the one that is supposed to be problematic for the iterative conception — is not part of our default theory. Nonetheless, I want to argue that the conclusion may in any case be too pessimistic: although we have no need to do so here in order to defend our default theory, I think that we may be able to justify replacement on a ground that it is open to the dependency theorist to hold. To do this, we split the weak limitation-of-size principle which we considered in the last section into two parts. The weak limitation-of-size principle is evidently equivalent to the conjunction of the following two principles.

Generalized separation principle. *If every F is a G and the Gs form a collection, then the Fs form a collection.*

Size principle. *If there are just as many Fs as Gs, then the Fs form a collection if and only if the Gs do.*

The first of these principles generalizes the separation principle stated in §3.5 in that it applies to collections generally, rather than just to sets. Of course, if the internal platonist argument for the well-foundedness of dependency sketched in §3.3 is correct, this generalization is harmless, since all collections are sets anyway. If it is not, then some further argument is required in order to justify it. But in that circumstance more would in any case have to be said about the metaphysical nature of ungrounded collections in order to justify whatever other axioms were assumed about them: whether the generalized separation principle is correct would presumably depend on that.

So let us leave that question on one side and focus instead on what I am calling the size principle. I wish to suggest that this can be given a plausible motivation based on the idea that what is essential to a collection is how many members it has. To put the idea in context, let us consider a notion that is nowadays a commonplace in mathematics. A property is group-theoretic if two isomorphic groups cannot differ with respect to it; a property is *topological* if two homeomorphic metric spaces cannot differ with respect to it; and so on. Now the analogue of isomorphism or homeomorphism in the case of collections is the notion of a one-to-one correspondence. By extension, then, a property is *collection-theoretic* if a class has it if and only if all equinumerous classes have it. If, as seems reasonable, we take the property a class may have of there being a collection with just those members to be collection-theoretic, we arrive straightforwardly at the size principle.

This argument does little more than spell out the simple thought that a collection is *barely* composed of its members: no further structure is imposed on them than they have already. So, the thought runs, what *else* could there be to determine whether some objects form a collection than how many there are of them? What else could even be relevant?

13.7 Higher still

Recent research in set theory has been dominated by the study of what are known as *large cardinal axioms*. These are cardinal existence claims that are independent of ZFC but are nevertheless believed to be consistent with it. In order to qualify for the title, however, it has usually been taken that they must also exhibit mathematical fruitfulness, for instance by generating new combinatorial results, as well as set-theoretic stability, by being fixed points of certain thinning procedures studied by set theorists. Each large cardinal axiom is in particular an axiom of infinity in the sense of this chapter, since the existence of a large cardinal entails the existence of the levels in the hierarchy needed to arrive at it. There is now a huge literature on large cardinals — for instance (in increasing order of size) strongly inaccessible, strongly Mahlo, measurable, Woodin, and supercompact cardinals — and on the corresponding axioms of infinity.

The study of these axioms has brought to light further close connections with the structure of sets of real numbers: we have noted already that the proof of determinacy for Borel sets requires the axiom of ordinals; and this result has turned out to have analogues at higher levels. The pattern is thus that proofs of the determinacy of more and more complex subsets of the Baire line appeal to larger and larger cardinals.

Perhaps the most striking result of this sort has its origins in a famous error of Lebesgue (1905, pp. 191–2). He claimed that a continuous image of a Borel set is in turn a Borel set, but his argument was, as he later acknowledged, 'simple, short, but false' (Lusin 1930, p. vii): he had mistakenly assumed that

$$f^{-1}(\bigcap_{n\in\omega} A_n) = \bigcap_{n\in\omega} f^{-1}(A_n),$$

which need not be true if f is not one-to-one. The error was noticed by Souslin (1917), who gave an example to show that an *analytic* set, i.e. one that is the continuous image of a Borel set, need not be Borel.[2] This discovery led naturally to the study of the *projective* sets, i.e. the sets obtainable from closed sets if we are allowed to form continuous images as well as countable unions and complements arbitrarily often. The projective sets form a much more inclusive class than the Borel sets, and the definition of a projective set may be much more complex. In studying such sets the key idea is the realization by Kuratowski and Tarski (1931) that methods for obtaining projective sets correspond to logical operations. This has the consequence that whereas the axiom of ordinals is sufficient to prove the determinacy of the game on any Borel set, to generalize this result to an arbitrary projective set it is not enough to assume even the existence of a measurable cardinal; the proof of this determinacy claim that *has* been found makes the very much stronger assumption that there exist infinitely many Woodin cardinals (Martin and Steel 1988). And with a yet stronger axiom of infinity the result can be generalized still further, to what are sometimes called the *quasi-projective* sets, an even more inclusive class consisting of sets which occur in a constructible hierarchy starting

[2] Ironically, Lebesgue had constructed just such a set for another purpose in his original paper.

from the set of real numbers: if we assume that there are not only infinitely many Woodin cardinals but also a measurable cardinal above them, we can prove that the game on every quasi-projective set is determined.[3]

We shall have a little more to say about these results in §15.7. For the moment, however, we confine ourselves to comments on the motivation for large cardinal axioms. Gödel's theorems warn us, of course, that we cannot expect a *proof* that any such axiom is consistent. Intuitive arguments for the plausibility of large cardinal axioms have been attempted, e.g. Reinhardt's (1974) argument for the existence of supercompact cardinals,[4] but those that have been given so far have a notably sketchy and provisional character. When reading about large cardinals, it is hard to avoid a nagging feeling that their size makes them literally incredible. Having swallowed the notion of a measurable cardinal, for instance, we turn the page and read that a supercompact cardinal is larger still — so much larger that the set of measurable cardinals less than the least supercompact cardinal is itself of measurable cardinality.

It is notable that the formalist mode of speech, always popular among mathematicians, seems especially so when large cardinals are in question. Indeed it is tempting to sympathize with Wittgenstein's suggestion that it is a mistake to regard infinite cardinals as *big* at all.

> $\aleph_0$ is not an enormous number. ... The child who has learnt $\aleph_0$ multiplications hasn't learnt anything huge. ... "I bought something infinite and carried it home." You might say, "Good lord! How did you manage to carry it?" — A ruler with an infinite radius of curvature. (1976, pp. 32, 142)

Part of what makes large cardinals so hard to accept is precisely the *Cantorian finitism* which has sometimes been used to motivate them. This is the idea that infinite sets are, as far as possible, just like finite sets. If we adopt this idea, we are encouraged to regard the claim that a large cardinal is much bigger than, say, $\aleph_0$ as having the same sort of import as the claim that 10^{20} is much bigger than 12. But if we do that, do we lose our grip on reality? Do we, as Boolos has suggested,

> suspect that, however it may have been at the beginning of the story, by the time we have come thus far the wheels are spinning and we are no longer listening to a description of anything that is the case? (2000, p. 268)

[3] This assertion is often abbreviated as $AD^{L(\mathbb{R})}$ in the literature.

[4] Maddy has declined to call arguments such as Reinhardt's intuitive on the ground that 'they extend beyond anything that could plausibly be traced to an underlying perceptual, neurological foundation' (1990, p. 141). I shall not follow her in this because I do not expect an intuitive argument for accepting an axiom to depend on a perceptual, neurological foundation any more than any other argument does.

13.8 Speed-up theorems

So far, our discussion of abstract (i.e. higher-order) methods has focused on their utility in proving results not provable without them. But we should not ignore another use they have in application to results which *are* provable without them but which can be given much shorter, more elegant or more perspicuous proofs by means of them.

Notice straightaway that although they are all desirable in their way, these three criteria — brevity, elegance and perspicuity — are undoubtedly distinct: the shortest proof is often not the most perspicuous, nor sometimes the most elegant. When it comes to studying the matter systematically, we have to recognize that elegance and perspicuity are of course much less objective than mere length and hence less amenable to formal study. In relation to length, however, the phenomenon we are alluding to — that there is a trade-off between the abstractness of a proof and its brevity — is a familiar fact of mathematical experience: when learning mathematics we sometimes struggle through a long but (at least in an informal sense) elementary proof of a result, only to discover much later that it can be proved more quickly by applying an abstract theory. So far this is just an informal observation, though: in making it precise, we face the difficulty that the size of a proof is highly sensitive to the form in which the theory is presented. If we measure the size of a proof by the number of lines it contains, every axiomatizable first-order theory is equivalent to one in which every theorem has a 3-line proof. This brevity is only achieved, however, by sacrificing a feature of all the theories we actually deal with, namely that they are axiomatized by means of a finite number of axioms or schemes, and for such theories we have the following theorem: if U' is strictly stronger than U, there exists a number N such that for any m there is a sentence provable in no more than N lines in U' whose shortest proof in U has more than m lines (Buss 1994).

On the other hand, the number of lines in a proof is not a good measure of feasibility, since each line might be unfeasibly long. But even if we measure the size of a proof by the number of characters it contains, small variations of presentation may have an enormous effect on size. A dramatic example of this is provided by Mathias (2002), who has shown that the term used to express the cardinal number 1, which in many formal systems has a few dozen characters, in Bourbaki's (1954) formal system has about 10^{12} characters; and an apparently trivial emendation of the system in the 4th edition of the book, by which the ordered pair is given the Kuratowski definition rather than being treated as a primitive, makes the term explode to about 10^{54} characters.[5] Despite this, however, it turns out that the analogue of Buss's theorem for this

[5] More precisely, Mathias claims that the term in question has 2,409,875,496,393,137,472,149, 767,527,877,436,912,979,508,338,752,092,897 characters, but I have not checked his arithmetic myself.

measure of size is available (Mostowski 1952): if U' is strictly stronger than U, there exists a number N such that for any m there is a sentence with a proof shorter than N characters in U' whose shortest proof in U has more than m characters.

Nonetheless, as with the incompleteness phenomenon discussed earlier, what we are interested in here is not so much these wholly general results as genuinely mathematical instances of them. After all, the speed-up results of Buss and Mostowski would be of little relevance to mathematical practice if the length of proof N for which arbitrary speed-up occurs were 10^{100}, since it would then be inconceivable that we could ever encounter the phenomenon ourselves.

In fact, however, we can obtain genuinely mathematical examples of speed-up by restriction from mathematical incompletenesses. Thus, although the claim that *every* Goodstein sequence terminates cannot be proved in PA, the instance of it for a particular starting point m can; but we should expect the proof in PA that the Goodstein sequence starting at m terminates to be very long indeed for some values of m, whereas in ZU it is, of course, a trivial particular instance of Goodstein's theorem, of which we gave in §13.1 an evidently feasible proof.

Another, even simpler and more intuitive example of speed-up has been given by Boolos (1987). Consider the first-order language containing a constant 0, a one-place function symbol s, a two-place function symbol f, and a one-place predicate symbol D. Let T be the theory whose axioms are

(1) $(\forall x) f(x, 0) = s(0)$

(2) $(\forall y) f(0, sy) = s(s(f(0, y)))$

(3) $(\forall x)(\forall y) f(s(x), s(y)) = f(x, f(s(x), y))$

(4) $D(0)$

(5) $(\forall x)(D(x) \Rightarrow D(s(x)))$.

Now $D(f(ssss0, ssss0))$ is a theorem of T. This can easily be shown in ZU as follows. Take any set-theoretic model of T and write $\overline{D}$, $\overline{f}$, $\overline{s}$ and $\overline{0}$ for the interpretations of D, f, s and 0 in the model. Let $\overline{N} = \mathrm{Cl}_{\overline{s}}(\overline{0})$. Then we can prove by induction on y that $(\forall x \in \overline{N})(\forall y \in \overline{N}) f(x, y) \in \overline{N}$. To do this, note first that $\overline{f}(\overline{0}, \overline{0}) = \overline{s}(\overline{0}) \in \overline{N}$; and if $y \in \overline{N}$ and $\overline{f}(\overline{0}, y) \in \overline{N}$, then $\overline{f}(\overline{0}, \overline{s}(y)) = \overline{s}(\overline{s}(\overline{f}(\overline{0}, y))) \in \overline{N}$, so by induction on y $\overline{f}(\overline{0}, y) \in \overline{N}$ for all $y \in \overline{N}$. Now suppose that $\overline{f}(x, y) \in \overline{N}$ for all $y \in \overline{N}$. Now $\overline{f}(\overline{s}(x), \overline{0}) = \overline{s}(\overline{0}) \in \overline{N}$; and if $y \in \overline{N}$ and $\overline{f}(\overline{s}(x), y) \in \overline{N}$, then $\overline{f}(\overline{s}(x), \overline{s}(y)) = \overline{f}(x, \overline{f}(\overline{s}(x), y)) \in \overline{N}$ by hypothesis. So by induction on y, $\overline{f}(s(x), y) \in \overline{N}$. By induction on x, therefore, $\overline{f}(x, y) \in \overline{N}$ for every x and y in $\overline{N}$ as required. But $\overline{0} \in \overline{D}$ by (4), and $\overline{D}$ is s-closed by (5). So $\overline{N} \subseteq \overline{D}$, from which it follows immediately that $\overline{f}(\overline{4}, \overline{4}) \in \overline{D}$. Thus the sentence $D(f(ssss0, ssss0))$ is true in every set-theoretic model of T and hence by the completeness of first-order logic it is provable in T as required.

Yet Boolos shows in his paper that in a fairly standard system of first-order logic the shortest formal proof of this theorem has (using the notation defined in §5.4) more than $2_{65,536}$ symbols. We thus have a simple illustration of how set theory makes arguments feasible which are out of reach if we are restricted to using wholly elementary methods.

Now the example quoted above of a term in Bourbaki's system with 10^{54} characters shows that we need to exercise a little care if we are to be sure that the unabbreviated formal version of the set-theoretic proof just sketched is feasible. Moreover, our examples earlier show that we must be cautious lest this is merely a feature of the particular first-order formal system being studied: there are examples showing, for instance, that adding the cut rule to the logic effects considerable speed-ups on proofs. However, these examples come nowhere near to the degree of speed-up involved in this case, so we may be tolerably confident that it is not feasible to write down a proof of Boolos' 'curious inference' in any familiar first-order system.

Of course, Boolos devised his example specifically to provide an illustration of speed-up. Examples of speed-up that we meet in practice may not be so striking. Moreover, perspicuity is just as important as length: it is not much help that a short proof exists if we cannot find it. Examples of these phenomena occur at various levels in the hierarchy. We mentioned in chapter 5 that Dirichlet's theorem in arithmetic was shown to have a proof in PA only a century after an analytic one was discovered. Adjoining infinitesimals to the theory of real numbers is capable of speeding up proofs. And the theorem that the game on any Borel set is determined was first proved assuming the existence of measurable cardinals; Martin's later proof (1975) which requires just the axiom of ordinals is significantly more complicated and depends on an entirely new idea.

Notes

We have been concerned here with two related phenomena: there are results provable in a strong set theory U' whose proofs in a weaker theory U are very much longer; and there are results provable in U' which are not provable in U at all. The general theorems showing that these phenomena are possible are due to Gödel, but many natural examples of both, some of them arguably mathematical, are now known. For some instances consult Friedman (1986; 1998). Many further examples obtained by Friedman still await publication at the time of writing.

The details of the theory of large cardinals are expounded by Kanamori (1997). A beautifully clear analysis of set theorists' reasons for accepting large cardinal axioms has been given by Maddy (1988). She distinguishes between intrinsic and extrinsic reasons, but, consistently with her decidedly naturalistic leanings, her distinction does not coincide with the one between intuitive and regressive arguments which I have emphasized here. Maddy's extrinsic reasons include not only the regressive arguments of the realist but reasons for studying particular theories that can be given by formalists, such as simplicity, elegance or the ability to generate hard and interesting problems.

There are treatments of von Neumann's theory of ordinals and Mostowski's collapsing lemma in very many set theory textbooks: Drake 1974 is especially clear. The philosophical basis for various reflection principles is discussed by Wang (1977). The best discussion of the history of limitation of size is Hallett 1984. The abstraction principle known as New V is discussed by Shapiro and Weir (1999).

Chapter 14

The axiom of choice

In §9.4 we introduced a principle — the axiom of countable choice — which differed from the axioms of our default theory because it asserted the existence of a set of a particular sort (actually, in this case, a sequence) without supplying a condition that characterizes it uniquely. In this chapter we shall investigate some generalizations of the axiom of countable choice that share this feature, and enquire a little further into whether the lack of uniqueness in such specifications should be regarded as troubling.

14.1 The axiom of countable dependent choice

Consider the following attempt to prove that a partially ordered set is partially well-ordered iff it contains no strictly decreasing sequences. Certainly one direction of this equivalence is straightforward: the range of a strictly decreasing sequence is a non-empty set with no minimal element. To prove the reverse implication, suppose that A is not partially well-ordered, so that it has a non-empty subset B without a minimal element. Now choose an element x_0 in B and define a sequence (x_n) in B as follows: once x_n has been chosen, let x_{n+1} be any element of B less than x_n. (Such an element exists because B has no minimal element.) The sequence (x_n) is clearly strictly decreasing.

The difficulty with this argument is that it requires a countable infinity of choices to be made in order to generate the sequence (x_n). However, it cannot be justified by appeal to the axiom of countable choice, because that axiom licenses only *independent* choices. One way of putting the point might be by appeal to a temporal metaphor: the choices involved are not simultaneous, since x_{n+1} cannot be chosen until the value of x_n is known. The necessity for making choices in this way arises sufficiently often in mathematics for it to be worthwhile to single out the set-theoretic principle which licenses the procedure.

Axiom of countable dependent choice. *If r is a relation on a set A such that $(\forall x \in A)(\exists y \in A)(x\ r\ y)$, then for any $a \in A$ there exists a sequence (x_n) in A such that $x_0 = a$ and $x_n\ r\ x_{n+1}$ for all $n \in \boldsymbol{\omega}$.*

(14.1.1) **Proposition.** *The axiom of countable dependent choice implies the axiom of countable choice.*

Proof. Suppose that (A_n) is a sequence of disjoint non-empty sets. Choose an element $a \in A_0$ and define a relation r on $\bigcup_{n\in\omega} A_n$ by letting $x\ r\ y$ iff there exists $n \in \omega$ such that $x \in A_n$ and $y \in A_{n+1}$. Now clearly $\operatorname{dom}[r] = \bigcup_{n\in\omega} A_n$ and so by the axiom of countable dependent choice there exists a sequence (x_n) such that $x_0 \in A_0$ and $x_n\ r\ x_{n+1}$ for all $n \in \omega$. It follows easily by induction that $x_n \in A_n$ for all $n \in \omega$. We have thus proved the special case of the axiom of countable choice where the sets from which elements are to be chosen are pairwise disjoint. It is now an easy exercise to show that the general case follows. □

(14.1.2) **Theorem.** *These three assertions are equivalent:*

(i) *The axiom of countable dependent choice;*

(ii) *If r is a relation on a non-empty set B such that* $\operatorname{dom}[r] = B$, *i.e.* $(\forall x \in B)(\exists y \in B)(x\ r\ y)$, *then there exists a sequence (y_n) in B such that $y_n\ r\ y_{n+1}$ for all $n \in \omega$ (the value of y_0 is not stipulated);*

(iii) *Every partially ordered set which does not contain the image of a strictly decreasing sequence is partially well-ordered.*

(i) ⇒ (ii). Trivial.

(ii) ⇒ (iii). Suppose that $(A, \leqslant)$ is a partially ordered set which is not partially well-ordered. So it has a non-empty subset B without a minimal element. Now $\operatorname{dom}[>_B] = B$ (since otherwise B would have a minimal element), and hence by hypothesis there exists a sequence (y_n) in B such that $y_n > y_{n+1}$ for all $n \in \omega$.

(iii) ⇒ (i). Suppose that r is a relation on a non-empty set A such that $(\forall x \in A)(\exists y \in A)(x\ r\ y)$ and $a \in A$. Let $\mathcal{A}$ be the set of all strings s in String(A) such that $a = s(0)\ r\ s(1)\ r\ \dots\ r\ s(n-1)$ (where n is the length of the string s). The *opposite* of the inclusion relation does not partially well-order $\mathcal{A}$, and therefore by hypothesis there exists a strictly decreasing sequence (s_n) in $\mathcal{A}$. Now $\{s_n : n \in \omega\}$ is a chain, and so if $s = \bigcup_{n\in\omega} s_n$, then s is a sequence in A [proposition 4.8.1], $s(0) = a$, and $s(n)\ r\ s(n+1)$ for all $n \in \omega$. □

The axiom of countable dependent choice was first stated explicitly by Bernays in 1942, although mathematicians (especially analysts) had been using it informally for many years before that. It is in fact stronger than countable choice; i.e. it cannot be proved from it (Mostowski 1948), even if we assume the axiom of purity (Jensen 1966). The stability of dependent choice as a mathematical principle has been emphasized by the discovery that it is equivalent to various mathematically interesting statements in other parts of mathematics — most notably, perhaps, the Baire category theorem (Blair 1977).

The formalist might, I suppose, treat results of this sort as reasons to be interested in the axiom. For the realist, on the other hand, the equivalence of dependent choice with propositions in diverse parts of mathematics says nothing as to its truth unless there is independent reason to believe these other propositions. However, the constructivist motivation we sketched for countable choice in §9.4 does seem to apply equally to countable dependent choice. What it hinges on, after all, is just the coherence of the notion of a countably infinite supertask, a succession of operations performed in a finite time by an idealized being: it does not seem to place any greater load on *that* notion to require that the operations to be performed should depend on one another.

Exercises

1. Do the 'easy exercise' mentioned in the proof of proposition 14.1.1.

2. Assuming the axiom of countable dependent choice, show that if $(A, \leqslant)$ is an infinite partially ordered set, then A has either an infinite totally ordered subset or an infinite totally unordered subset. [Suppose that all totally unordered subsets of A are finite: show that every infinite subset B of A has a maximal totally unordered subset and therefore has an element b which is comparable with infinitely many elements of B.]

3. Show that the axiom of countable dependent choice is equivalent to the assertion that if $(A, \leqslant)$ is a partially ordered set and (D_n) is a sequence of cofinal subsets of A, then there is an increasing sequence (x_n) in A such that $\{x_n : n \in \omega\}$ intersects every D_n.

14.2 Skolem's paradox again

One instance in which dependent choices are needed arises in model theory.

Löwenheim/Skolem theorem (submodel form). *The axiom of countable dependent choice entails that every structure has a countable elementarily equivalent substructure.*

For the proof of this result consult a model theory textbook such as Hodges 1993. It is stronger than the version of the Löwenheim/Skolem theorem mentioned in §6.6, which claimed only that every structure is elementarily equivalent to a countable structure, not that it could be chosen to be a substructure of the given structure. The stronger theorem certainly requires some version of choice for its proof in general[1] (as can be seen from the fact that it can be used

[1] However, McIntosh's (1979, n. 3) assertion that the submodel form implies the axiom of choice is incorrect.

to give a quick proof that every set is finite or infinite). Some authors have thought that this dependence on choice could be used to neutralize the submodel version as a philosophical weapon. Indeed the reason Skolem (1922) gave for proving the weaker version of the theorem which we stated in §6.6, when he had already proved the submodel version two years earlier, was that he wanted to derive philosophical consequences from it without depending on choice.

But in fact the issue about the use of choice here is a red herring, for in the case which is used to generate Skolem's paradox the theorem is applied to a model of set theory itself, and in this case the use of dependent choice to prove the theorem is eliminable. That is to say, the following theorem is provable in ZU *without* any use of choice.

Theorem. *Every [transitive] model of* ZF *has a countable [transitive] submodel.*

It is this theorem that provides us with a more precise (and, some have thought, more troubling) version of Skolem's paradox. To see why, let us once more shift our perspective to that of a metalanguage. Applying the submodel version of the Löwenheim/Skolem theorem, we can deduce that there is a countable transitive set M such that $(M, \in)$ is a model of set theory. Moreover, since M is countable and transitive, every member of M is also countable. Yet all the theorems of ZF are true in M: in particular, M has members, such as the set which acts in M as the power set of the natural numbers, which are not countable-in-M. So countability is not equivalent to countability-in-M. Set theorists express this by saying that countability is not an *absolute* property.

What we saw when we first discussed Skolem's paradox in §6.6 was that if we keep the meaning of the logical vocabulary fixed but leave the domain of interpretation and the extension of the membership relation unconstrained, we cannot tie down the cardinality of the domain. What we can now do on the basis of the stronger version of the theorem is to extend this even to the case where the extension of the membership relation is kept fixed. What varies between interpretations is now *only* the range of the quantifiers, but this degree of variation is sufficient to change the extension of the predicate 'countable'.

Thus set theory has (assuming always that it is consistent) a model M which is uncountable from within and countable from without. It is easy to see how we have allowed this to happen: the first-order scheme of separation is weak because at each level in the construction of a model it forces us to include $\{x \in V : \Phi\}$ only for formulae Φ in the (countable) language of the theory. The second-order variant of the theory has no such weakness.

So for some models of first-order set theory there are perspectives from which they are countable. This does *not* show that for every model of set theory there is such a perspective. It certainly does not compel us to believe that every set is really countable. The point is obvious, but the literature on Skolem's paradox has repeatedly been marred by misunderstandings of it.

14.3 The axiom of choice

Definition. A choice function *is a function* f *such that* $f(A) \in A$ *for all* $A \in \text{dom}[f]$. *The set of all choice functions* f *such that* $\text{dom}[f] \subseteq \mathcal{A} \setminus \{\emptyset\}$ *is denoted* choice$(\mathcal{A})$.

If $\text{dom}[f] = \mathcal{A} \setminus \{\emptyset\}$, we shall say that f is a choice function *for* $\mathcal{A}$.

(14.3.1) **Lemma.** *The maximal elements of* choice$(\mathcal{A})$ *are the choice functions for* $\mathcal{A}$.

Proof. If f is a choice function for $\mathcal{A}$, then $\text{dom}[f] = \mathcal{A} \setminus \{\emptyset\}$, and so f is evidently maximal in choice$(\mathcal{A})$. If, on the other hand, f is a member of choice$(\mathcal{A})$ which is not a choice function for $\mathcal{A}$, then $\text{dom}[f] \subset \mathcal{A} \setminus \{\emptyset\}$ and there exists $A \in (\mathcal{A} \setminus \{\emptyset\}) \setminus \text{dom}[f]$: if $a \in A$, then $f \cup \{(A, a)\}$ is a choice function which strictly contains f and so f is not maximal in choice$(\mathcal{A})$. □

(14.3.2) **Proposition.** *Every finite set has a choice function ('The principle of finite choice').*

Proof. If $\mathcal{A}$ is a finite set, then choice$(\mathcal{A})$ is finite and non-empty: it therefore has a maximal element [theorem 6.4.5], which must be a choice function for $\mathcal{A}$ [lemma 14.3.1]. □

In §9.4 we discussed the axiom of countable choice, which asserts that every countable set has a choice function. What we shall consider now is an obvious generalization of this principle.

Axiom of choice. *For every set* $\mathcal{A}$ *of disjoint non-empty sets there is a set* C *such that for each* $A \in \mathcal{A}$ *the set* $C \cap A$ *has exactly one member.*

Many authors treat the axiom of choice as part of the default theory, but we shall not do so here. It is customary to write ZFC for the theory obtained by adding the axiom of choice to ZF, and to use corresponding notations for the theories obtained by adding it to other set theories.

(14.3.3) **Proposition.** *The axiom of choice is equivalent to the assertion that every set has a choice function.*

Necessity. Assume the axiom of choice and let $\mathcal{B}$ be any set of non-empty sets. We want to define a choice function whose domain is $\mathcal{B}$. To do this, let $\mathcal{A} = \{B \times \{B\} : B \in \mathcal{B}\}$. Then $\mathcal{A}$ is obviously pairwise disjoint, and every element of it is non-empty. So by the axiom of choice there is a set C intersecting each $B \times \{B\}$ in a unique ordered pair: call the first element of that ordered pair $f(B)$. Then the function f thus defined is evidently a choice function for $\mathcal{B}$.

Sufficiency. If $\mathcal{A}$ is a set of disjoint non-empty sets and f is a choice function for $\mathcal{A}$, the set $\{f(A) : A \in \mathcal{A}\}$ intersects each $A \in \mathcal{A}$ in just the one element $f(A)$. □

The axiom of choice obviously implies the axiom of countable choice; let us now show that, slightly less obviously, it implies the stronger axiom of countable *dependent* choice.

(14.3.4) **Proposition.** *The axiom of choice implies the axiom of countable dependent choice.*

Proof. Suppose that a, A and r satisfy the hypotheses of the axiom of countable dependent choice. Then $r[x] \neq \emptyset$ for all $x \in A$. So by the axiom of choice there exists a function f from A to itself such that for each $x \in A$ we have $f(x) \in r[x]$, i.e. $x\ r\ f(x)$. Now let $x_0 = a$ and once x_n has been defined let $x_{n+1} = f(x_n)$. Evidently $x_n\ r\ x_{n+1}$ for all $n \in \boldsymbol{\omega}$. □

The axiom of choice certainly does not introduce inconsistency into set theory, even if we assume the axiom of purity (Gödel 1938); moreover, this relative consistency result continues to hold whichever of the higher axioms of infinity discussed in chapter 13 we add to the theory. On the other hand, the axiom of choice cannot be proved in any of the theories we have considered so far — even if we assume countable dependent choice (Mostowski 1948) and purity (Feferman 1965). It is by far the most important of the choice axioms, not least because it is the most stable in the sense of the last section: a very large number of significant assertions in apparently unconnected parts of mathematics can be shown to be equivalent to it.

Exercise

Show that the axiom of choice is equivalent to the assertion that if $(B_i)_{i \in I}$ is a family of non-empty sets, then $\prod_{i \in I} B_i$ is non-empty.

14.4 The well-ordering principle

(14.4.1) **Proposition.** *Every countable set is well-orderable.*

Proof. Trivial. □

Once Cantor had discovered that $\mathbf{R}$ is uncountable, it was natural that he should ask whether it is well-orderable: he soon claimed, in fact, not only that $\mathbf{R}$ is well-orderable but that *every* set is. Cantor made this claim, which has come to be known as the *well-ordering principle*, in 1883, describing it as 'a law of thought which appears to me to be fundamental, rich in consequences, and particularly remarkable for its general validity' (p. 550). By 1895, however, he had retreated from regarding the well-ordering principle as an axiom and thought of it only as a conjecture requiring proof; indeed he attempted such a proof in a letter to Hilbert of 1896, but what he says in the letter is very far from convincing. Even so, in his 1900 lecture to the International Congress of Mathematicians, Hilbert rather generously referred to the well-ordering principle as a 'theorem of Cantor', but instead posed the problem of finding a *definite* well-ordering on the real line. Soon Zermelo (1904) formulated the axiom of choice (*Auswahlaxiom*) explicitly and showed how to deduce the well-ordering principle from it.

(14.4.2) **Lemma (Zermelo 1904).** *For every well-ordering on a set A there exists a definite choice function for $\mathfrak{P}(A)$, and conversely.*

Necessity. Suppose that $\leqslant$ is a well-ordering relation on A. For each $B \in \mathfrak{P}(A) \setminus \{\emptyset\}$ let $f(B)$ be the least element of B with respect to $\leqslant$. It is clear that this defines a choice function for $\mathfrak{P}(A)$.

Sufficiency. If f is a choice function for $\mathfrak{P}(A)$, then by the general principle of transfinite recursion and Hartogs' theorem 11.4.2 there exist a unique ordinal α and function g from $\boldsymbol{\alpha}$ to A such that $g(\beta) = f(A \setminus g[\boldsymbol{\beta}])$ for all $\beta < \alpha$ and $g[\boldsymbol{\alpha}] = A$; this function is one-to-one and therefore gives rise to a well-ordering on A. □

(14.4.3) **Theorem.** *The axiom of choice is equivalent to the assertion that every set is well-orderable.*

Proof. Immediate from the lemma. □

The more demanding problem posed by Hilbert of *defining* a well-ordering on **R** is not soluble even if we assume the axiom of choice: there is no term σ in the first-order language of set theory for which we can prove in ZFC that σ is a well-ordering of **R** (Feferman 1965). Put more informally, this means that the axiom of choice guarantees the existence of a well-ordering on **R** without giving us the means to define one.

The axiom of choice was used a number of times by mathematicians in the first few years of the 20th century — for details see the notes at the end of the chapter — but it was certainly Zermelo's use of it in his 1904 proof of the well-ordering principle that pushed the axiom of choice to centre stage and made it a matter of active controversy among mathematicians. This was even more the case after König (1905) published a purported proof of a result *contradicting* the axiom of choice: there then followed a flurry of discussion of Zermelo's proof in *Mathematische Annalen* (Borel 1905; Bernstein 1905*a*; Jourdain 1905; Schönflies 1905), *Bulletin de la Société Mathématique de France* (Borel, Baire, Hadamard and Lebesgue 1905), *Proceedings of the London Mathematical Society* (Hobson 1905; Dixon 1905; Hardy 1906; Jourdain 1906; Russell 1906*b*), and elsewhere.

One motive driving some of the critics seems to have been that they already believed the well-ordering principle to be false and therefore needed to find fault with Zermelo's proof somehow. Now his 1904 proof made use of transfinite induction on the ordinals (as does the one we have just given), and any use of ordinals was still tainted by Burali-Forti's paradox, so some of the objectors focused on this rather than on the use of the axiom of choice. Hobson (1905, p. 185), for instance, complained that 'the non-recognition of the existence of "inconsistent" aggregates, which existence, on the assumption of Cantor's theory cannot be denied, introduces an additional element of doubt as regards this proof'. It was in order to answer *this* criticism that Zermelo published another proof (1908*a*) which eliminated the previous use of ordinals by means of an extension of Dedekind's *Kettentheorie*. This left the axiom of choice as the only principle used in the new proof that the critics could reasonably object to.

The well-ordering principle is often a useful way of applying the axiom of choice: here is a more or less typical illustration.

(14.4.4) **Proposition.** *Assuming the axiom of choice, every partial ordering $\leqslant$ on a set A can be extended to a total ordering.*

Proof. The key to the proof is that by the well-ordering principle we can express $A \times A$ as the range of a transfinite sequence $A \times A = \{(x_\alpha, y_\alpha) : \alpha < \beta\}$. We then define recursively a family $(<_\alpha)_{\alpha \leqslant \beta}$ of partial orderings on A containing $<$ as follows. Start by letting $<_0$ be $<$. Once $<_\alpha$ has been defined, let $<_{\alpha^+}$ be if possible the smallest partial ordering containing both $<_\alpha$ and the ordered pair (x_α, y_α); if there is no such partial ordering, let $<_{\alpha^+} = <_\alpha$. Finally, if λ is a limit ordinal, let $<_\lambda = \bigcup_{\alpha<\lambda} <_\alpha$. The final element of this transfinite sequence of partial orderings contains $<$ and is obviously a maximal element of the set of all partial orderings on A, i.e. a total ordering on A. □

14.5 Maximal principles

It gradually became apparent that Zermelo's (1908*a*) elimination of ordinals from his proof of the well-ordering principle was a case of a much more general method for applying the axiom of choice without invoking ordinals: very many applications of the axiom of choice in mathematics use it to obtain maximal elements of particular partially ordered sets. In this section we shall isolate the properties which these uses of the axiom depend on. This material is principally of mathematical interest and is not used in the remainder of the book.

Definition. *A partially ordered set $(A, \leqslant)$ is said to be* inductively ordered *(and $\leqslant$ is said to be an* inductive ordering *on A) if every totally ordered subset of A has a supremum in A.*

Every inductively ordered set is non-empty and has a least element (since $\varnothing$ is totally ordered). For a set $\mathcal{A}$ to be inductively ordered by inclusion, it is sufficient (but not necessary) that $\bigcup \mathcal{B} \in \mathcal{A}$ for every chain $\mathcal{B} \subseteq \mathcal{A}$.

(14.5.1) **Lemma (Bourbaki 1949*b*).** *If $(A, \leqslant)$ is inductively ordered, then every function f from A to itself such that $f(x) \geqslant x$ for all $x \in A$ has a definite fixed point.*

Proof. Suppose that f is such a function and let α be the least ordinal such that $|\alpha| \nleqslant \text{card}(A)$ [Hartogs' theorem 11.4.2]. Then [simple principle of transfinite recursion] there exists a unique function g from $\boldsymbol{\alpha}$ to A such that

$$\begin{aligned} g(0) &= \bot; \\ g(\beta^+) &= f(g(\beta)) \text{ if } \beta^+ < \alpha; \\ g(\lambda) &= \sup_{\beta<\lambda} g(\beta) \text{ for every limit ordinal } \lambda < \alpha. \end{aligned}$$

Evidently g is normal [proposition 12.1.1], but it is not strictly normal (since if it were, it would in particular be one-to-one and we would have $|\alpha| \leqslant \text{card}(A)$). So there exists $\beta < \alpha$ such that $g(\beta^+) = g(\beta)$ [proposition 12.1.1]. If we choose the least such β (to be definite) and let $b = g(\beta)$, then $f(b) = f(g(\beta)) = g(\beta^+) = g(\beta) = b$. □

(14.5.2) **Theorem.** *If $(A, \leqslant)$ is a well-orderable inductively ordered set, then $(A, \leqslant)$ has a maximal element.*

Proof. Suppose on the contrary that A has no maximal element with respect to this ordering. Let $\preccurlyeq$ be some well-ordering on A. Now for each $x \in A$ there exist elements $y \in A$ such that $y > x$: to be definite, let $f(x)$ be the least such y with respect to the well-ordering $\preccurlyeq$. Then $f(x) > x$ for all $x \in A$, contradicting lemma 14.5.1. □

Note, incidentally, that in the proof of theorem 14.5.2 $f(x)$ is chosen to be at a minimum with respect to $\preccurlyeq$, not with respect to $\leqslant$; the well-ordering $\preccurlyeq$ is being exploited here only to give us a way of defining f without using the axiom of choice directly.

In practice almost all of the inductive orderings one comes across are cases of the inclusion relation. In particular, the following more restrictive condition occurs frequently (especially in algebra).

Definition. *A set $\mathcal{A}$ is said to have* finite character *if any set A belongs to $\mathcal{A}$ iff every finite subset of A belongs to $\mathcal{A}$.*

(14.5.3) **Proposition.** *If $\mathcal{A}$ is a non-empty set of finite character, then it is inductively ordered by inclusion.*

Proof. Let $\mathcal{B}$ be a chain in $\mathcal{A}$ and let $B = \bigcup \mathcal{B}$. Now certainly $\emptyset$ is contained in an element of $\mathcal{A}$ since $\mathcal{A}$ is non-empty. So consider a non-empty finite subset $\{b_0, \ldots, b_{n-1}\}$ of B. For $0 \leqslant r \leqslant n-1$ there exists $B_r \in \mathcal{B}$ such that $b_r \in B_r$. Now $\{B_0, B_1, \ldots, B_{n-1}\}$ is a finite non-empty chain and therefore has a greatest element B_{r_0} [theorem 6.4.5]. So $\{b_0, b_1, \ldots, b_{n-1}\} \subseteq B_{r_0}$, i.e. $\{b_0, b_1, \ldots, b_{n-1}\}$ belongs to $\mathcal{A}$. What we have now shown is that every finite subset of B belongs to $\mathcal{A}$. So B itself belongs to $\mathcal{A}$ and hence is the supremum of $\mathcal{B}$ in $\mathcal{A}$. □

(14.5.4) **Proposition.** *Every countable inductively ordered set has a maximal element.*

Proof. Immediate [proposition 14.4.1 and theorem 14.5.2]. □

(14.5.5) **Proposition.** *If $\mathcal{B}$ is a non-empty set of finite character such that $\bigcup \mathcal{B}$ is countable, and if $A \in \mathcal{B}$, then $\mathcal{B}$ has a maximal element with respect to inclusion containing A.*

Proof. If $\bigcup \mathcal{B} = \emptyset$, then $\mathcal{B} = \{\emptyset\}$ and the result is trivial. If not, then there exists a sequence (b_n) whose range is $\bigcup \mathcal{B}$. Now let $A_0 = A$. Once A_n has been defined, let A_{n+1} be the intersection of the elements of $\mathcal{B}$ containing $A_n \cup \{b_n\}$ if there are any; otherwise let $A_{n+1} = A_n$. In this way we recursively define an increasing sequence (A_n) in $\mathcal{B}$ whose range $\{A_n : n \in \boldsymbol{\omega}\}$ therefore has a supremum B since $\mathcal{B}$ is inductively ordered by inclusion [proposition 14.5.3]. B is evidently a maximal element of $\mathcal{B}$ containing A. □

(14.5.6) **Theorem.** *The following are equivalent:*

(i) *The axiom of choice;*

(ii) *Every inductively ordered set has a maximal element* (Zorn 1935);

(iii) *If $\mathcal{B}$ is a set of finite character and $A \in \mathcal{B}$, then $\mathcal{B}$ has a maximal element with respect to inclusion containing A* (Teichmüller 1939; Tukey 1940).

(i) ⇒ (ii). Theorem 14.5.2.

(ii) ⇒ (iii). Suppose that $\mathcal{B}$ is a set of finite character and $A \in \mathcal{B}$. Then $\mathcal{B}$ is inductively ordered by inclusion [proposition 14.5.3], hence so also is $\mathcal{A} = \{B \in \mathcal{B} : A \subseteq B\}$, which therefore by hypothesis has a maximal element.

(iii) ⇒ (i). Suppose that $\mathcal{A}$ is a set. Then it is easy to check that choice($\mathcal{A}$) is of finite character and therefore by hypothesis has a maximal element f with respect to inclusion, which must be a choice function for $\mathcal{A}$ [lemma 14.3.1]. □

The two maximal principles stated in this theorem provide us with a very powerful tool for deriving mathematical consequences of the axiom of choice without using ordinals. In order to be able to compare the two methods in use, let us consider again the claim that every partial ordering on a set can be extended to a total ordering: we proved this in the last section by using the well-ordering principle to label the members of the set with ordinals; let us now prove it again by our new method.

Proposition. *Assuming the axiom of choice, every partial ordering $\leqslant$ on a set A can be extended to a total ordering.*

Proof. Let $\mathcal{B}$ be the set of all the relations r such that $r^{\mathbf{t}}$ is a strict partial ordering on A. This is a set of finite character [corollary 6.2.5], and so by the Teichmüller/Tukey principle each element of it is contained in an element which is maximal with respect to inclusion, hence is a strict total ordering. □

The earliest statement of what is recognizably a maximal principle of this kind is in Hausdorff 1909, p. 301, where it is asserted that if the union of every well-ordered chain in $\mathcal{A}$ is an element of $\mathcal{A}$, then $\mathcal{A}$ has a maximal element with respect to inclusion. But it is one thing to prove a result and quite another to appreciate its utility. Hausdorff's widely read *Mengenlehre* (1914) does not even mention this result, and although it does contain the closely related result that every partially ordered set has a maximal totally ordered subset (see exercise 6 below), which is often referred to as 'Hausdorff's maximality principle', neither he nor anyone else seems to have appreciated its usefulness then. Whitehead and Russell came rather close to stating the maximal principle in their presentation of Zermelo's theorem in *Principia Mathematica* (1910–13, ∗258), but once again they failed to see its usefulness.

When Kuratowski rediscovered the maximal principle in 1922, on the other hand, it was quite explicitly as part of a reductionist programme stemming from Zermelo's (1908*b*) axiomatization of set theory. As we have already noted, the form this axiomatization took was strongly influenced by Zermelo's desire to provide a basis for his new proof of the well-ordering principle: crudely speaking, he chose the weakest natural-seeming axioms which would justify his proof. As a result the system was not strong enough to contain ordinal arithmetic without an additional postulate on the existence of transfinite numbers. But in the following years numerous important results were proved from the axiom of choice by transfinite induction. For example, Steinitz (1910) demonstrated by this means the existence and uniqueness up to isomorphism of the algebraic closure of an arbitrary field. Then Kuratowski turned Zermelo's method for eliminating ordinals into a general procedure and hence obtained the maximal principle: the result was a method by which proofs which use transfinite induction could be transformed into proofs which do not and which could therefore be formalized in Zermelo's system.

But just when Kuratowski was demonstrating how uses of ordinals could be uniformly eliminated from proofs, the adoption of the axioms for set theory strong enough to deliver von Neumann's theory of ordinals was making this manoeuvre formally unnecessary.[2] So although after Kuratowski the maximal principle continued to be used occasionally, it did not achieve widespread fame because the advantages he claimed for it were axiomatic and aesthetic rather than practical: mathematicians have always been loath to give up convenient tools for the sake of logical purity. The second rediscovery of the maximal principle by Zorn (1935) was decisive in ensuring its lasting popularity as a mathematical tool, partly no doubt because he gave convincing evidence of its usefulness rather than merely its elegance, but also because he had recently emigrated from Hamburg to New England: it was taken up by the active research communities there, among whom it became known as 'Zorn's lemma', the name by which it is universally known today (see Campbell 1978).

The final step in popularizing this maximal principle was taken by Bourbaki, who not only stated both the version for abstract partial order relations (due to Bochner 1928) and the Teichmüller/Tukey principle (in 1939), but — more importantly — went on to exploit these principles systematically in the subsequent parts of the treatise. Bourbaki's presentation is in this regard the fulfilment of Kuratowski's reductionist programme of eliminating ordinals: Bourbaki does not even trouble to define the notion of an ordinal in the text of his work but relegates it to an exercise, and he avoids using ordinals in his proofs. However, Bourbaki's reason for proceeding in this manner cannot have been foundational in quite the way that Kuratowski's had been, since Bourbaki's formal system is quite strong enough to define the ordinals if required; instead the reason was presumably aesthetic, stemming from a preference for what were seen as purely algebraic methods.

Exercises

1. Show that the set $\mathfrak{P}(A, B)$ of functions f such that $\text{dom}[f] \subseteq A$ and $\text{im}[f] \subseteq B$ is inductively ordered by inclusion.

2. If $(A, \leqslant)$ is inductively ordered and $a \in A$, show that $\{x \in A : x \leqslant a\}$ is also inductively ordered.

3. Is $\mathfrak{F}(\boldsymbol{\omega})$ inductively ordered by inclusion?

4. Let $\text{Well}(A)$ be the set of all relations on A which are well-orderings. If $r, r' \in \text{Well}(A)$, define $r \leqslant r'$ iff $r \subseteq r'$ and $\text{dom}[r]$ is an initial subset of the well-ordered set $(\text{dom}[r'], r')$. Show that $(\text{Well}(A), \leqslant)$ is an inductively ordered set. Hence deduce the well-ordering property directly from Zorn's lemma.

5. Show by an example that a function of the kind referred to in lemma 14.5.1 need not have a *least* fixed point.

[2]For the details see appendix A.

6. Show that the axiom of choice is equivalent to the assertion that every partially ordered set has a maximal totally ordered subset. [*Necessity.* Use the Teichmüller/Tukey property. *Sufficiency.* Prove Zorn's lemma.]

7. Assuming the axiom of choice, prove that every partially ordered set $(A, \leqslant)$ has a maximal totally unordered subset. [*First method.* Use the well-ordering principle. *Second method.* Show that the set of all totally unordered subsets of A is of finite character and then use the Teichmüller/Tukey property.]

8. Assuming the axiom of choice, prove that a relation is a partial ordering [resp. partial well-ordering] on the set A iff it is the intersection of a set of total orderings [resp. well-orderings] on A.

9. Assuming the axiom of choice, prove that every partially ordered set $(A, \leqslant)$ has a cofinal partially well-ordered subset. [Let $\mathcal{A}$ be the set of partially well-ordered subsets of A. Apply Zorn's lemma to $\mathcal{A}$ with the partial ordering 'is an initial subset of'.]

14.6 Regressive arguments

We have seen how the axiom of choice emerged as a new mathematical tool around the turn of the century, and how it soon became apparent that its consequences were not restricted to set theory but cropped up in many disparate areas of pure mathematics. However, we have not yet considered whether the axiom is true. Certainly many mathematicians have doubted it. For instance, Littlewood (1926, p. 25): 'Reflection makes the intuition of its truth doubtful, analysing it into prejudices derived from the finite case, and short of intuition there seems no evidence in its favour.' What is clear is that the sort of temporal motivation which we gave for the axiom of countable dependent choice is not available. A quasi-temporal argument can perhaps be given in favour of the axiom of well-ordered choice, first proposed by Hardy (1906), which asserts that the range of every transfinite sequence $(A_\alpha)_{\alpha<\beta}$ of non-empty sets has a choice function, but even here the temporal analogy seems rather far-fetched when β is uncountable. And in the general case, where we want to choose one element from each of an arbitrary family of non-empty sets, whatever is left of the temporal idea evaporates, as the sets are not presented in any particular order, temporal or otherwise.

The arguments that are given by mathematicians for believing the axiom of choice are often quite weak. One common argument generalizes from the finite case on the basis that there is no reason to suppose that infinite collections behave any differently. The difficulty with this is that we have been given nothing except a wing and a prayer to support the view that they do *not* behave differently. Another variant proceeds more cautiously by generalizing first from the finite to the countable case in the constructive manner already outlined, and then generalizing to the uncountable case by appeal to the idea that an ideal being could achieve the choices required of him (or perhaps Him). The main difficulty with this is that the most convincing argument for the extension from the finite to the countable case, namely that it depends merely on performing the supertask of making an infinite number of choices in a finite time, does not extend to the uncountable case. If we say, on the other hand, that the choices involved are merely logical choices and do not actually have to be *made*, hence do not occur in time at all, then it is very hard to see what is left in the metaphor of 'choice' that is doing any work. In short, this sort of argument for the axiom of choice appears to rest on a version of the limiting case platonism which we were so suspicious of in §3.2.

Another line that has often been taken is to justify the axiom of choice by appealing to qualities of the theory which results from assuming it. We have had cause to mention already a few of the consequences of the axiom of choice, but it has many more. A common pattern in many parts of mathematics is that a theorem provable for a restricted class of cases without the axiom becomes provable without restriction if we assume it. Here are a few examples:

(1) Every finite-dimensional vector space has a basis.
(AC) *Every* vector space has a basis.

(2) Every countable field has an algebraic closure.
(AC) *Every* field has an algebraic closure.

(3) Every separable Hilbert space contains a complete orthonormal sequence.
(AC) Every Hilbert space contains a complete orthonormal set.

(4) Every consistent set of sentences in a countable first-order language has a model.
(AC) Every consistent set of sentences in a first-order language of arbitrary cardinality has a model.

If the criterion on which we judge mathematical theories is their elegance, then, the axiom of choice may be counted a success. Many parts of pure mathematics attain a more elegant form if we assume the axiom of choice than if we do not. Another criterion on which the axiom of choice scores highly is fruitfulness. There are many problems in diverse parts of mathematics which can be solved only with its aid; and many authors have taken the fruitfulness of an axiom as an argument for its truth.[3]

For the genuine formalist, of course, there is little more to be said: even if the elegance and fruitfulness of the resulting theories are reason enough to accept the axiom of choice — which for the formalist means 'treat its consequences as worthy of attention' — this does not preclude the possibility that

[3]Curiously, though, one occasionally finds quite the opposite view expressed: 'The more problems a new axiom settles, the less reason we have for believing the axiom is true.' (Shoenfield 1977, p. 344)

other set-theoretic principles contradicting the axiom of choice might also be worthy of study. The difficulty with assessing how attractive these competing theories might be, however, is that very little work has been done on finding out their properties: a few mathematicians have studied the consequences of the axiom of determinacy, which contradicts the axiom of choice (see §15.7 below), but at the moment this is a rather isolated case. If the study of axioms contradicting the axiom of choice became more common, moreover, it is not evident that this would simply lead to competing versions of the whole of mathematics, one version assuming the axiom of choice and the other assuming the other principle. What seems more likely is that different sorts of mathematicians might settle on different set-theoretic principles as appropriate to their own disciplines.

It is still not clear, however, whether a split of this sort — different additional axioms for different parts of mathematics — could usefully be accommodated within a single theory of sets. Such an accommodation would be possible only if the objects used as proxies in embedding the various theories into the theory of sets were set-theoretically distinguishable in some principled manner, but this does not at the moment seem at all plausible. Although mathematicians do not seem to have articulated the point in quite this form, it may be one of the reasons why some of them eschew the idea that any single theory can act as a foundation for the whole of mathematics.

None of these considerations, however, is of much direct help to the *realist* in deciding whether the axiom of choice is true, unless there is some general reason to think that the truth is always pretty. On the contrary, there is some reason to suspect that, in mathematics at least, the truth, while not perhaps downright ugly, is at any rate not always optimally beautiful. So for the consequences of the axiom of choice to give the realist a regressive reason to believe that it is true, it is not enough that they should form an elegant theory: there needs to be some reason to believe that they are true, independent of the fact that they follow from it. In this respect, of course, the axiom is in just the same position as any other candidate for extending the default theory, such as the higher axioms of infinity of the last chapter.

14.7 The axiom of constructibility

There is one striking difference, though: in contrast to the axioms of infinity, the consequences of the axiom of choice which might bear on the question of its truth are never number-theoretic. To explain this point, we need to examine in more detail the method by which Gödel proved that the axiom of choice is consistent with set theory. At the opposite extreme to the maximal conception of the formation of power sets, mentioned earlier, is the *minimal* conception according to which the only sets created at each stage are those forced on us by the axioms of ZU. Fraenkel (1922*b*) suggested adding an axiom of restriction (*Axiom der Beschränktheit*) to achieve this, but did not succeed in formalizing the notion. The first satisfactory formulation of an axiom of this broad sort is due to Gödel (1938), who defined in the language of set theory a much more restrictive hierarchy consisting (roughly) of sets which can be defined by means of a formula which refers only to sets which have already been created.

In §3.5 we briefly canvassed the idea of a hierarchy formed by a wholly predicative process. If L_α is a level in such a hierarchy, the following level $L_{\alpha+1}$ will consist only of sets of the form $\{x \in L_\alpha : \Phi^{(L_\alpha)}\}$, i.e. sets definable by means of formulae whose quantifiers are restricted to range only over L_α. Gödel showed that the definition of this notion of constructibility, which *prima facie* quantifies metalinguistically over Φ, can in fact be formalized within the theory of sets, and hence that the constructible hierarchy **L** consisting of the subsets of the constructible levels L_α is a well-defined subclass of the universe of sets **V**.[4] The assertion that *every* set is constructible, i.e. that $\mathbf{V} = \mathbf{L}$, can then, somewhat surprisingly, be expressed as a single sentence in the language of set theory. We shall call it the *axiom of constructibility*.

Axiom of constructibility. *Every set is constructible.*

Gödel showed that not only the axiom of constructibility but all the axioms of ZF hold when all the quantifiers in them are restricted to **L**. It follows at once, of course, that if ZF is consistent, then so is ZF together with the axiom of constructibility. This is significant because of the following result.

Theorem (Gödel 1938). *The axiom of constructibility entails the axiom of choice.*

If we combine this with the relative consistency result just mentioned, we reach the conclusion that if ZF is consistent, then it remains so when we add the axiom of choice.

Now we noted earlier that predicative set theory on its own is rather weak: if we replace the axiom scheme of separation with its predicative weakening, we cannot, for instance, prove the existence of any uncountable sets. So Gödel's demonstration that all the axioms of set theory, including impredicative separation, hold in the constructible hierarchy is at first sight surprising. The point to realize, however, is that what gives Gödel's constructible hierarchy its strength is its appeal to a *prior* theory of ordinals. What Gödel discovered was that a predicative process of the formation of levels can generate impredicative sets if the number of iterations of the process is given by an impredicatively specified ordinal: each level L_α in the constructible hierarchy contains only sets specifiable predicatively in terms of the lower levels, but this hierarchy is

[4]Here once again we use for convenience the language of classes. See appendix C.

parasitic on the full, impredicative hierarchy in which it is embedded, since it needs this to supply the ordinal α.

For Gödel's proof to work, though, it is necessary to assume a moderately strong axiom of infinity. This bears on the discussion in the last chapter of the technical advantages of such axioms. We noted there that set theorists themselves have a reason to assume the axiom of ordinals because it ensures that every well-founded set-theoretic model of set theory itself is isomorphic to a standard model, i.e. a model in which '$\in$' is interpreted as membership (Mostowski's collapsing lemma). The axiom scheme of reflection supports the study of models even more strongly because it gives the hierarchy room for other operations to close out with fixed points. One example is provided by Gödel's proof of the consistency of the axiom of choice just described: Gödel discovered the proof in 1935, but one reason for the delay in publication was that he spent a long time trying to make it work in such a way that it would apply to a theory like Z which does not assume anything as strong as reflection.

When we discussed the submodel form of the Löwenheim/Skolem theorem, we touched briefly on one consequence of reflection that is relevant here: if ZF is consistent, there is a countable transitive class which is a standard model of it; and reflection shows that this class is a *set*. In fact, Gödel's constructible hierarchy allows us to describe one particular countable model much more precisely. This is because whether a set is an inner model of ZF is only a matter of whether it satisfies certain closure conditions, so that the intersection of all the transitive standard models of ZF is itself a model, called the *minimal model*. Now this minimal model can be shown to be equal to L_{ξ_0} for some countable ordinal ξ_0. This provides us with a graphic illustration of the relativity of cardinality that we discussed earlier: the von Neumann ordinals which belong to L_{ξ_0} are precisely those $< \xi_0$, and yet, since L_{ξ_0} is a model of ZF, many of these von Neumann ordinals are, relative to L_{ξ_0}, uncountable. What makes this possible, of course, is just that the power-set operation has been interpreted in the constructible hierarchy as thinly as possible within the constraints of the first-order theory ZF. We might be tempted to think of L_{ξ_0} as realizing a sort of contrary of the principle of plenitude — a principle of paucity, if you will — in respect both of the thinness of each level and of the total number of levels.

We have called the assumption that every set is constructible an 'axiom', but is there any reason to think it is true? At first sight the principle of ontological parsimony which encourages some authors to eliminate individuals and un-well-founded classes makes it an attractive assumption, since it asserts that every set occurs in a highly restrictive hierarchy in which only those sets essential to the theory are created. Gödel himself initially flirted with the thought that it gives 'a natural completion of the axioms of set theory, in so far as it determines the vague notion of an arbitrary infinite set in a definite way' (1938, p. 557). However, the picture of the set-theoretic universe which this forces on us has seemed very implausible to many subsequent writers (including Gödel himself in later life): it is difficult to find a reason for believing that the predicative creation process represented by the constructible hierarchy and the impredicative process represented by the traditional hierarchy should both result in the same sets being created, as the axiom of constructibility would have us believe.

So the axiom of constructibility has little direct support. Does it then have any regressive support? It certainly gives a neat theory which settles not just the axiom of choice and the continuum hypothesis (see below) but various otherwise problematic questions in the theory of sets of real numbers. However, this has been thought by most set theorists not to give regressive support to the axiom, because it is felt to settle these questions in the 'wrong' way. It would take us too far afield to examine here the intuitions they appeal to in reaching this conclusion: I shall note only that it is not shared universally. Friedman (2000, p. 437), for instance, regards the intuition that the axiom of constructibility is false as dubious: 'I don't have it, and mathematicians in general disclaim it.' And Jensen (1995, p. 398) has even said, 'I personally find [constructibility] a very attractive axiom.'

In any case, the axiom of constructibility is of no help to us in deciding whether we should believe the axiom of choice if we do not already believe the axiom of purity, for if the set of individuals is formless — and hence not well-orderable — no amount of care in limiting the construction of the hierarchy can change that. But even if constructibility is not a plausible hypothesis, the method of proving relative consistency results by forming inner models of the theory such as **L** generates useful information about the strength of the axioms that hold in it. The most striking result of this sort is that the axiom of constructibility (and hence the axiom of choice) makes no difference whatever in the sphere of first-order arithmetic (Ax and Kochen 1965). For suppose we have a proof of a first-order arithmetical sentence which uses the axiom of choice. This proof is not correct as it stands if the axiom of choice is false. But the key fact to note is that if we relativize all the quantifiers in our definition of the set of natural numbers to **L**, the set that is picked out does not change. (In the set theorists' jargon, $\boldsymbol{\omega}$ is *absolute* for **L**.) This is significant because the axiom of choice is certainly true in **L**, whether or not it is true in the whole set-theoretic universe **V**. So if we now relativize all the quantifiers in our proof to **L**, what we obtain is a correct proof, not assuming the axiom of choice, of a conclusion concerning the set of natural numbers of **L**. But because $\boldsymbol{\omega}$ is absolute for **L**, the conclusion of the relativized proof is the *same* as the conclusion of the original one. Thus, to repeat, any first-order number-theoretic sentence provable with the axiom of choice is provable without it.

Similar remarks apply even more directly to elementary geometry, because it can be given a complete first-order axiomatization (Tarski 1959) and so *any* consistent extension of it will be trivially conservative. Therefore adding the axiom of constructibility (or *a fortiori* the axiom of choice) to set theory will make no difference to what can be proved in elementary geometry either. Indeed Putnam (1980) has extended the idea even further by observing that for any given countable set S of real numbers there is a model of set theory which satisfies the axiom of constructibility and contains the given set S as well as a standard copy of the natural numbers. By applying this result to the case in which the set S contains all the 'operational constraints' — correct assignments of values to all magnitudes which sentient beings in this physical universe can actually measure — Putnam draws the conclusion that the truth-value of the axiom of constructibility (and *a fortiori* of the axiom of choice) *cannot* be determined by these operational constraints.

These facts are important partly because they indicate that whether the axiom of choice is true is a question to which many mathematicians, not just number theorists, may safely remain indifferent. But they are important, too, because they entail that the only propositions which the regressive theorist could use to test the axiom of choice belong to the parts of mathematics whose application to the world might be thought to be already theory-laden. We cannot, in other words, expect to find results which could provide simple empirical tests of the axiom, such as 'If the axiom of choice is true, the Forth Bridge will not fall down.' The consequences of the axiom of choice which the regressive theorist has to work on belong to relatively abstract branches of mathematics where our intuitions are already stretched taut.

14.8 Intuitive arguments

So if regressive arguments for the truth of the axiom are likely to remain inconclusive, we must fall back on intuitions bearing directly on the axiom itself. An argument one finds quite frequently is that the axiom of choice can be derived from the first principle of plenitude which we stated in §3.5. The idea in outline is that if a level in the hierarchy really does contain *all* possible subsets of the previous levels, it will in particular contain all the choice sets. 'For the fat (or "full") hierarchy, the axiom of choice is quite evident.' (Kreisel 1980, p. 192) This sort of argument goes back to Ramsey (1926): he advanced a conception of sets as wholly extensional entities not dependent for their existence on there being any means of specifying their members, and claimed that on this conception the axiom of choice is 'an obvious tautology'.

The argument can be spelt out as follows. Consider the following second-order logical principle:

$$(\forall x)(\exists y)\Phi(x, y) \Rightarrow (\exists F)(\forall x)\Phi(x, F(x)). \qquad (1)$$

Hintikka (1998, pp. 39–48) and others have argued that it is hard to see how we could deny the truth of this principle for all formulae Φ except by reading the first-order existential quantifier classically and the second-order one constructively; when both are read classically, the principle is quite unobjectionable.

Let us suppose for the sake of argument, then, that (1) *is* a logical truth. In that case we can deduce the axiom of choice from the second-order separation principle. For suppose that $\mathcal{A}$ is a set of disjoint non-empty sets. Then

$$(\forall A \in \mathcal{A})(\exists x)(x \in A),$$

from which it follows — in classical, but not in intuitionistic, logic (see Tait 1994) — that

$$(\forall A)(\exists x)(A \in \mathcal{A} \Rightarrow x \in A).$$

So by (1) there is a (logical) function F such that

$$(\forall A \in \mathcal{A})(F(A) \in A).$$

The set $C = \{x \in \bigcup \mathcal{A} : (\exists A \in \mathcal{A})(x = F(A))\}$ exists by the second-order separation principle. And because the members of $\mathcal{A}$ are disjoint, for each $A \in \mathcal{A}$ the set $C \cap A$ has just one element $F(A)$.

Now it is important to see that this does nothing to threaten the independence result quoted earlier to the effect that the axiom of choice is not provable in ZU. Since the formalization of first-order logic is complete, that result holds good whatever second-order logical principles we manage to persuade ourselves of. What the argument just given does, rather, is to draw attention to the fact that the axiomatizations of set theory from which the axiom of choice has been shown to be independent are all first-order. Specifically, the axiom of choice is not derivable from any instance of the *first-order* separation scheme: the sign 'F' in the argument for the axiom of choice stands for a *logical* function, i.e. a second-order entity, not a set of ordered pairs. So the explanation is simply that the instance of separation used to obtain the set C is not expressible in the first-order language.

The effect of this is to narrow down the options for anyone wishing to reject the axiom of choice. Short of denying the second-order principle (1), one is forced to adopt some argument for first-order separation other than the one we gave in §3.5 which justified it as an approximation to the second-order axiom. This might well not bother the constructivists, who are unlikely to have been impressed by that argument in the first place, but their arguments fall well short of justifying the impredicative separation scheme, and so they can be expected to have stopped reading long ago. The question of interest here, therefore, is not whether the constructivists should believe the axiom of choice but whether there is a moderate platonist argument that grounds all the instances of the separation scheme expressible in the first-order language (even the impredicative ones) but does not extend to the instances of separation involved in the axiom of choice. If there is not, we will have reached a substantial conclusion, because we will have shown that ZU is conceptually unstable: the argument we used to justify will also be an argument for the stronger system ZCU.

Before we leap to that conclusion, however, it is worth noting that there is at any rate a significant logical difference between the axiom of choice and the axioms of ZU. All of the latter can easily be transformed into a form in which the existential claims they make are claims of *unique* existence. We could, that is to say, have stated the axioms of ZU as follows.

Axiom of infinity. *There exists a* unique *earliest limit level.*

Axiom of creation. *For each level there exists a* unique *next level.*

Axiom scheme of separation. *For every level V there exists a* unique *collection a such that $a = \{x \in V : \Phi\}$.*

Because of the well-foundedness of the hierarchy the same also applies to ZFU.

Axiom scheme of reflection. *For all $x_1, \ldots, x_n$ there is a* unique *earliest level V such that $\Phi \Rightarrow \Phi^{(V)}$.*

The axiom of choice, on the other hand, has no such equivalent.

One reason why this difference is worth taking seriously is that it is quite stable under minor perturbations of the background logic. It is no doubt because of this that even mathematicians who believe that the axiom of choice is true nevertheless regard proofs which do not use it as providing more information than those that do. However, it is a large step from there to saying that sets which are not fully specified in this manner not only encode less information but do not even exist. For that we would require a further argument. And it presumably could not simply be a general argument against non-unique existence claims, since the second-order principle (1), which we have agreed to accept as uncontroversial for the moment, evidently makes just such a claim. It seems that it would instead have to be a quite specific argument limited in its applicability to collections, or if not to collections alone then at any rate to the objects of mathematics.

Perhaps at this point the discussion might return to the perspective adopted at the end of §3.3. I put forward there what I called an internal platonist argument for the well-foundedness of the hierarchy of sets that avoided outright constructivism while accepting as a premise that mathematics is part of our attempt to represent the world. It is hard to see how the conception of mathematics thus invited could support the axiom of choice. One rather vague way of putting the point would be to say that the argument for the axiom of choice depends crucially on the coherence of the notion of a wholly arbitrary subset, but that even if this notion is coherent, it cannot participate in our attempt to represent the world, and hence is not part of the mathematics we use to help us do so.

Perhaps what this debate about whether to accept the axiom of choice indicates is that the disjunction between regularity and randomness is as fundamental to our conception of the world as that between discreteness and continuity. Even the uncritical platonist ought not to deny the distinction, but should claim only that in order for us to comprehend the world we represent, we *must* see it as a limited whole — see it, that is to say, as part of a more inclusive whole containing things that we do not and cannot represent directly, such as sets we cannot explicitly define. The internal platonist, on the other hand, maintains that the attempt to sit astride this divide is illusory.

Notes

Cantor made frequent use of the axiom of choice in his work on cardinal arithmetic. Indeed there is no evidence to suggest that Cantor ever doubted the validity of the axiom for a moment: it was a principle which, in Zermelo's words, he 'unconsciously and instinctively used everywhere and expressly stated nowhere' (Cantor 1932, p. 451). We noted in §9.4 how implicit uses of the axiom of countable choice became common in the last quarter of the 19th century. The unrestricted axiom of choice, by contrast, was hardly used by anyone other than Cantor until after the appearance of his *Beiträge* (1895; 1897). Felix Bernstein, a pupil of Cantor working in Germany, used a consequence of the axiom of choice called the partition principle in his 1901 doctoral thesis on cardinal arithmetic (published in 1905*b*) and was immediately criticized for doing so by Levi (1902); the axiom of choice was also used in Italy by Burali-Forti (1896), even though he elsewhere expressed antagonism to the axiom of countable choice. The axiom of choice was also used implicitly by the Cambridge mathematicians Whitehead (1902) and Hardy (1904). Russell came rather close to an explicit statement of the axiom in the work he contributed to Whitehead's 1902 paper when he postulated that every non-finite set is a disjoint union of countably infinite sets (which is equivalent to the axiom of choice). But it was only later that Russell came to see that Whitehead had implicitly assumed the axiom of choice in his proof (in the same paper) that any family of cardinal numbers has a product; it was Russell's (1906*b*) attempt to prove this assumption that led him to formulate explicitly what he called the 'multiplicative axiom'. Meanwhile in Germany Zermelo (1904) had also stated the axiom of choice explicitly, deciding that he needed it if he was to prove that every set is well-orderable.

At a purely descriptive level the best source for more on the history of the emergence of the axiom of choice is G. H. Moore (1982). The idea that a plenitudinous conception of the hierarchy makes it a triviality is essentially due to Ramsey (1926), although the manner in which I have developed that idea here owes more to later writers.

The role which the axiom of choice plays in mathematics is now rather well understood. We have done no more here than touch on the large number of statements in diverse parts of mathematics that are equivalent to it. This information is exhaustively catalogued by Rubin and Rubin (1985). Many branches of abstract mathematics are very much streamlined by the assumption of the axiom of choice. A good example is general topology, which becomes decidedly disconcerting in its absence (see Good and Tree 1995).

The axiom of constructibility and the large topic of inner models, of which the constructible hierarchy is only the most famous example, are discussed in many textbooks, e.g. Devlin 1984 and Kunen 1980. The reasons for the tendency of mathematicians to reject this axiom are discussed by Maddy (1993); a dissenting voice is Jensen (1995). The technique of forcing by which Cohen (1963) proved the independence of the axiom of choice from ZF has been much refined subsequently. Kunen 1980 is once again a good introduction.

Chapter 15

Further cardinal arithmetic

The axiom of choice leads to a considerable simplification of the arithmetic of cardinals, but even so it leaves some questions in this domain unsettled. Our aim in this chapter is to focus on these issues.

15.1 Alephs

Let us (by a minor abuse of language) say that a cardinal $\mathfrak{a} = \text{card}(A)$ is *well-orderable* if A is well-orderable; this definition is independent of the choice of representative set A because whether a set is well-orderable depends only on its cardinality. The well-orderable cardinals are thus precisely those of the form $|\alpha|$ for some ordinal α.

(15.1.1) **Proposition.** *Every set of well-orderable cardinals has a least element.*

Proof. This follows at once from the corresponding fact for ordinals since the function given by $\alpha \mapsto |\alpha|$ is increasing. □

In particular, any two well-orderable cardinals are comparable. Therefore every well-orderable cardinal is either finite or infinite (and of course every finite cardinal is well-orderable).

Definition. *An infinite well-orderable cardinal is called an* aleph.

(15.1.2) **Proposition.** *The alephs do not form a set.*

Proof. By Hartogs' theorem 11.4.2 there is no cardinal which is an upper bound for the alephs: it follows that they do not form a set [proposition 9.2.5]. □

The smallest alephs are $\aleph_0$ and $\aleph_1$. We now generalize this and write $\aleph_\alpha$ to denote the αth aleph. $\aleph$ is the Hebrew letter 'aleph', which explains the terminology introduced a moment ago. The least element of the set $\{\beta : |\beta| = \aleph_\alpha\}$ is denoted ω_α: again this conforms with our previous usage since the least infinite ordinal is ω_0 and the least uncountable ordinal is ω_1. By analogy with the terminology for cardinals, one might call ordinals of the form ω_α *omegas*, but in practice no one does.

If we assumed the axiom of ordinals, we could prove that $\aleph_\alpha$ exists for every ordinal α. In ZU, however, the only alephs whose existence we can be sure of are the $\aleph_n$ for all $n \in \omega$.

Definition. *If* $\mathfrak{a}$ *is a cardinal, then we let* $\mathfrak{a}^+$ *denote the least well-orderable cardinal* $\mathfrak{b}$ *such that* $\mathfrak{b} \nleqslant \mathfrak{a}$.

The fact that $\mathfrak{a}^+$ exists follows at once from Hartogs' theorem. If $\mathfrak{a}$ is finite, then of course $\mathfrak{a}^+ = \mathfrak{a} + 1$. If $\mathfrak{a}$ is not finite, then $\mathfrak{a}^+$ is an aleph; in particular, $\aleph_\alpha^+ = \aleph_{\alpha+1}$.

15.2 The arithmetic of alephs

The arithmetic of alephs is much simpler than the arithmetic of other infinite cardinals: it turns out, in fact, that addition and multiplication collapse into triviality, leaving only exponentiation as a way of obtaining different cardinals. The clue to this came when we saw earlier that $2\aleph_0 = \aleph_0$ and $\aleph_0^2 = \aleph_0$. What we shall show now is that both these results generalize to all the alephs, leading at once to the aforementioned triviality of addition and multiplication.

(15.2.1) **Proposition.** *If* $\mathfrak{a}$ *is an aleph, then* $2\mathfrak{a} = \mathfrak{a}$.

Proof. We shall prove by transfinite induction that $2|\alpha| = |\alpha|$ for every infinite ordinal α. This is certainly true for $\alpha = \omega$ since $2\aleph_0 = \aleph_0$. If it is true for α, then

$$\begin{aligned} |2(\alpha + 1)| &= |2\alpha + 2| = 2|\alpha| + 2 \\ &= |\alpha| + 2 = |\alpha| + 1 = |\alpha + 1|, \end{aligned}$$

and so it is true for $\alpha + 1$. Finally, if λ is a limit ordinal and the hypothesis is true for $\omega \leqslant \alpha < \lambda$, then $\lambda = \omega\beta$ for some $\beta < \lambda$ [corollary 12.3.6], so that

$$|2\lambda| = |2\omega\beta| = 2\aleph_0|\beta| = \aleph_0|\beta| = |\omega\beta| = |\lambda|,$$

and hence it is true for λ. This completes the proof. □

(15.2.2) **Corollary.** *If* $\mathfrak{a}$, $\mathfrak{b}$ *are alephs, then* $\mathfrak{a} + \mathfrak{b} = \max(\mathfrak{a}, \mathfrak{b})$.

Proof. Either $\mathfrak{a} \leqslant \mathfrak{b}$ or $\mathfrak{b} \leqslant \mathfrak{a}$ [proposition 15.1.1]; suppose for the sake of argument that $\mathfrak{a} \leqslant \mathfrak{b}$. Then $\mathfrak{b} \leqslant \mathfrak{a} + \mathfrak{b} \leqslant \mathfrak{b} + \mathfrak{b} = 2\mathfrak{b} = \mathfrak{b}$ [proposition 15.2.1], and so $\mathfrak{a} + \mathfrak{b} = \mathfrak{b} = \max(\mathfrak{a}, \mathfrak{b})$. □

(15.2.3) **Proposition.** *If* $\mathfrak{a}$ *is an aleph, then* $\mathfrak{a}^2 = \mathfrak{a}$.

Proof. It will be sufficient to prove that $\text{card}(\boldsymbol{\alpha} \times \boldsymbol{\alpha}) = |\alpha|$ for every infinite ordinal α. Suppose for a contradiction that this is false and that α is the least ordinal for which it fails. Note that if σ is an ordinal, then

$$\sigma < \alpha \Leftrightarrow |\sigma| < |\alpha|. \tag{1}$$

Note also that $|\alpha| > \aleph_0$ [proposition 10.3.2].

Define an ordering on $\boldsymbol{\alpha} \times \boldsymbol{\alpha}$ by writing $(\beta, \gamma) \leqslant (\delta, \epsilon)$ iff **either** $\beta+\gamma < \delta+\epsilon$ **or** $\beta + \gamma = \delta + \epsilon$ and $\beta < \delta$. It is easy to check that this is a well-ordering on $\boldsymbol{\alpha} \times \boldsymbol{\alpha}$.

Now for each ordinal $\sigma < \alpha$ let $A(\sigma) = \{(\beta, \gamma) : \beta + \gamma < \sigma\}$. It is clear that $A(\sigma) \subseteq \boldsymbol{\sigma} \times \boldsymbol{\sigma}$ and that $A(\sigma)$ is an initial subset of $\boldsymbol{\alpha} \times \boldsymbol{\alpha}$. So

$$\begin{aligned} \text{card}(A(\sigma)) &\leqslant \text{card}(\sigma \times \sigma) \\ &= |\sigma| \\ &< |\alpha| \text{ by (1)}, \end{aligned}$$

and therefore $\text{ord}(A(\sigma)) < \alpha$.

If $(\beta, \gamma) \in \boldsymbol{\alpha} \times \boldsymbol{\alpha}$, then $|\beta|, |\gamma| < |\alpha|$ by (1), so that

$$\begin{aligned} |\beta + \gamma| &= |\beta| + |\gamma| \text{ [proposition 12.2.2]} \\ &< |\alpha| \text{ [corollary 15.2.2]}. \end{aligned}$$

Therefore $\beta + \gamma < \alpha$, so that $(\beta, \gamma) \in A(\sigma)$ for some $\sigma < \alpha$. In other words $\boldsymbol{\alpha} \times \boldsymbol{\alpha} = \bigcup_{\sigma<\alpha} A(\sigma)$. So

$$\begin{aligned} \text{ord}(\boldsymbol{\alpha} \times \boldsymbol{\alpha}, \leqslant) &= \sup_{\sigma<\alpha} \text{ord}(A(\sigma), \leqslant) \text{ [proposition 11.2.3]} \\ &= \alpha. \end{aligned}$$

Hence $\text{card}(\boldsymbol{\alpha} \times \boldsymbol{\alpha}) = |\alpha|$, which is what we wanted. □

(15.2.4) **Corollary.** *If* $\mathfrak{a}$ *and* $\mathfrak{b}$ *are alephs, then* $\mathfrak{a}\mathfrak{b} = \max(\mathfrak{a}, \mathfrak{b})$.

Proof. Suppose for the sake of argument that $\mathfrak{a} \leqslant \mathfrak{b}$. Then

$$\mathfrak{b} \leqslant \mathfrak{a}\mathfrak{b} \leqslant \mathfrak{b}\mathfrak{b} = \mathfrak{b}^2 = \mathfrak{b} \text{ [proposition 15.2.3]},$$

so that $\mathfrak{a}\mathfrak{b} = \mathfrak{b} = \max(\mathfrak{a}, \mathfrak{b})$. □

15.3 Counting well-orderable sets

(15.3.1) **Theorem.** *If A is an infinite well-orderable set, then*

$$\text{card}(\mathfrak{F}(A)) = \text{card}(A).$$

Proof. Suppose not. So there exists an infinite ordinal α such that $\mathrm{card}(\mathfrak{F}(\boldsymbol{\alpha})) \neq |\alpha|$: choose α as small as possible. Note that α is the least element of $\{\beta : |\beta| = |\alpha|\}$ and hence

$$\beta < \alpha \Leftrightarrow |\beta| < |\alpha|. \tag{2}$$

Now if $X \in \mathfrak{F}(\boldsymbol{\alpha})$, then we can let $X = \{\gamma_0, \gamma_1, \ldots, \gamma_{n-1}\}$ with $\alpha > \gamma_0 > \gamma_1 > \cdots > \gamma_{n-1}$ and define

$$f(X) = 2^{(\gamma_0)} + 2^{(\gamma_1)} + \cdots + 2^{(\gamma_{n-1})}$$

(unless $X = \emptyset$, in which case let $f(X) = 0$). Now if $0 \leqslant r \leqslant n-1$, then either γ_r is finite, in which case

$$|2^{(\gamma_r)}| < |\omega| \leqslant |\alpha|,$$

or γ_r is infinite, in which case

$$\begin{aligned} |2^{(\gamma_r)}| &= \mathrm{card}(\mathfrak{F}(\boldsymbol{\gamma}_r)) \\ &= |\gamma_r| \text{ by the induction hypothesis} \\ &< |\alpha| \text{ since } \gamma_r < \alpha. \end{aligned}$$

So

$$\begin{aligned} |f(X)| &= |2^{(\gamma_0)}| + |2^{(\gamma_1)}| + \cdots + |2^{(\gamma_n)}| \text{ [proposition 12.2.2]} \\ &< |\alpha| \text{ [corollary 15.2.2]}, \end{aligned}$$

and therefore $f(X) < \alpha$ by (2). In other words, f is a function from $\mathfrak{F}(\boldsymbol{\alpha})$ to $\boldsymbol{\alpha}$. Since this function is a one-to-one correspondence [theorem 12.5.1], it follows that $\mathrm{card}(\mathfrak{F}(\boldsymbol{\alpha})) = |\alpha|$. Contradiction. □

(15.3.2) **Proposition.** *If A is an infinite well-orderable set, then*

$$\mathrm{card}(\mathrm{String}(A)) = \mathrm{card}(A).$$

Proof. Each element of String(A) is a function from $\boldsymbol{n}$ to A, hence a finite subset of $\boldsymbol{\omega} \times A$. So $\mathrm{String}(A) \subseteq \mathfrak{F}(\boldsymbol{\omega} \times A)$. Now $\boldsymbol{\omega}$ and A are both well-orderable, hence so is $\boldsymbol{\omega} \times A$ [lemma 12.3.1]. Therefore

$$\begin{aligned} \mathrm{card}(\mathrm{String}(A)) &\leqslant \mathrm{card}(\mathfrak{F}(\boldsymbol{\omega} \times A)) \\ &= \mathrm{card}(\boldsymbol{\omega} \times A) \text{ [theorem 15.3.1]} \\ &= \aleph_0 \,\mathrm{card}(A) \\ &= \mathrm{card}(A) \text{ [corollary 15.2.4]}. \end{aligned}$$

The result follows, since the opposite inequality

$$\mathrm{card}(A) \leqslant \mathrm{card}(\mathrm{String}(A))$$

is obvious. □

(15.3.3) **Proposition.** *If* $(A, \leqslant)$ *and* $(B, \leqslant)$ *are infinite well-ordered sets, then*

$$\operatorname{card}({}^{(A)}B) = \max(\operatorname{card}(A), \operatorname{card}(B)).$$ [1]

Proof. If $f \in {}^{(A)}B$, let $\{x_0, \dots, x_{n-1}\}$ be $\{x \in A : f(x) \neq \bot\}$ arranged in order, and let

$$g(f) = (\{x_0, \dots, x_{n-1}\}, (f(x_r))_{r \in \mathbf{n}}).$$

The function g from ${}^{(A)}B$ to $\mathfrak{F}(A) \times \operatorname{String}(B)$ thus defined is evidently one-to-one. Hence

$$\begin{aligned}
\operatorname{card}({}^{(A)}B) &\leqslant \operatorname{card}(\mathfrak{F}(A) \times \operatorname{String}(B)) \\
&= \operatorname{card}(\mathfrak{F}(A)) \operatorname{card}(\operatorname{String}(B)) \\
&= \operatorname{card}(A) \operatorname{card}(B) \text{ [theorem 15.3.1 and proposition 15.3.2]} \\
&= \max(\operatorname{card}(A), \operatorname{card}(B)) \text{ [corollary 15.2.4]}.
\end{aligned}$$

The converse inequality is obvious, whence the result. □

Exercises

1. If $\mathfrak{b}$ is an aleph, show that $\mathfrak{a} = 2^{\mathfrak{b}}$ iff $\mathfrak{a} \geqslant \mathfrak{b}$ and $\mathfrak{a} + \mathfrak{b} = 2^{\mathfrak{b}}$.

2. If $\mathfrak{a} = \operatorname{card}(A)$ and $\mathfrak{b} = \operatorname{card}(B)$, let us write $\mathfrak{a} \leqslant^* \mathfrak{b}$ if either $A = \varnothing$ or there exists a function from B onto A. Establish the following results.

   (a) $\mathfrak{a} \leqslant \mathfrak{b} \Rightarrow \mathfrak{a} \leqslant^* \mathfrak{b}$.

   (b) The converse holds if $\mathfrak{b}$ is well-orderable.

   (c) $\mathfrak{a} \leqslant^* \mathfrak{b} \Rightarrow 2^{\mathfrak{a}} \leqslant 2^{\mathfrak{b}}$.

   (d) $\aleph_{\alpha+1} \leqslant^* 2^{\aleph_\alpha}$.

   (e) $\aleph_{\alpha+1} < 2^{2^{\aleph_\alpha}}$.

3. (a) Given a well-orderable cardinal $\mathfrak{b} \neq 0$, find an infinite $\mathfrak{a}$ such that $\mathfrak{a}^{\mathfrak{b}} = \mathfrak{a}$.

   (b) Can we choose $\mathfrak{a} \leqslant \mathfrak{b}$?

4. If A is an infinite well-orderable set of cardinal $\mathfrak{a}$, show that each of the following sets has cardinal $2^{\mathfrak{a}}$:

   (a) the set of infinite subsets of A;

   (b) the set of subsets of A equinumerous with A;

   (c) the set of equivalence relations on A;

   (d) the set of well-ordering relations on A.

[1] This is the result whose proof we have owed since we used it in establishing proposition 12.4.2.

15.4 Cardinal arithmetic and the axiom of choice

(15.4.1) **Proposition.** *These three assertions are equivalent:*

(i) *The axiom of choice.*

(ii) *Every cardinal that is not finite is an aleph.*

(iii) *Every infinite cardinal is an aleph.*

(i)⇒(ii). Assume the axiom of choice and let $\mathfrak{a}$ be any non-finite cardinal. Then $\mathfrak{a}$ is infinite. Moreover, every set, and therefore every cardinal, is well-orderable. So $\mathfrak{a}$ is an aleph.

(ii)⇒(iii). Trivial.

(iii)⇒(i). Let $\mathfrak{a}$ be any cardinal. If it is finite, it is trivially well-orderable, so suppose that it is not finite. Then $\mathfrak{a} + \mathfrak{a}^+ \geqslant \mathfrak{a}^+ \geqslant \aleph_0$, so $\mathfrak{a} + \mathfrak{a}^+$ is infinite and hence by hypothesis an aleph. But $\mathfrak{a} \leqslant \mathfrak{a}+\mathfrak{a}^+$, and so $\mathfrak{a}$ is well-orderable. Thus every cardinal is well-orderable, and the axiom of choice follows [theorem 14.4.3]. □

So if the axiom of choice is true, the simplifying results we proved in §15.2 about the arithmetic of alephs apply to all infinite cardinals: addition and multiplication of infinite cardinals become completely trivial, exponentiation is the only arithmetical operation that generates anything new, and the partial ordering of cardinals becomes a total ordering. In fact we can go further: these simplifications of cardinal arithmetic are equivalent to the axiom of choice.

(15.4.2) **Proposition (Hartogs 1915).** *The axiom of choice is equivalent to the assertion that any two cardinals are comparable.*

Necessity. If the axiom of choice holds, then every set is well-orderable, so every cardinal that is not finite is an aleph, and it follows from proposition 15.1.1 that any two cardinals, and hence in particular any two infinite cardinals, are comparable.

Sufficiency. If $\mathfrak{a}$ is an infinite cardinal, then $\mathfrak{a}^+ \nleqslant \mathfrak{a}$, so $\mathfrak{a} < \mathfrak{a}^+$ by hypothesis: as $\mathfrak{a}^+$ is an aleph, it follows that $\mathfrak{a}$ is one too. The axiom of choice follows [proposition 15.4.1]. □

(15.4.3) **Lemma.** *If* $\mathfrak{a} + \mathfrak{a}^+ = \mathfrak{a}\mathfrak{a}^+$, *then* $\mathfrak{a}$ *is an aleph.*

Proof. Let A and B be sets such that $\operatorname{card}(A) = \mathfrak{a}$ and $\operatorname{card}(B) = \mathfrak{a}^+$. By hypothesis there exist disjoint sets A' and B' equinumerous with A and B respectively such that $A \times B = A' \cup B'$. Suppose first that $(\exists x \in A)(\forall y \in B)((x, y) \in A')$. Then there is a one-to-one function from B to A' given by $y \mapsto (a, y)$. So $\mathfrak{a}^+ \leqslant \mathfrak{a}$. Contradiction. So $(\forall x \in A)(\exists y \in B)((x, y) \in B')$. Now choose a well-ordering of B and for each $x \in A$ let $f(x)$ be the least $y \in B$ such that $(x, y) \in B'$. Then $x \mapsto (x, f(x))$ is a one-to-one function from A into B' and therefore $\mathfrak{a} \leqslant \mathfrak{a}^+$. Consequently $\mathfrak{a}$ is well-orderable since $\mathfrak{a}^+$ is. □

(15.4.4) **Proposition (Tarski 1924).** *The axiom of choice is equivalent to the assertion that* $\mathfrak{a} + \mathfrak{b} = \mathfrak{a}\mathfrak{b}$ *for any infinite cardinals* $\mathfrak{a}$ *and* $\mathfrak{b}$.

Necessity. If we assume the axiom of choice, then any infinite cardinals $\mathfrak{a}$ and $\mathfrak{b}$ must be alephs, so that

$$\begin{aligned}\mathfrak{a} + \mathfrak{b} &= \max(\mathfrak{a}, \mathfrak{b}) \text{ [corollary 15.2.2]} \\ &= \mathfrak{a}\mathfrak{b} \text{ [corollary 15.2.4]}.\end{aligned}$$

Sufficiency. If $\mathfrak{a}$ is any infinite cardinal, then $\mathfrak{a} + \mathfrak{a}^+ = \mathfrak{a}\mathfrak{a}^+$ by hypothesis, whence $\mathfrak{a}$ is an aleph [lemma 15.4.3]. The axiom of choice follows [proposition 15.4.1]. □

(15.4.5) **Theorem (König 1905).** *If* $(A_i)_{i\in I}$ *and* $(B_i)_{i\in I}$ *are families of sets such that* $\operatorname{card}(A_i) < \operatorname{card}(B_i)$ *for all* $i \in I$ *and* $\bigcup_{i\in I} B_i$ *is well-orderable, then* $\operatorname{card}(\bigcup_{i\in I} A_i) \neq \operatorname{card}(\prod_{i\in I} B_i)$.

Proof. Suppose on the contrary that f is a function from $\bigcup_{i\in I} A_i$ onto $\prod_{i\in I} B_i$. Choose first some well-ordering of $\bigcup_{i\in I} B_i$. For each $i \in I$ the set $B_i \setminus \{f(a)_i : a \in A_i\}$ is non-empty, since otherwise $a \mapsto f(a)_i$ would be a function from A_i onto B_i, contrary to hypothesis; so we can let b_i be the member of $B_i \setminus \{f(a)_i : a \in A_i\}$ which is least with respect to the chosen well-ordering of $\bigcup_{i\in I} B_i$. In this way we define a family $(b_i)_{i\in I}$, and because f is onto, $(b_i)_{i\in I} = f(a)$ where $a \in A_j$ for some $j \in I$. But then $b_j = f(a)_j$. Contradiction. □

Note, incidentally, that by putting $A_i = \{i\}$ and $B_i = \{0, 1\}$ for all $i \in I$ we retrieve Cantor's theorem

$$\operatorname{card}(I) \neq \operatorname{card}({}^I\{0, 1\}) = \operatorname{card}(\mathfrak{P}(I))$$

as a special case (since $1 < 2$).

(15.4.6) **Corollary.** *The axiom of choice holds iff for any families* $(A_i)_{i\in I}$ *and* $(B_i)_{i\in I}$ *such that* $\operatorname{card}(A_i) < \operatorname{card}(B_i)$ *for all* $i \in I$ *we have* $\operatorname{card}(\bigcup_{i\in I} A_i) < \operatorname{card}(\prod_{i\in I} B_i)$.

Necessity. Suppose that $\operatorname{card}(A_i) < \operatorname{card}(B_i)$ for all $i \in I$. It is easy to use the axiom of choice to show that $\operatorname{card}(\bigcup_{i\in I} A_i) \leqslant \operatorname{card}(\prod_{i\in I} B_i)$. But the axiom of choice also entails that $\bigcup_{i\in I} B_i$ is well-orderable, and so it follows by König's theorem that $\operatorname{card}(\bigcup_{i\in I} A_i) < \operatorname{card}(\prod_{i\in I} B_i)$.

Sufficiency. If $B_i \neq \varnothing$ for all $i \in I$, then (putting $A_i = \varnothing$ for all $i \in I$) we obtain $\prod_{i\in I} B_i \neq \varnothing$ (since $0 < 1$). This is equivalent to the axiom of choice. □

15.5 The continuum hypothesis

In §11.3 we briefly mentioned the hypothesis, first conjectured by Cantor, that every uncountable set of reals has the power of the continuum.

Continuum hypothesis. *There is no cardinal* $\mathfrak{b}$ *such that* $\aleph_0 < \mathfrak{b} < 2^{\aleph_0}$.

If we assume the axiom of choice, then $\aleph_0 < \aleph_1 \leqslant 2^{\aleph_0}$, and the continuum hypothesis is therefore equivalent to the equation $2^{\aleph_0} = \aleph_1$: indeed it is often stated in this form. If we do not assume the axiom of choice, however, the version we have stated above is strictly weaker (Solovay 1970), and the equation $2^{\aleph_0} = \aleph_1$ is then equivalent to the conjunction of the continuum hypothesis and the claim that $2^{\aleph_0}$ is an aleph (i.e. that the real numbers are well-orderable).

Cantor devoted a great deal of time to investigating whether or not the continuum hypothesis is true and on several occasions believed briefly that he had proved it. Indeed he first stated it (in 1878, p. 258) not as a conjecture but as something he claimed to have proved ('by a process of induction which we do not describe further at this point'). One approach Cantor used in trying to tackle the problem was to study the properties of perfect sets (closed subsets of the real line without isolated points). We showed in §10.4 that every non-empty perfect set has the power of the continuum. It follows that the continuum hypothesis is entailed by the following stronger claim.

Perfect set hypothesis. *Every uncountable subset of the real line has a non-empty perfect subset.*

The Cantor/Bendixson theorem, first proved in the 1880s, establishes this hypothesis for *closed* subsets of **R**, but it is far from easy to extend this result to other more inclusive classes (not least because the property is not preserved by complementation). Eventually, though, new methods enabled Alexandroff (1916) to prove it for Borel sets, and Souslin (see Lusin 1917) for analytic sets. But the more ambitious project of establishing the continuum hypothesis *via* the perfect set hypothesis was stymied, at least for those such as Cantor who accepted the axiom of choice, by Bernstein's (1908) discovery that the perfect set hypothesis *contradicts* the axiom of choice.

(15.5.1) **Proposition.** *If* $2^{\aleph_0}$ *is an aleph, there is a subset of the real line with the power of the continuum which neither contains nor is disjoint from any non-empty perfect set.*

Proof. Suppose that $2^{\aleph_0} = \aleph_\beta$. We noted earlier that the number of perfect sets is $2^{\aleph_0}$; so it follows that there is a transfinite sequence $(P_\alpha)_{\alpha<\omega_\beta}$ enumerating all the non-empty perfect sets. Let us now try to choose two transfinite sequences

(a_α) and (b_α) recursively so that

$$a_\alpha \in P_\alpha \setminus (\{a_\gamma : \gamma < \alpha\} \cup \{b_\gamma : \gamma < \alpha\})$$
$$b_\alpha \in P_\alpha \setminus (\{a_\gamma : \gamma \leqslant \alpha\} \cup \{b_\gamma : \gamma < \alpha\}).$$

At each stage the choice of a_α is indeed possible, since

$$\operatorname{card}(\{a_\gamma : \gamma < \alpha\} \cup \{b_\gamma : \gamma < \alpha\}) = 2|\alpha| = |\alpha| < |\omega_\beta| = 2^{\aleph_0} = \operatorname{card}(P_\alpha),$$

and so $P_\alpha \setminus (\{a_\gamma : \gamma < \alpha\} \cup \{b_\gamma : \gamma < \alpha\}) \neq \emptyset$; similarly for b_α. Moreover, this does not require the axiom of choice, since we are supposing that $2^{\aleph_0}$ is an aleph, and hence that the real line is well-orderable. The ranges of the two transfinite sequences $\{a_\alpha : \alpha < \omega_\beta\}$ and $\{b_\alpha : \alpha < \omega_\beta\}$ are evidently disjoint sets with the power of the continuum and every non-empty perfect set intersects both of them. □

(15.5.2) **Corollary.** *The axiom of choice and the perfect set hypothesis cannot both be true.*

Proof. If the axiom of choice is true, then $2^{\aleph_0}$ is an aleph and hence by proposition 15.5.1 there is an uncountable set of real numbers with no non-empty perfect subset, contradicting the perfect set hypothesis. □

So the axiom of choice refutes the perfect set hypothesis. The much stronger axiom of constructibility refutes even the special case of the perfect set hypothesis for projective sets (Gödel 1938). But the continuum hypothesis itself is sufficiently weaker than the perfect set hypothesis to escape this stricture: it is entailed by the axiom of constructibility, but it is standardly used nowadays as an example of a proposition that is *independent* of the ordinary axioms of set theory: even if we assume the whole of ZFC, we can prove neither the continuum hypothesis (Cohen 1963) nor its negation (Gödel 1938), provided only that ZFC itself is consistent.

So is there anything we can prove about the size of $2^{\aleph_0}$ without assuming further axioms? It turns out that there is.

Definition. *A cardinal* $\mathfrak{a}$ *is said to be of* countable cofinality *if there is a sequence* $(A_n)_{n\in\omega}$ *of sets such that* $\operatorname{card}(A_n) < \mathfrak{a}$ *for all* $n \in \omega$ *but* $\operatorname{card}(\bigcup_{n\in\omega} A_n) = \mathfrak{a}$.

(15.5.3) **Theorem.** $2^{\aleph_0}$ *is not an aleph of countable cofinality.*

Proof. Suppose on the contrary that there is a well-orderable set B such that $B = \bigcup_{n\in\omega} A_n$ with $\operatorname{card}(A_n) < 2^{\aleph_0} = \operatorname{card}(B)$. In that case

$$\operatorname{card}({}^{\omega}B) = (2^{\aleph_0})^{\aleph_0} = 2^{\aleph_0} = \operatorname{card}(\bigcup_{n\in\omega} A_n),$$

contradicting König's theorem. □

(15.5.4) **Corollary.** $2^{\aleph_0} \neq \aleph_\omega$.

Proof. The cardinal $\aleph_\omega$ is of countable cofinality, since $\boldsymbol{\omega}_\omega = \bigcup_{n\in\boldsymbol{\omega}} \boldsymbol{\omega}_n$. □

But this is the only restriction: any value for $2^{\aleph_0}$ in the hierarchy of alephs which is not of countable cofinality is consistent with ZFC (Solovay 1964). It is thus consistent (although, one somehow feels, rather unlikely) that $2^{\aleph_0} = \aleph_{4049}$, for instance, or $2^{\aleph_0} = \aleph_{\omega^2+61}$.

Many mathematicians conclude on the basis of these independence results not only that the continuum hypothesis is undecided but that it is undecidable. Let us pause now to consider whether this is the right conclusion to draw.

Exercise

(Sierpinski 1924) Show that $2^{\aleph_0} = \aleph_1$ iff $\aleph_2^{\aleph_0} > \aleph_1^{\aleph_0}$.

15.6 Is the continuum hypothesis decidable?

Note first that the continuum hypothesis is obviously equivalent to the statement

$$(\forall A \subseteq \mathfrak{P}(\boldsymbol{\omega}))(A \sim \boldsymbol{\omega} \text{ or } A \sim \mathfrak{P}(\boldsymbol{\omega})),$$

where $\sim$ expresses equinumerosity, and that *this* statement, even when it is expressed fully without abbreviations, quantifies only over the first few infinite levels of the hierarchy of sets. (The exact number of levels involved depends on just how the ordered pair is defined, but by careful use of coding we could if it mattered reduce the quantification to the third infinite level.) This is significant because it shows that the continuum hypothesis — in stark contrast, for instance, to the higher axioms of infinity considered in chapter 13 — is *decided* by second-order set theory Z2. This contrast, although relatively trivial in itself, is certainly not always appreciated by mathematicians. It is well known to set theorists, of course, and is a theme of Scott's foreword to Bell (1977). 'There are any number of contradictory set theories, all extending the Zermelo-Fraenkel axioms,' he observes (p. xiv), 'but the models are all just models of the first-order axioms, and first-order logic is weak.' The point has also been made repeatedly by Kreisel, but usually in works unlikely to be read by mainstream mathematicians (e.g. Kreisel 1967*a*); it is rare for books aimed at a general mathematical audience to give the point any prominence.

Two consequences of this second-order decidability result should be noted. The first is that the uncritical platonist who accepts the argument given in the last chapter that the axiom of choice follows from the second-order logical principle must correspondingly accept that the truth or falsity of the continuum hypothesis might be settled by second-order logic. It is, indeed, easy enough to formulate a sentence in the language of pure second-order logic which is a logical truth iff the continuum hypothesis is true, and another sentence (not, of course, the negation of the first) which is a logical truth iff the continuum hypothesis is false (see Shapiro 1985, p. 741). The difficulty is evidently that in contrast to the case of the axiom of choice we do not seem to have any intuitions about whether these second-order principles that could settle the continuum hypothesis are themselves true or false. So this observation does not seem especially likely to be a route to an argument that will actually settle the continuum hypothesis one way or the other. And even if we did find such an argument (for example, a mathematical argument from some new set-theoretic principle), although we could then work back from that to knowledge of the corresponding second-order logical truth, this would not automatically make the continuum hypothesis itself logical, since it would follow from the second-order logical truth in question only via the axioms of second-order set theory.

All of this, of course, applies only to the sort of platonist who accepts the second-order separation principle. But I also want to mention another consequence of the second-order decidability of the continuum hypothesis, this time one that does not seem to depend so directly on accepting second-order separation. The point I want to mention is that there is a difference in character between the continuum hypothesis and other sentences undecided by ZFC, such as the various large cardinal axioms. This is admittedly somewhat vaguer than the preceding point, and it is correspondingly more obscure how much it depends on the platonist commitment to the second-order system, but at the very least it shows that the analogy that has often been casually drawn (e.g. Errera 1952, A. Robinson 1968) between the position of the continuum hypothesis in set theory and that of the parallel postulate in geometry is much too hasty: the undecidability of the parallel postulate has nothing to do with the weakness of first-order systems.

Kreisel (1971) has urged that a much better analogy would be with the proven insolubility in elementary geometry of the classical problems of squaring the circle and trisecting the angle: what is shown is not that an angle cannot be trisected but only that it cannot be done with a straightedge and compasses. But even if this analogy is apposite, it is not clear that it helps us to solve the continuum problem, since it does not give us much of a clue where to look for the new methods that we need.

One superficially appealing strategy would be to bring higher infinities to our aid. We saw in chapter 13 how one property (determinacy) can be proved successively for closed sets, Borel sets and projective sets, but only by invoking higher and higher infinities at each stage. By analogy one might conjecture, as Gödel did (1947), that the continuum hypothesis, which we have proved for closed sets already, could be extended to more inclusive categories of set in something like the same manner.

However plausible this conjecture may have seemed when Gödel made it, subsequent work in set theory has shown that it is very unlikely to be true. What Gödel had presumably not expected was that the method of forcing devised by Cohen (1963) to prove the independence of the continuum hypothesis from ZF would turn out to be even more robust than Gödel's inner model construction when confronted with large cardinal axioms (Levy and Solovay 1967). Broadly stated, every large cardinal axiom so far proposed is known not to settle the continuum hypothesis.

This marks another way of differentiating between independence claims. We have already noted that the independence of the continuum hypothesis is different from that of the parallel postulate in geometry, because it is distinctively a first-order result; what we can now see is that it is also different from the sort we considered in chapter 13, such as the independence of the Gödel sentence of a theory or the independence of Borel determinacy from ZU, since these claims can be decided by ascent to a higher level in the hierarchy, whereas the continuum hypothesis cannot.

With this distinction in mind let us call a sentence *strongly undecidable* if it is independent of set theory even if any axiom of infinity, however strong, is adjoined to it. We should recognize at once, of course, that this is not a formalizable notion, since Gödel's theorem shows that no formal characterization is possible of what should count as an axiom of infinity. The best we could hope for would be, as Gödel (1965, p. 85) suggested, 'a characterization of the following sort: An axiom of infinity is a proposition which has a certain (decidable) formal structure and which in addition is true.' In any such characterization truth would of course remain as the inherently non-formal notion involved. It must be said, however, that neither Gödel nor anyone else has yet offered a plausible candidate for a formal characterization of the required sort.

In lectures he gave in 1939 or 1940, shortly after proving the consistency of the continuum hypothesis, Gödel speculated that it might be strongly undecidable whether every real number is constructible (see Gödel 1986–2003, vol. III, pp. 175 and 185). Later, though, he seems to have had a change of heart and not only suggested that the continuum hypothesis might be decided by a suitably strong axiom of infinity but even briefly speculated in 1946 that there might be *no* strongly undecidable propositions in set theory.

> It is not impossible that ... some completeness theorem would hold which would say that every proposition expressible in set theory is decidable from the present axioms plus some true assertion about the largeness of the universe of sets. (Gödel 1965, p. 85)

But even if the continuum hypothesis is strongly undecidable in the sense just outlined, it does not automatically follow that it is *absolutely* undecidable — undecidable, that is to say, by *any* true principles about sets, whether or not they count as axioms of infinity. We are surely not entitled to make this stronger claim until the *reason* for the robustness of the continuum hypothesis is well understood. And in any case large cardinal axioms are not the only way of extending ZF. Indeed Gödel himself, only a year after he had speculated that there might be no strongly undecidable statements, pointed out another way in which a proposition could be decided. Not only do there probably exist new axioms of infinity based on unknown principles, he said, but also

> there may exist, besides the ordinary axioms [and] the axioms of infinity ... other (hitherto unknown) axioms of set theory which a more profound understanding of the concepts underlying logic and mathematics would enable us to recognize as implied by these concepts. (Gödel 1947, pp. 520–1)

In line with Gödel's suggestion, various authors have offered arguments which aim to settle the continuum hypothesis on the basis of more or less intuitively appealing principles. Hilbert, who in 1900 regarded settling it as one of the most important challenges in mathematics, sketched in 1925 a purported proof of the continuum hypothesis based on a classification of the elements of Baire space into orders of recursive definability; but he never completed the details, and Zermelo is reported as saying that 'no one understood what he meant' (P. Levy 1964, p. 89). Gödel himself quite late in his life believed he had a proof that $2^{\aleph_0} = \aleph_2$ on the basis of several new set-theoretic axioms, but he withdrew the paper before publication (see Gödel 1986–2003, vol. III, pp. 405–25). More recently, various other set-theoretic principles have been shown to entail that $2^{\aleph_0} = \aleph_2$: e.g. the principle known as 'Martin's maximum' (Foreman, Magidor and Shelah 1988) or an axiom proposed by Woodin (2001*b*).

What should we make of these new axioms? Formalists, of course, will as usual regard universes in which the new axioms hold and those in which they fail as equally valid (although whether they are equally *interesting* will depend on how the mathematics develops). But there does not seem to be any reason for them to regard the continuum hypothesis as special in this regard. For the realist, on the other hand, there is always the possibility that intuitive principles will settle the continuum hypothesis one way or the other. One statement which entails it is the axiom of constructibility, which may be thought of as a minimizing principle whose approximate effect is to make each level of the hierarchy as thin as is permitted by the other axioms. As we suggested in the last chapter, part of the reason why few realist mathematicians are willing to regard the axiom of constructibility as true is that it seems to contravene the first principle of plenitude which guides the formation of the hierarchy.

The reason for repeating this point here is that if it is right, it is natural to wonder whether a converse argument can be mounted to the effect that the first principle of plenitude requires $2^{\aleph_0}$ to be as large as possible. This idea has been urged by Cohen.

A point of view which the author feels may eventually come to be accepted is that CH is *obviously* false. The main reason one accepts the Axiom of Infinity is probably that we feel it absurd to think that the process of adding only one set at a time can exhaust the entire universe. Similarly with the higher axioms of infinity. Now $\aleph_1$ is the set of countable ordinals and this is merely a special and the simplest way of generating a higher cardinal. The set $\mathfrak{P}(\boldsymbol{\omega})$ is, in contrast, generated by a totally new and more powerful principle, namely the Power Set Axiom. It is unreasonable to expect that any description of a larger cardinal which attempts to build up that cardinal from ideas deriving from the Replacement Axiom can ever reach $\mathfrak{P}(\boldsymbol{\omega})$. Thus $\mathfrak{P}(\boldsymbol{\omega})$ is greater than $\aleph_n$, $\aleph_\omega$, $\aleph_{\omega_\omega}$, etc. This point of view regards $\mathfrak{P}(\boldsymbol{\omega})$ as an incredibly rich set given to us by one bold new axiom, which can never be approached by any piecemeal process of construction. (1966, p. 151, modified)

This is a radical argument, and it is hard to make sense of it in conventional terms. For it is certainly provable in ZFC that $2^{\aleph_0} = \aleph_\alpha$ for some α. One possibility, of course, would be simply to deny that $\mathfrak{P}(\boldsymbol{\omega})$ is a set, but that is not Cohen's intention: his proposal is not that an ordinal α such that $2^{\aleph_0} = \aleph_\alpha$ does not exist, but only that it cannot be *described* in any other terms already available to us in the first-order theory. Cohen (1973) followed up his suggestion by proposing one axiom which attempts to give expression to the idea that $\mathfrak{P}(\boldsymbol{\omega})$ is large, and Takeuti (1971) has suggested others. Scott, meanwhile, went further, speculating (in Bell 1977, p. xiv) that 'we would be pushed in the end to say that all sets are *countable* (and that the continuum is not even a set!) when at last all cardinals are absolutely destroyed'.

But it is by no means clear why the maximal conception of the power-set operation should deliver the sort of conclusion these authors want. If we enrich the power set at each level, we enlarge not only $\mathfrak{P}(\boldsymbol{\omega})$ but also the set of non-isomorphic well-orderings of $\boldsymbol{\omega}$ and hence (in one sense) the size of $\aleph_1$.

While the property of *being an ordinal* is invariant or absolute, the property (of ordinals) of *being [uncountable]* is not. The point is often overlooked in the (popular) 'debate' on the continuum hypothesis, where the *orderliness* of the ordinals (in V_κ or L_κ) is contrasted with the *mess* of $\mathfrak{P}(\boldsymbol{\omega})$ (in V_κ): a similar mess is involved in the collection of maps (in V_κ) of $\boldsymbol{\omega}$ onto initial segments of the ordinals. It does not seem at all surprising that we have not (yet) decided whether the two 'messes' match. (Kreisel 1980, p. 198)

But the matter is even harder to resolve for the regressivist, because it is difficult to come by consequences of the continuum hypothesis to use as data. Indeed, one reason for the tendency of mathematicians to regard the continuum hypothesis as absolutely undecidable may well be that it receives so little regressive support from its consequences. For we saw in the last chapter that the axiom of constructibility does not enable us to prove any new theorems in first-order arithmetic; and since the continuum hypothesis is entailed by the axiom of constructibility, the same will be true of it. But by a more elaborate argument we can show still more: even in second-order arithmetic there is nothing provable using the continuum hypothesis that is not already provable using only the axiom of choice (Platek 1969). It follows that anyone who wishes to justify the continuum hypothesis on the basis of its consequences must claim independent knowledge, i.e. knowledge acquired by some other route, of truths of at least the third order. But, as we noted at the beginning of this section, the continuum hypothesis is itself of the third order; so at any rate the regressive approach does not seem to effect any logical simplification of the problem.

Feferman's view is that the continuum hypothesis is 'inherently vague' (2000, p. 405). This, or something like it, is a common view among mathematicians. Not all of those who hold it, though, are clear about its consequences. For whether a sentence is vague or not is presumably a function of its meaning. So we cannot corral the undecided sentences and leave the others untainted: if we admit the continuum hypothesis as vague, we shall be hard pressed to resist the conclusion that all other sentences involving quantification at the third infinite level of the hierarchy are more or less vague as well. The concern has been well expressed by Steel.

> There may be something in the idea that the language of third order arithmetic is vague, but the suggestion that it is inherently so is a gratuitous counsel of despair. If the language of third order arithmetic permits vague or ambiguous sentences, then it is important to trim or sharpen it so as to eliminate these. ... In his argument that the concept of an arbitrary set of reals is inherently vague, Feferman likens it to the 'concept' of a feasible number. This analogy is far-fetched at best. The concept of an arbitrary set of reals is the foundation for a great deal of mathematics, and has never led into contradiction. The first two things of a general nature one is inclined to say about feasible numbers will contradict each other. (2000, p. 432)

15.7 The axiom of determinacy

The continuum hypothesis may be thought of as making a general claim about arbitrary sets of real numbers — that they are all countable or have the power of the continuum. As we have seen, Cantor succeeded in showing, assuming the axiom of choice, that every closed set has this property, and he hoped to extend this to all sets in due course. In this project we now know that he was doomed to failure. There is a striking contrast, though, between the property Cantor was investigating and several others which arise naturally in the study of the real line. Consider, for instance, the property of measurability central to the theory of integration: every closed set is measurable, and if we assume the axiom of countable choice, we can prove that every Borel set is measurable. But the axiom of choice entails the existence of non-measurable sets. The existence of such sets may be thought unwelcome on naive grounds of simplicity, but in fact matters are somewhat worse.

Theorem (Banach and Tarski 1924). *The axiom of choice implies that there is a decomposition of the surface of the unit sphere into a finite number* *of pieces which can by rigid motions of three-dimensional Euclidean space be reassembled to form the surfaces of two spheres of unit radius.*[2]

Now of course it is trivially the case that the decomposition described in the theorem is impossible if we require the pieces into which the sphere is decomposed all to be *measurable,* since rigid motions preserve area, and the surface area of each of the spheres in question is 4π. So what the theorem claims must be a decomposition into non-measurable pieces.

The Banach/Tarski theorem has sometimes been used in an attempt to *refute* the axiom of choice: the conclusion of the theorem is intuitively false, it is said, and therefore the axiom of choice cannot be true. In order to use it in this way, though, we would need to have an intuitive argument not depending on the concept of area for disbelieving in the possibility of the decomposition mentioned in the theorem, and it is by no means clear that such an argument exists. The point is one we came across when we were considering real analysis in chapter 8. We have already seen that in testing the axiom of choice geometrical intuitions derived from *elementary* geometry — the geometry of straightedge and compasses — are irrelevant. So the geometrical intuitions involved here cannot be elementary in this sense, but must depend on our general grasp of properties of transcendental functions. But experience already suggests that our intuitions concerning such functions need to be educated before much reliance can be placed on them.

This is a common phenomenon in mathematics. The ancient Greeks apparently regarded their discovery of the existence of irrational numbers as paradoxical (whether or not one of them drowned because of it, as myth claims); if no trained mathematician would have this reaction today, that is precisely because by studying the phenomenon we have reached an understanding of the reasons for it, and hence, far from seeming paradoxical, it comes to be just what we intuitively expect. In much the same way, those who have received the appropriate education seem generally disinclined to regard the conclusion of the Banach/Tarski theorem as false. (What is harder to judge, of course, is whether they are influenced in this view by also having been educated to believe the axiom of choice.)

The impression that the Banach/Tarski theorem does not show the axiom of choice to be false is further strengthened if it is compared to the following result.

Theorem (Mazurkiewicz and Sierpinski 1914). *There is a non-empty subset E of the Euclidean plane which has two disjoint subsets each of which can be split into finitely many parts which can be rearranged isometrically to form a partition of E.*

[2] R. M. Robinson (1947) has shown that the number of pieces in the decomposition can be made as small as four.

This theorem is certainly surprising, but this time we cannot blame the axiom of choice since the proof does not require it: the sets involved in the decomposition are measurable.[3] It may indeed be that this is not *quite* as surprising as the previous result, but it surely weakens one's confidence that the conclusion of the Banach/Tarski theorem is intuitively false.

Nonetheless, the Banach/Tarski theorem has led some mathematicians to speculate on the idea of abandoning the axiom of choice and putting in its place an axiom which ensures that every set is measurable and hence rules out the decompositions of the sphere which they find paradoxical. The prime candidate for such an axiom is one that was first proposed by Mycielski and Steinhaus (1962).

Axiom of determinacy. *The game on every subset of the Baire line is determined.*

The axiom of determinacy entails that every set of real numbers is measurable (Mycielski and Swierczkowski 1964). This is welcome news to anyone who finds the Banach/Tarski decomposition paradoxical, since it shows that determinacy rules this result out. It follows, of course, that determinacy must be incompatible with the axiom of choice, but in fact this is something we can easily prove directly.

(15.7.1) **Proposition.** *The axiom of determinacy entails that $2^{\aleph_0}$ is not an aleph.*

Proof. Suppose that $2^{\aleph_0} = \aleph_\beta$. The set of first player strategies has cardinal $2^{\aleph_0}$ and hence can be enumerated as the range of a transfinite sequence $\{\sigma_\alpha : \alpha < \omega_\beta\}$. In the same way we can let $\{\tau_\alpha : \alpha < \omega_\beta\}$ enumerate the strategies available to the second player. Suppose now that $\alpha < \omega_\beta$. The function $t \mapsto \sigma_\alpha * t$ is one-to-one, and so the set $\{\sigma_\alpha * t : t \in {}^{\boldsymbol{\omega}}\boldsymbol{\omega}\}$ of all the possible games in which the first player follows the strategy σ_α has the same cardinal as ${}^{\boldsymbol{\omega}}\boldsymbol{\omega}$, i.e. $2^{\aleph_0}$; similarly the set $\{s * \tau_\alpha : s \in {}^{\boldsymbol{\omega}}\boldsymbol{\omega}\}$ has cardinal $2^{\aleph_0}$ as well. It is therefore possible recursively to choose $a_\alpha, b_\alpha \in {}^{\boldsymbol{\omega}}\boldsymbol{\omega}$ so that $b_\alpha = \sigma_\alpha * t$ for some t but $b_\alpha \notin \{a_\gamma : \gamma < \alpha\}$, and $a_\alpha = s * \tau_\alpha$ for some s but $a_\alpha \notin \{b_\gamma : \gamma < \alpha\}$. It is simple to check that $A = \{a_\alpha : \alpha < \omega_\beta\}$ and $B = \{b_\alpha : \alpha < \omega_\beta\}$ are disjoint and neither player has a winning strategy for the game on A. □

In particular, therefore, the axiom of determinacy is incompatible with the axiom of choice and also entails that $2^{\aleph_0} \neq \aleph_1$. Rather more elaborate methods can be used to strengthen proposition 15.7.1 to the following.

Theorem (Davis 1964). *The axiom of determinacy entails the perfect set hypothesis.*

[3] It is easy to deduce that the area of the set E mentioned in the theorem must be zero.

As an immediate corollary it follows that the axiom of determinacy entails the continuum hypothesis (which, as we noted earlier, does not contradict $2^{\aleph_0} \neq \aleph_1$ in the absence of choice).

The principal reason for the interest set theorists have taken in the axiom of determinacy is the connection between determinacy and large cardinal axioms. We saw in chapter 13 how successively stronger axioms had to be added to ZF to prove the determinacy of all Borel sets and then of all projective sets. Mimicking Gödel's failed programme for settling the continuum hypothesis, we might even wonder whether a still stronger axiom of infinity would prove the axiom of determinacy. Indeed, the axiom of determinacy is already in a sense an axiom of infinity, as it was shown by Solovay in 1967 to entail that $\aleph_1$ and $\aleph_2$ are measurable cardinals; and it is possible to convert the proof in ZfU of any proposition of a sufficiently simple syntactic form into a proof in ZU plus Borel determinacy.[4]

An axiom that entailed determinacy would have to contradict the axiom of choice, of course, and few suitable candidates are known. One that might have been a candidate was stated by Reinhardt in his 1967 doctoral thesis (see Reinhardt 1974). His proposal was, in effect, that we add to set theory an operator j which permutes the members of the hierarchy while leaving their first-order properties unchanged, i.e.

$$(\forall x_1, \ldots, x_n)\, \Phi(x_1, \ldots, x_n) \Leftrightarrow \Phi(jx_1, \ldots, jx_n)$$

for every formula Φ. Kunen (1971) showed that if we assume the axiom of choice, then the only such operator is the identity, i.e. $jx = x$ for all x, and hence that Reinhardt's axiom asserting the existence of a non-trivial permutation is inconsistent with ZFC. Since then, Reinhardt's axiom has been regarded by set theorists as an upper limit to their invention of large cardinal axioms when working under the constraint of the axiom of choice. What remains unknown, however, is whether this limit applies in the absence of the axiom of choice — whether, for instance, Reinhardt's axiom is inconsistent even with ZF.

In the absence of much work on the consequences of Reinhardt's proposal, it is hard to speculate, but it does seem to throw up the intriguing possibility that the axiom of choice might act in some way as a barrier to the free construction of the hierarchy. In other words, the axiom of choice, which was billed as an expression of the platonist's desire to maximize the number of sets at each level, might conflict with the desire to maximize the number of levels. Something of this sort is already known to hold for the much stronger axiom of constructibility, which has been shown by Scott (1961) to contradict the existence of a measurable cardinal.

[4] Bizarrely, though, the axiom of determinacy also entails that $\aleph_n$ is *not* measurable for any $n > 2$.

Of course, in the unlikely event that considerations of this sort led mathematicians to give up the axiom of choice, that would still not in itself settle matters in favour of determinacy. Since the axiom of determinacy entails the existence of large cardinals, no proof of its consistency relative to ZF is possible. It is natural, then, to look for an intuitive argument that the axiom of determinacy is true. But proponents of the axiom have generally stopped short of offering this. When they first proposed the axiom, for example, Mycielski and Steinhaus (1962) were equivocal. They did offer what they called an 'intuitive justification' for it.

> Suppose that both players I and II are infinitely clever and that they know perfectly well what [the game] is, then owing to the complete information during every play, the result of the play cannot depend on chance. [The axiom of determinacy] expresses exactly this. (p. 1)

However, they went on to deny that they wished to

> depreciate classical mathematics with its fundamental 'absolute' intuitions on the universum of sets (to which belongs the axiom of choice), [but] only to propose another theory which seems very interesting.

The axiom of determinacy, they said,

> can be considered as a restriction of the classical notion of a set leading to a smaller universum, say of determined sets, which reflect some physical intuitions which are not fulfilled by the classical sets (e.g. paradoxical decompositions of the sphere are eliminated). (p. 2)

The idea that determinacy should simply be adopted as an axiom instead of choice has not found much favour among mathematicians since then. Moschovakis (1980, p. 379), for instance, called the axiom of determinacy 'blatantly false'. Recently set theorists have been more inclined to consider as an axiom candidate the weaker *projective determinacy*, an axiom asserting that every projective set is determined. This weaker claim has the advantage that it is known to be consistent with the axiom of choice unless the axiom of determinacy is itself inconsistent. Moreover, as we have already noted in §13.7, it is provable from a large cardinal axiom asserting the existence of infinitely many Woodin cardinals. On the other hand, projective determinacy does not refute the Banach/Tarski theorem, nor does it settle the continuum hypothesis (Levy and Solovay 1967).

However, the general notion of a projective set of real numbers is so remote from geometrical intuition that it is hard to see what direct intuitive reason might be given for believing a proposition about them such as projective determinacy. So any argument for accepting it as an axiom (as opposed to treating it as a theorem in set theory with a suitable large cardinal axiom) is likely to be wholly regressive. Martin (1977, p. 814) regards it as 'an hypothesis with a status similar to that of a theoretical hypothesis in physics'. Because of its 'pleasing consequences', he suggests, 'it is not unreasonable to suspect that it may be true' (Martin 1976, p. 90). This view is echoed by Woodin (1994, p. 34).

There is little a priori evidence that [projective determinacy] is a plausible axiom, or even that it is a consistent axiom. However, the theory that follows from the assumption of [projective determinacy] is so rich that, a posteriori, the axiom is both consistent and true. The lesson here is an important one. Axioms need not be a priori true.

And more recently he has re-affirmed his view that there is 'compelling evidence that [projective determinacy] is the "right axiom" for the projective sets' (2001*a*, p. 571). Perhaps the position has been summarized best by Moschovakis (1980, pp. 610–11).

At the present state of knowledge only few set theorists accept [projective determinacy] as highly plausible and none is quite ready to believe it beyond a reasonable doubt; and it is certainly possible that someone will refute [it] in ZFC. On the other hand, it is also possible that the web of implication involving determinacy hypotheses and relating them to large cardinals will grow steadily until it presents such a natural and compelling picture that more will succumb to its beauty.

15.8 The generalized continuum hypothesis

Once mathematicians became interested in the continuum hypothesis, it was natural that they would also wish to study the following natural generalization of it.

Definition. *The* generalized continuum hypothesis *is the proposition that for no infinite cardinal* $\mathfrak{a}$ *is there a cardinal* $\mathfrak{b}$ *such that* $\mathfrak{a} < \mathfrak{b} < 2^{\mathfrak{a}}$.

(15.8.1) **Lemma.** $2^{\mathfrak{a}^+} \leqslant 2^{2^{\mathfrak{a}^2}}$.

Proof. Let A be a set such that $\text{card}(A) = \mathfrak{a}$, and let β be the least ordinal such that $|\beta| = \mathfrak{a}^+$. It is easy to check that the function f from $\mathfrak{P}(\beta)$ to $\mathfrak{P}(\mathfrak{P}(A \times A))$ given by $f(X) = \{r \subseteq A \times A : \text{ord}(\text{dom}[r], r) \in X\}$ is one-to-one. So

$$2^{\mathfrak{a}^+} = \text{card}(\mathfrak{P}(\beta)) \leqslant \text{card}(\mathfrak{P}(\mathfrak{P}(A \times A))) = 2^{2^{\mathfrak{a}^2}}. \qquad \square$$

(15.8.2) **Theorem (Sierpinski 1924).** *The generalized continuum hypothesis entails the axiom of choice.*

Proof. Suppose not. So the generalized continuum hypothesis holds, but there is a non-finite cardinal $\mathfrak{a}$ which is not an aleph. Let $\mathfrak{b} = 2^{\mathfrak{a}+\aleph_0}$. We intend to show first that $\mathfrak{b}^+ = (2^{\mathfrak{b}})^+$. For suppose not. Then $\mathfrak{b}^+ < (2^{\mathfrak{b}})^+$ and so $\mathfrak{b}^+ \leqslant 2^{\mathfrak{b}}$. Now $\mathfrak{b} \leqslant \mathfrak{b} + \mathfrak{b}^+ \leqslant \mathfrak{b}\mathfrak{b}^+ + \mathfrak{b}^+ = (\mathfrak{b} + 1)\mathfrak{b}^+ = \mathfrak{b}\mathfrak{b}^+ \leqslant (2^{\mathfrak{b}})^2 = 2^{2\mathfrak{b}} = 2^{\mathfrak{b}}$. But if $\mathfrak{b} + \mathfrak{b}^+ = \mathfrak{b}$, then $\mathfrak{b}^+ \leqslant \mathfrak{b}$, which is absurd. Hence $\mathfrak{b} + \mathfrak{b}^+ =$

$2^{\mathfrak{b}}$ by the generalized continuum hypothesis and therefore $\mathfrak{b} + \mathfrak{b}^+ = \mathfrak{b}\mathfrak{b}^+$. Consequently $\mathfrak{b}$ is an aleph [lemma 15.4.3]. But $\mathfrak{a} < \mathfrak{b}$ and so $\mathfrak{a}$ is an aleph. Contradiction. So $\mathfrak{b}^+ = (2^{\mathfrak{b}})^+$. Similar arguments show that $(2^{\mathfrak{b}})^+ = (2^{2^{\mathfrak{b}}})^+$ and that $(2^{2^{\mathfrak{b}}})^+ = (2^{2^{2^{\mathfrak{b}}}})^+$. It follows that $\mathfrak{b}^+ = (2^{2^{2^{\mathfrak{b}}}})^+ \not\leqslant 2^{2^{2^{\mathfrak{b}}}}$.

But $2\mathfrak{b} = 2^{\mathfrak{a}+\aleph_0+1} = 2^{\mathfrak{a}+\aleph_0} = \mathfrak{b}$, so that $\mathfrak{b}^2 \leqslant (2^{\mathfrak{b}})^2 = 2^{2\mathfrak{b}} = 2^{\mathfrak{b}}$, and therefore

$$\begin{aligned} \mathfrak{b}^+ &< 2^{\mathfrak{b}^+} \text{ [Cantor's theorem]} \\ &\leqslant 2^{2^{\mathfrak{b}^2}} \text{ [lemma 15.8.1]} \\ &\leqslant 2^{2^{2^{\mathfrak{b}}}}. \end{aligned}$$

Contradiction. □

(15.8.3) **Corollary.** *These four assertions are equivalent:*

(i) *the generalized continuum hypothesis;*

(ii) $\mathfrak{a}^+ = 2^{\mathfrak{a}}$ *for every infinite cardinal* $\mathfrak{a}$;

(iii) *the axiom of choice holds and* $\mathfrak{a}^+ = 2^{\mathfrak{a}}$ *for every aleph* $\mathfrak{a}$;

(iv) V_0 *is well-orderable and* $\mathfrak{a}^+ = 2^{\mathfrak{a}}$ *for every aleph* $\mathfrak{a}$.

(ii) ⇒ *(i)*. There is no cardinal between $\mathfrak{a}$ and $\mathfrak{a}^+$. So if $\mathfrak{a}^+ = 2^{\mathfrak{a}}$, there is no cardinal between $\mathfrak{a}$ and $2^{\mathfrak{a}}$.

(i) ⇒ *(iv)*. Assume the generalized continuum hypothesis. We have shown above [theorem 15.8.2] that this implies the axiom of choice. So if $\mathfrak{a}$ is an aleph, then $\mathfrak{a} < \mathfrak{a}^+ \leqslant 2^{\mathfrak{a}}$ and so $\mathfrak{a}^+ = 2^{\mathfrak{a}}$ by the generalized continuum hypothesis.

(iv) ⇒ *(iii)*. The method is to show by transfinite induction that there is a well-ordering on every level V. To do this, we let σ be the least ordinal such that $|\sigma| \not\leqslant \text{card}(V)$ [Hartogs' theorem 11.4.2], so that

$$\gamma < \sigma \Leftrightarrow |\gamma| \leqslant \text{card}(V).$$

By hypothesis there exists a well-ordering $\prec$ on $\mathfrak{P}(\sigma)$. We now define recursively a well-ordering $<$ on V as follows. First define $a < b$ whenever $\rho(a) < \rho(b)$. Suppose now that the well-ordering has been defined for all members of V of rank $< \beta$. Let $\gamma = \text{ord}(\mathrm{V}_\beta, <)$ and let g_β be the unique isomorphism of $(\mathrm{V}_\beta, <)$ onto $\boldsymbol{\gamma}$. Now $\mathrm{V}_\beta \subset V$, so that $|\gamma| \leqslant \text{card}(V)$ and therefore $\gamma < \sigma$. So for any a and b of rank β we can define $a < b$ iff $g_\beta[a] \prec g_\beta[b]$.

(iii) ⇒ *(ii)*. It follows from the axiom of choice that every infinite cardinal is an aleph. □

If V_0 is well-orderable, therefore, the generalized continuum hypothesis is equivalent to the claim that $2^{\mathfrak{a}} = \mathfrak{a}^+$ for every aleph $\mathfrak{a}$. This equivalence holds in particular in **Z**, since in that theory the set of individuals is empty and therefore trivially well-orderable. In **Zf**, where the aleph $\aleph_\alpha$ exists for every α, the generalized continuum hypothesis thus takes the form

$$\aleph_{\alpha+1} = 2^{\aleph_\alpha} \text{ for every ordinal } \alpha;$$

this is how it was expressed when first conjectured by Hausdorff (1908, p. 494), and how it has most often been expressed in the literature of the subject since. In the notation due to Peirce that we introduced in §11.5 we can write it even more compactly as $\beth_\alpha = \aleph_\alpha$.

The generalized continuum hypothesis is consistent with **ZF** (Gödel 1938), and this result is stable under the addition of many large cardinal axioms. The generalized continuum hypothesis is redundant in the elementary parts of mathematics (arithmetic in particular). It is also known to be independent of **ZFC** even if we assume the continuum hypothesis as well.

From a logical point of view, however, the most significant point of difference with the continuum hypothesis is that the generalized continuum hypothesis is not obviously decided by the second-order set theories **Z2**, **ZF2**, etc. This is because we are not able to rule out the possibility that the behaviour of the operations concerned might be different at different levels in the hierarchy.

Very little is known about the relationship between the generalized continuum hypothesis and large cardinal axioms. The one striking exception is a theorem of Solovay (1974): the existence of a strongly compact cardinal entails that there is no upper bound to the cardinals $\mathfrak{a}$ such that $2^{\mathfrak{a}} = \mathfrak{a}^+$.

As one might expect, the generalized continuum hypothesis leaves its trace on ordinary mathematics rather more faintly than the continuum hypothesis, but generally in the more abstract parts of the same branches. Its character is overwhelmingly that of a simplifying assumption: it simplifies cardinal arithmetic in the same sort of way as, but more radically than, the axiom of choice. For that reason it is not uncommon for mathematicians to use the generalized continuum hypothesis as an assumption in proving theorems even when it is not strictly needed (sometimes, indeed, when any competent set theorist would be able to see immediately that it is not needed).

This is emphatically not the speed-up phenomenon we encountered in §13.8: a proof that avoids the generalized continuum hypothesis is typically not vastly longer than the one that does not. It seems rather to be a case of *psychological* speed-up: the proof using the generalized continuum hypothesis is easier to find, especially if one is not expert in the niceties of cardinal arithmetic. In any case, a proof assuming the generalized continuum hypothesis is better than no proof at all: it entails relative consistency and hence demonstrates the futility of searching for a counterexample in **ZFC**; but also, because the generalized continuum hypothesis holds in the constructible hierarchy, it entails that the result holds if the quantifiers are restricted to constructible sets, and may therefore be seen as proving a sort of restricted case of the fully general result aimed for. This is not dissimilar to the perfectly ordinary mathematical practice, when attempting to prove a general theorem, of proving a restricted case first.

Exercise

Show that the generalized continuum hypothesis holds iff for any non-finite cardinals $\mathfrak{a}$ and $\mathfrak{b}$ either $\mathfrak{a} \leqslant \mathfrak{b}$ or $2^{\mathfrak{b}} \leqslant \mathfrak{a}$.

Notes

The axiom of choice and the generalized continuum hypothesis simplify cardinal arithmetic substantially, but it nevertheless remains a rich subject, well expounded by Bachmann (1955).

The early history of the continuum hypothesis is recounted by G. H. Moore (1989). The technical position is summarized by Martin (1976). Sierpinski (1934) lists a great many equivalent formulations. The status of the continuum hypothesis has been much discussed. Gödel 1947 is a good introduction. Woodin (2001*a*) gives arguments against believing it, while Feferman (2000) argues that the question does not have a determinate answer.

For more on the merits of assuming projective determinacy as an axiom, see Jensen 1995 and Maddy 1988.

Conclusion to Part IV

In the first part of this book we presented a theory of sets ZU; in the second we showed how to embed mathematics in this theory; and in the third we developed the theory of cardinals and ordinals within it. This work certainly made a strong case for the practical virtues of ZU as a theory: it is elegant and simple; and its axioms can be justified on the basis of the conception of sets as subject, through their intrinsic nature, to a primitive relation of dependence.

In this last part of the book, however, the neat picture has begun to fragment. The unitary account we were building up has split into several competing strands as we have come across different ways, some of them mutually inconsistent, of extending the system. The best-known point of bifurcation, of course, concerns the truth-value of the continuum hypothesis, and it is as good a case as any to focus on here. We have already cautioned against taking too seriously the popular analogy between Cohen's (1963) proof that the continuum hypothesis is independent of ZFC and the proof discovered a century earlier of the independence of the parallel postulate in geometry. But even if that point is disputed, there is in both cases reason to be cautious about the significance of purely formal results: the independence of the parallel postulate does not in itself show that non-Euclidean geometry describes a way that *space* could be; nor does Cohen's result show on its own that there are two competing theories of *sets*.

But saying that is of itself no help whatever in working out how we might be able to settle whether the continuum hypothesis is true. Naturally enough, attempts to make progress with this question have generally focused on the two principles of plenitude which we formulated in part I. The thought, in broad terms, is this. We are aware that first-order separation is a pale approximation to the full intended import of the first principle; and infinity and creation are inadequate to express the second. So there is plenty of scope for further axioms which simply express somewhat more of the originally intended meaning of one or other of the two principles.

As we have presented them, the two principles of plenitude are distinct, and it has sometimes been assumed that they are independent of each other. Many accounts of the matter nonetheless seem implicitly to give some sort of primacy to the first principle. The constructivist account, for instance, conceives of each stage in the hierarchy as being constructed *after* the preceding one, so the question of the richness of each level is prior (even in a temporal sense) to the question of how many levels there are in total. The most popular sort of platonist explanation, on the other hand, has been the one which is a limiting case of the constructivist account, and it seems to have been tacitly assumed that the priority just mentioned of the first principle over the second is preserved in the progress to this limit.

On the dependency account which we have favoured here, on the other hand, it is surely dubious whether the first principle is prior (even conceptually) to the second. The only reason to think that it is would be if we had distinguished between the sorts of possibility invoked in the two principles. For instance, if the first principle is represented as saying that for any level V the set $\{x \in V : Xx\}$ exists for any property X, one way we might try to express this is by saying that all subsets of V exist that are logically possible. When we come to express the second principle, on the other hand, we might do so by saying that all the levels exist that are possible, where the type of possibility invoked might now be narrower (e.g. metaphysical or conceptual possibility).

There is a difficulty at this point, of course, which we raised when we first stated this principle of plenitude. To say that there are as many levels as possible seems contradictory, because however many there are, there could have been more: just take the union of all of them. If we simply deny that there *is* such a union, we are left struggling to explain why not.

There is no denying that this *is* a real difficulty for the platonist. For the platonist, unlike the constructivist, conceives of the universe of sets as static. There are just the sets there are: they do not depend on my, or anyone else's, construction of them, and hence are not subject to any of the ordinary modalities of logical possibility or conceivability. And yet it seems to be of the *essence* of the conception that the second principle of plenitude urges on us to keep on trying to destabilize this static picture.

Now the standard platonist view invites us to see a striking asymmetry here. For there is, it is claimed, no similar instability in the first principle of plenitude. This asymmetry emerges very clearly in the second-order formulation, where it shows itself as a precise technical result, Zermelo's categoricity theorem. This tells us that once it is determined what the individuals are, the only variation between models of second-order set theory is their rank, i.e. how many levels there are. No variation is possible in the constitution of each level because we are forced by the second-order quantifier to include $\{x \in V : Xx\}$ for *all* properties X, and the range of this last quantifier is a matter for logic to settle, not set theory.

But even if this is right, and the second principle is *somehow* consequent upon the first, it does not follow automatically that its meaning is affected by the first. The standard conception, at any rate of platonists, has been that in *this* sense the two principles are independent. In other words, a difference in the richness of low levels in the hierarchy should make no difference to how many levels in the hierarchy there can be.

But notice how the technical results here concerning first-order set theory point in quite the opposite direction. The continuum hypothesis is third-order, i.e. it makes a claim about the first three infinite levels of the hierarchy. So at first sight its truth or falsity is answerable only to the first principle, which governs how rich each level is relative to the previous one, not to the second principle, which governs the total number of levels. Gödel discovered that there are sentences of the same logical shape as the continuum hypothesis which *are* settled by appealing to new axioms justified by the second principle. Now this does not yet *establish* any interaction between the two principles: it can be explained simply by a feature of the system we have *already* accepted, namely that the impredicativity of the axiom scheme of separation has the consequence that any new axiom of infinity increases the power of the scheme.

If the effect that strong axioms of infinity have on the low levels of the hierarchy does not quite *prove* that the second principle affects the first, it does nonetheless seem suggestive. The uncritical platonist can stamp his foot, of course, and insist that the range of the second-order quantifier is determinate: if so, there is probably nothing that can be done to dissuade him. But it is striking how little anyone can think of to say about how to determine this range more fully. We mentioned earlier that there is a sentence of pure second-order logic which is a logical truth iff the continuum hypothesis is true. But no one has suggested even the outline of a research programme to find out, by reflection within the domain of pure logic alone, whether this sentence is true. What this suggests (to me, at least) is that we ought to consider whether the second-order concept *all* might be indefinitely extensible in the *same* way that the concept *set* is. This thought is already familiar to constructivists: it has been advanced by Dummett (1978), for instance. But although *some* retreat from a wholly external platonism is no doubt necessary if this view is to seem at all attractive, I do not think it needs to be a retreat all the way to constructivism. A platonist with modestly internalist leanings might well feel cautious about the idea that logic can comprehend the notion of a *wholly* random property. Logicians are inclined to think of this on the model of a mathematical function, so that in general a property F is a sort of function taking an object as its argument. But what should it return as its value? Ramsey (1926) proposed an understanding of the second-order variables as ranging over what he called propositional functions in extension, which were functions from objects to *propositions*. But, as Sullivan (1995) has pointed out, that is not what we want, for if F is such a function and a is an object, 'Fa' will be a term referring to a proposition, whereas what we need if we are to use it in logic is a *sentence*.

Sullivan ends his article by declining to generalize beyond Ramsey's proposal and the Wittgensteinian framework within which it was intended to operate. But that seems wrong. It is no doubt true that the Wittgensteinian argument expounded by Sullivan does not constitute a wholly general refutation of any attempt to understand the range of second-order quantifiers extensionally. However, it takes only a modest internalism to see our failure to settle the continuum problem solely by some sort of second-order logical contemplation as a symptom of the fact that we have no extensional understanding of the range of second-order quantifiers as circumscribed from without. It would follow that our only understanding of this range is from within, as exhausted by more and more general sorts of predication.

If this is correct, the asymmetry which the external platonist sees between the two principles of plenitude disappears. The first principle is on this view every bit as problematic as the second, since it exhibits indefinite extensibility in just the same way. Moreover, the (rather weak) reasons we canvassed earlier for thinking of the first principle as conceptually prior to the second have disappeared.

The possibility which we are now free to countenance is that on an internal platonist view the symptoms of impredicativity which we have come across repeatedly, according to which higher axioms of infinity force the enrichment of low infinite levels of the hierarchy, might be not parochial features of the first-order formulation, but symptoms of the real state of things.

Appendices

Appendix A

Traditional axiomatizations

In chapter 13 I suggested that the regressive support for strengthening **ZU** by adding reflection or replacement is rather weak because almost all of current mathematics can be represented within **ZU** (and even without much use of the artifice of coding). But the theories that were current in the 1920s, when the proposal to adopt a theory equivalent to **ZFU** was first mooted, were significantly weaker than **ZU**, and so the regressive support for the strengthening was somewhat greater. To explain this point, we shall have to trace how these axiomatizations developed.

A.1 Zermelo's axioms

The centrepiece of Zermelo's original axiomatization was the axiom of choice, which we have discussed in chapter 14. If we put it to one side, his remaining axioms were approximately as follows.

Axiom of extensionality. *If* a, b *are sets*, $(\forall x)(x \in a \Leftrightarrow x \in b) \Rightarrow a = b$.

Axiom of individuals. *There is a set of all individuals.*

Axiom of separation. *For any definite property* X *and any set* a, $\{x \in a : Xx\}$ *is a set.*

Axiom of power sets. *If* a *is a set*, $\mathfrak{P}(a)$ *is a set.*

Axiom of union. *If* a *is a set*, $\bigcup a$ *is a set.*

Axiom of pairs. *If* a, b *are sets*, $\{a, b\}$ *is a set.*

Axiom of infinity″. *There is a set* U *such that* $\emptyset \in U$ *and* $(\forall a \in U)(\{a\} \in U)$.

The overwhelming majority of axiomatizations to be found in subsequent textbooks have been recognizably variants or extensions of this one.

Critics of these axioms at first focused their attention principally on separation: they objected to Zermelo's appeal to (what he took to be) the primitive notion of a 'definite' property on the ground of vagueness. The first workable proposal to correct the perceived deficiency was made by Weyl. Only some time later was it realized that Weyl's formulation amounted to the same thing as restricting separation to those properties expressible by a formula in the first-order language. The axiom of separation was therefore replaced by the following axiom scheme.

Axiom scheme of separation. *If $\Phi(x)$ is a formula, then the following is an axiom:*

> *If a is a set, $\{x \in a : \Phi(x)\}$ is a set.*

I shall, at the risk of some historical inexactitude, call the system with this adjustment 'Zermelo's theory': it is formal (or at any rate straightforwardly formalizable). Not only that but it is also weaker than **ZU**, in the sense that every axiom of Zermelo's theory can be read as a theorem of **ZU**.

A.2 Cardinals and ordinals

The converse is false, however: there are proofs in **ZU** that cannot be validly replicated in Zermelo's theory. One instance of this which was especially important for the historical development arises because Zermelo's theory is still, in the terminology I have been using in this book, a theory of *collections* rather than of sets: there is no axiom which rules out non-grounded collections. As a consequence we cannot in this theory use the Scott/Tarski method to define cardinals and ordinals. And in fact no other device will do instead: if in order to obtain a theory of cardinals we add Hume's principle as an axiom, what we get is (provided that we let the new operator 'card' occur in instances of the axiom of separation) a non-conservative extension of Zermelo's theory (Levy 1969). In fact, the salience of Zermelo's theory was in any case temporary. Soon the interest of set theorists focused on the *grounded* collections, and they were then led to add the following restrictive assumption.

Axiom of foundation. $a \neq \emptyset \Rightarrow (\exists x \in a)(x \text{ is an individual or } x \cap a = \emptyset)$.

Yet even with this axiom our theory still is not strong enough to support the efficient exploitation of its hierarchical structure. The definition of the cardinal of a set applied foundation to the *class* of all sets equinumerous with the given set, and this is not licensed by the axiom of foundation in the first-order form we have stated (Jensen and Schröder 1969; Boffa 1969).

So what should we do? For Zermelo himself at the time of 1908*b* no further axiom was needed: his purpose was to give a system of axioms that would ground his new proof (1908*a*) of the well-ordering principle from the axiom of choice. The key feature of this proof was that, unlike his earlier proof (1904) of the same result, it does not make use of ordinals. So in this case at least Zermelo had shown that ordinals could be regarded as ideal elements whose use in his proof was eliminable. For a time, indeed, there was considerable interest in the general project of eliminating uses of ordinals from proofs: Lindelöf showed in 1905 that Cantor's use of ordinals in proving the Cantor/Bendixson theorem was eliminable, and in 1922 Kuratowski gave a general method for eliminating them from a large class of mathematical contexts.

But mathematicians have never been inclined to give up fruitful techniques solely for reasons of foundational hygiene. The theory of ordinals is a convenient tool for the pure mathematician to have, so have it he will. What, then, can be done to give regressive support to this intransigence? One possibility would be to find an axiom that will allow the Scott/Tarski definitions of cardinals and ordinals to go through.

Axiom of transitive containment. *For each set there is a transitive set containing it.*

Call the resulting theory ZU_0''. On the obvious interpretation it becomes a subtheory of **ZU**, since all its axioms are recognizable as things we have already proved in **ZU**. However, it is in fact significantly weaker than **ZU**, but to see why, we need to look more closely at the form of Zermelo's axiom of infinity. Zermelo chose this form because it is equivalent to the claim that the set

$$\omega'' = \{\varnothing, \{\varnothing\}, \{\{\varnothing\}\}, \{\{\{\varnothing\}\}\}, \ldots\}$$

exists, and this is the set he intended to use as his surrogate for the natural numbers. If we were still working in the context of the theory of levels developed in chapter 3, the difference would be unimportant: Zermelo's axiom of infinity would imply the version of the axiom of infinity given in chapter 4. But this is because there we took the notion of a *level* as fundamental (although by using Scott's other trick we were able to avoid taking it as a primitive), whereas in Zermelo's theory the basic notion is that of a *set*. The consequence is that in a theory of the sort we set up in chapter 3, asserting the existence of one set guarantees the existence of a whole level, so that Zermelo's axiom of infinity would in that context have had the same effect as the axiom of infinity we actually stated there, namely to guarantee the existence of the *whole* level V_ω. But in the context of Zermelo's system his axiom does only what it says and not much more: it guarantees the existence of some sets of rank ω, such as Zermelo's set of natural numbers ω'' itself, but not of others such as the set

$$\omega' = \{\varnothing, \{\varnothing\}, \{\varnothing, \{\varnothing\}\}, \{\varnothing, \{\varnothing\}, \{\varnothing, \{\varnothing, \{\varnothing\}\}\}\}, \ldots\}$$

which von Neumann used as *his* set of natural numbers (Mathias 2001; Drabbe 1969). One consequence of this is that if we had wanted even a slightly

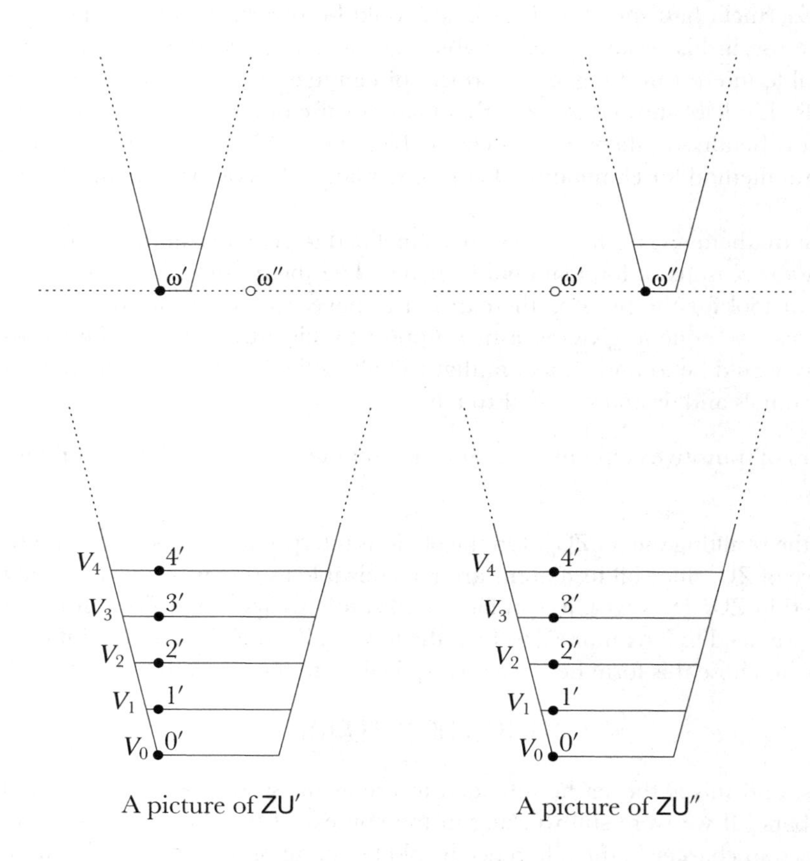

A picture of ZU′ A picture of ZU″

different theory of natural numbers we would have had to use a different axiom of infinity. If we had used the von Neumann definition, for example, we would have had to replace Zermelo's axiom of infinity with one ensuring that ω' exists, such as the following.

Axiom of infinity′. *There is a set U such that $\varnothing \in U$ and $(\forall a \in U)(a \cup \{a\} \in U)$.*

Call the theory that results if we make this substitution ZU'_0. As we have just seen, it is not equivalent to ZU''_0, but the significance of this fact depends to some extent on one's viewpoint. Someone whose belief in set theory is intuitive will deduce from this inequivalence that anyone who expresses a commitment to only one of the two theories is plainly telling us only part of what they believe, since there could hardly be any ground for the existence of ω' that is not an equally good reason to believe in ω''. From an intuitive perspective it would therefore be odd to regard either ZU'_0 *or* ZU''_0 as the whole of an axiomatization of set theory.

But consider now someone whose reason for believing the axioms is regressive. The body of knowledge which the regressive argument appeals to as its basis is presumably *mathematical* knowledge (some fragment of current mathematical practice) rather than set-theoretic knowledge. So the regressive argument can gain its grip only once we have shown how to embed mathematics (or at any rate the relevant fragment of it) in set theory. What the inequivalence of ZU'_0 and ZU''_0 does is merely to remind us that the regressive justification is relative to the embedding: if we had chosen a different embedding, we would have been justified in believing a slightly different set theory. Indeed the regressive justifier might even regard the inequivalence with approval, since it hints at a salutary parsimony in our choice of theory.

This cuts both ways, though: just because of its parsimony the theory is very inconvenient. Even if most mathematical reasoning *can* be embedded in ZU'_0, it will quite often take some knowledge of set theory to see how to do it. The restrictions ZU'_0 places on reasoning will seem to most mathematicians (who, after all, are not, and do not want to become, set theorists) to be arbitrary and unmotivated. And if we intend to use the Scott/Tarski definitions of cardinals and ordinals, we now have no reason to want ω' in our universe rather than ω'', and we might as well just assert the existence of *some* infinite set. The resulting theory will be adequate for embedding all the mathematics that can be represented in ZU, so there will be no regressive argument for preferring ZU to it.

In any case the historical development did not follow the course just outlined. We have already seen in part I that the iterative conception of set took some time to emerge, so that when the axiom of foundation was first mooted, it was regarded more as a convenient tool for metamathematical reasoning than as an evident truth. But even from this metamathematical perspective the iterative hierarchy was not well understood by most set theorists. In particular, the definitions of cardinals and ordinals that we have used in this book were not known until the 1950s. So the possibility of adding the axiom of transitive containment and using these definitions was not recognized.

Instead attention focused on adding axioms which allowed the von Neumann treatment of the ordinals to work. For this we would have to add some such axiom as the following.

Axiom of ordinals′. *Every well-ordered set is isomorphic to a von Neumann ordinal.*

Adding this axiom allows us to develop the theory of ordinals satisfactorily. But now what about the theory of cardinals? If we also assume the axiom of choice, we can use as a surrogate for the cardinal of a set the least von Neumann ordinal equinumerous with it (see chapter 14), but one of the points of interest, especially in the 1920s and 1930s, was to study cardinal arithmetic without the axiom of choice, and if that is our aim, we are still adrift. It seems that there is nothing for it but to add a new primitive operator 'card' and yet another axiom.

Axiom of cardinals. *For all sets A and B,* $\text{card}(A) = \text{card}(B)$ *iff A and B are equinumerous.*

A.3 Replacement

At this point we could be forgiven for thinking that things have got out of hand: our commitment to the regressive method will have to be quite unflinching if we are to find this jumble of axioms wholly acceptable. It is therefore easy to see the immediate appeal, on grounds of simplicity if on no others, of an axiom scheme which unifies this job lot of extra axioms into one form. So consider now what happens if, instead of transitive containment and the axiom of ordinals′, we add the following.

Axiom scheme of replacement. *If $\tau(x)$ is any term, this is an axiom:*

$$(\forall x \in a)(\tau(x) \text{ is a set}) \Rightarrow \{\tau(x) : x \in a\} \text{ is a set.}$$

We shall write ZFU′, for the system obtained from ZU_0' by deleting the axiom of transitive containment and adding this axiom scheme; and we shall write ZF′ for its pure variant. (The 'F' stands for Fraenkel, who was one of the devisors of replacement in the early 1920s.) Our choice of notation suggests that there is a connection between ZFU′ and ZFU, and indeed there is: ZFU′ is equivalent to ZFU as a theory of sets. In one direction we have already established this, of course: all the axioms of ZFU′ have already been shown to be theorems of ZFU. To establish the converse, we would have to work in ZFU′ and prove the axioms of ZFU. There is a certain amount of machinery we would have to develop again, mimicking the treatment already given in ZFU. But once we had done this, it would be possible to prove inductively in ZFU′ that the V_α are all levels and the axioms of creation and infinity follow at once from this.

The definition of what we have here been calling the von Neumann ordinals had already been identified by Zermelo in some unpublished work about 1915 (see Hallett 1984, pp. 270–80) and again by Mirimanoff (1917), but von Neumann was the first to see how the axiom scheme of replacement could be used to legitimate it. Fraenkel, who had stated the axiom scheme of replacement in 1922 before von Neumann, later recorded his surprise that there turned out to be this link between it and the theory of ordinals (Fraenkel 1967, p. 169).

What these mathematicians were discovering, and what made replacement so appealing, was in essence that it entails *all* the other axioms we have been considering as additions to Zermelo's theory with foundation, in particular the axiom of transitive containment and the axiom of ordinals′. The underlying reason for this, at least when viewed from the perspective of the iterative hierarchy, is that replacement acts as a hybrid: it asserts the existence of sets, such as von Neumann ordinals, of very high ranks, and hence has the effect of a higher axiom of infinity; but whenever a set exists it also entails the existence of its birthday. ZFU′ is thus the first of the traditional theories we have described in this Appendix that is strong enough to contain the whole of the theory of levels that we developed in part I.

The hybrid nature of replacement may be part of the explanation for the curious mixture of regressive enthusiasm and intuitive suspicion that it has engendered. Once we have seen how to treat cardinals and ordinals by the Scott/Tarski method, the regressive arguments in favour of replacement are overwhelmingly arguments for its ability to generate a hierarchy of levels rather than for its use as an axiom of infinity. Boolos (1971, p. 229) observes that the advantageous consequences of replacement 'include a satisfactory ... theory of infinite numbers, and a highly desirable result that justifies inductive definitions on well-founded relations'; but, as we have shown in this book, both of these are available in ZU, and hence they do not give any regressive support to replacement when viewed as a higher axiom of infinity.

Indeed it is striking, given how powerful an extension of the theory replacement represents, how thin the justifications for its introduction were. Skolem (1922) gives as his reason that 'Zermelo's axiom system is not sufficient to provide a complete foundation for the usual theory of sets', because the set $\{\omega, \mathfrak{P}(\omega), \mathfrak{P}(\mathfrak{P}(\omega)), \ldots\}$ cannot be proved to exist in that system; yet this is a good argument only if we have independent reason to think that this set does exist according to 'the usual theory', and Skolem gives no such reason. Von Neumann's (1925) justification for accepting replacement is only that

> in view of the confusion surrounding the notion 'not too big' as it is ordinarily used, on the one hand, and the extraordinary power of this axiom on the other, I believe that I was not too crassly arbitrary in introducing it, especially since it enlarges rather than restricts the domain of set theory and nevertheless can hardly become a source of antinomies. (In van Heijenoort 1967, p. 402)

But if the regressive argument for the axiom is not as strong as some have supposed, it is nevertheless given noticeably more weight here than in justifying the other axioms, even by authors otherwise inclined towards offering more intuitive justifications. Boolos (1971, p. 229), for instance, says explicitly that 'the reason for adopting the axioms of replacement is quite simple: they have many desirable consequences and (apparently) no undesirable ones.'

An emphasis on the regressive method of justification at the expense of the intuitive is often a sign of logical nervousness: Russell adopted the regressive method (1973*b*) only because he had been forced to include in his system an axiom (reducibility) which he saw no direct reason to believe was true. And when Frege (1893–1903, vol. II, p. 253) was told that his Basic Law V was contradictory, he admitted, in effect, that his reasons for assuming it had been largely regressive.

In the case of replacement there is, it is true, no widespread concern that it might be, like Basic Law V, inconsistent, but it is not at all uncommon to find expressed, if not by mathematicians themselves then by mathematically trained philosophers, the view that, insofar as it can be regarded as an axiom of infinity, it does indeed, as von Neumann (1925, p. 227) said, 'go a bit too far'. Putnam (2000, p. 24), for instance, admits, 'Quite frankly, I see no intuitive basis at all for ... the axiom of replacement. Better put, I do not see that a *notion* of set on which that axiom is clearly true has ever been *explained*.' And Boolos (2000) expresses at some length his discomfort with the ontological commitments of the theory that results if we assume it.

Notes

Mathias 2001 is a good account of the failure of systems such as ZU_0' and ZU_0'' to deliver a theory that can cope satisfyingly with transfinite recursions. Uzquiano (1999) shows that this is not wholly due to the weakness of first-order separation: the second-order versions of these theories do not deliver the hierarchy either.

Cantor's published works contain several assertions which are true only if the universe is rather large: for example, he claimed in (1883) that there is an aleph for every ordinal, although the assumptions which he made explicitly there do not imply this. It was only in work he chose not to publish that he stated (informally) a property approximating to the replacement principle. It re-appeared (again stated informally) in Mirimanoff 1917, Lennes 1922, and Fraenkel 1922*b*, and was given a precise first-order formulation by Skolem (1922); the name 'axiom of replacement' (*Ersetzungsaxiom*) under which it is nowadays known is due to Fraenkel.

Appendix B

Classes

An aggregate is something whose nature it is to be made up from other things. In §2.1 we distinguished two quite different conceptions of how this can happen, which we have called fusions and collections. But there does not seem to be much temptation to think of either fusions *or* collections as logical. This is not because we conceive of them as objects and logic is supposed not to be ontologically committed, but rather because the relation of constitution which is of their essence seems to be metaphysical rather than logical. But both these notions — fusions and collections — have been connected to (and at times confused with) other notions that do perhaps have title to be regarded as logical.

Consider fusions first. One reason they have often been thought to be ontologically innocent, and hence nominalistically acceptable, is that singular reference to a fusion is equivalent to plural reference to its parts. To say of my books that they are heavy is just to say of their fusion that it is. So the notion of a fusion is connected to the logical idea of plural quantification.

And in a rather similar manner the notion of a collection is connected to the logical idea of the *extension* of a property. This idea is standardly explained by means of tired examples such as the properties of having a heart and having a kidney. I am no physiologist, but I trust those logic textbooks which assure me that although they are different properties, they apply to just the same objects: apart from pathological cases such as people in the course of transplant surgery, any being with a heart also has a kidney and vice versa. Logicians express this by saying that the two predicates have different *intensions* but the same *extension*.

The correspondence between singular reference to a fusion and plural reference to its parts is not exact, however. We should therefore be cautious about too swiftly reducing one to the other. And we already know that there cannot be a wholly general correspondence between extensions of properties and collections because all properties have extensions but, as we saw in chapter 2, not all are collectivizing.

The fundamental property that makes extensions extensional is that properties have the same extensions just in case they have the same instances. If we use $[x : Xx]$ as a term to denote the extension of the property X, we can express extensionality by means of the second-order principle

$$(\forall x)(Xx \Leftrightarrow Yx) \Leftrightarrow [x : Xx] = [x : Yx].$$

We shall call this principle *Basic Law V* after Frege, who famously included it in the formal system of *Grundgesetze*.[1] It is an instance of a general method of introducing terms which is often known as *abstraction*, and for that reason it is sometimes referred to as an *abstraction principle*.

At this point we must tread carefully. For we need to be mindful of the reason for the lasting notoriety of Frege's principle. If we define

$$x \,\varepsilon\, y \Leftrightarrow (\exists X)(Xx \text{ and } y = [z : Xz]),$$

then

$$\begin{aligned} x \,\varepsilon\, [z : Fz] &\Leftrightarrow (\exists X)(Xx \text{ and } [z : Fz] = [z : Xz]) \\ &\Leftrightarrow (\exists X)(Xx \text{ and } (\forall z)(Fz \Leftrightarrow Xz)) \\ &\Leftrightarrow Fx, \end{aligned}$$

so that if we let $a = [x : x \not\varepsilon\, x]$, then

$$(\forall x)(x \,\varepsilon\, a \Leftrightarrow x \not\varepsilon\, x)$$

and therefore $a \,\varepsilon\, a \Leftrightarrow a \not\varepsilon\, a$, which is a contradiction.

What we have shown, therefore, is that the conjunction of the following three assumptions is contradictory:

(1) second-order logic;

(2) Basic Law V;

(3) the assumption that there is a single domain of objects over which all quantifiers range.

But we should not leap too hastily to judgement as to which of the three is guilty. The aim of this appendix is to explore how much of the content of the contradictory conjunction we should preserve.

B.1 Virtual classes

So let us begin again. To be definite, let us start with a prior theory U: the case which will interest us most is that in which U is a theory of sets such as

[1] To be strictly accurate, what Frege called Basic Law V was a somewhat more general principle, but the difference between the two is irrelevant to our current concerns.

ZU, but it will clarify the issues if for the moment we assume only that it is a conventionally formalized first-order system in a countable language. We introduce for each predicate Φ in the language of U a new term $[x : \Phi(x)]$, which we think of as denoting what we shall call a *class*. We want classes to be extensional entities obtained by a logical process of abstraction just as extensions were, so to express this idea we adjoin to U the following scheme.

Abstraction scheme. *If* Φ *and* Ψ *are formulae, then*

$$(\forall x)(\Phi(x) \Leftrightarrow \Psi(x)) \Leftrightarrow [x : \Phi(x)] = [x : \Psi(x)].$$

But mindful of our bad experience with Basic Law V we shall not assume as we did with extensions that classes are among the objects of which U already spoke. In order for the notation to be usable, though, we need a way of expressing the 'is' of predication as a relation between an object and an extension. This is conventionally done by means of the same sign '$\in$' that we have been using to express membership, but here we shall keep the two notions apart by using 'ε' for predication. We saw that in the second-order system this is definable, but in a first-order context that is no longer possible. So we shall for the moment treat '$y \; \varepsilon \; [x : \Phi(x)]$' merely as an abbreviation for $\Phi(y)$. General statements about classes can thus be represented at best schematically. We can if we wish use a single letter to stand for the class term $[x : \Phi(x)]$ here, but if we do so, it is important to remember that what such a letter stands for — the class term — is an incomplete symbol and not a name.

The point of using class terms in this way is really no more than notational convenience. A good example of this is that we might choose to adapt our notation for relativizing a sentence to a set and allow for relativizing to a class. Thus if $\mathbf{A}$ is a class term, we might write $(\forall x \; \varepsilon \; \mathbf{A})\Phi(x)$ for $(\forall x)(x \; \varepsilon \; \mathbf{A} \Rightarrow \Phi(x))$ and $(\exists x \; \varepsilon \; \mathbf{A})\Phi(x)$ for $(\exists x)(x \; \varepsilon \; \mathbf{A}$ and $\Phi(x))$; we could then write $\Phi^{(\mathbf{A})}$ for the result of relativizing all the quantifiers in Φ to $\mathbf{A}$ in this manner. The introduction of class terms thus permits us to mimic the notation $\Phi^{(a)}$ introduced in §2.6 for the case where a is a collection.

It is trivial to observe that extending our theory U by introducing class terms is *conservative* over U: that is to say, any sentence not mentioning classes that has a proof in the extended theory also has a proof in U. This is because all occurrences of class terms are trivially eliminable (see Quine 1969).

This much is wholly unsurprising. What is a little more unexpected, at least to anyone familiar with the inconsistency of Basic Law V which we demonstrated in the last section, is the following. I said that we did not *require* classes to be among the objects of the prior theory. In fact, if all that concerned us were logical consistency, there would be no need for us to be so permissive. The extension of U does not introduce an inconsistency even if we do require the class terms we have introduced to take their values in the domain of the prior theory U. Indeed it is even a conservative extension provided that U already entails the existence of infinitely many objects (Bell 1994).

This oddity is straightforwardly explained by the theory of cardinality we shall develop in part III. If the number of objects is $\mathfrak{a}$, the number of extensionally distinct properties is $2^{\mathfrak{a}}$. Since $\mathfrak{a} < 2^{\mathfrak{a}}$, there are not enough objects to go round and we obtain the familiar inconsistency. But if, as we are supposing here, U is a conventional first-order theory, there are only countably many formulae. So as long as our model is infinite, there is no difficulty about associating an object in its domain with each *formula* of the language so as to make all the instances of the scheme true. In other words, if Frege had restricted himself to a first-order language, his treatment of extensions of concepts as objects belonging to the domain over which the quantifiers range would not have been contradictory.

B.2 Classes as new entities

Suppose now that, emboldened by the harmlessness of speaking *as if* there were classes, we start using names as if referring to them. If we do this, we can no longer treat 'ε' as contextually defined, since we shall want to use it alongside class names, where its meaning cannot be unpacked contextually. We must therefore add 'ε' as a new logical primitive and add the following axiom scheme.

Epsilon scheme. *If Φ is any formula,*

$$x \mathrel{\varepsilon} [y : \Phi(y)] \Leftrightarrow \Phi(x).$$

Letters used to substitute for class terms are no longer metalinguistic as they were in the first proposal, but are an addition to the object language itself. We shall for the remainder of this chapter reserve upper-case bold type letters **A**, **B**, **C**, etc. for class names to keep them distinct from names of the entities referred to in the prior theory U. And we do need to keep them distinct: if we do not, and treat classes as invariably falling within the scope of the quantifiers of our prior theory U, then because ε is no longer defined contextually, the resulting theory fails to be conservative in the most spectacular way possible — by being inconsistent. The reason, of course, is that Russell's paradox applies to classes as much as to collections. If $\mathbf{A} = [x : x \not\mathrel{\varepsilon} x]$, then by the epsilon scheme

$$x \mathrel{\varepsilon} \mathbf{A} \Leftrightarrow x \not\mathrel{\varepsilon} x,$$

from which, if we do not place any restrictions on substitution, we obtain

$$\mathbf{A} \mathrel{\varepsilon} \mathbf{A} \Leftrightarrow \mathbf{A} \not\mathrel{\varepsilon} \mathbf{A}.$$

Now we should not be alarmed by this conclusion, but for a different reason from before. Russell's paradox for collections was unsurprising because it merely gave us an explicit example of a non-collectivizing property when we had no prior reason to expect that all properties *should* be collectivizing. Here, on the other hand, we do expect all properties to be class-forming, since classes are obtained from properties by a general logical process of abstraction; but precisely because they are obtained in this manner, they are to be conceived of as *new* entities not falling within the range of the quantifiers of our prior theory U, and Russell's argument merely demonstrates this explicitly.

Russell's paradox for classes is to be read, therefore, as demonstrating that if we treat classes as objects, they do not invariably fall within the range of the quantifiers of the prior theory. As long as we observe this restriction, the theory which introduces classes in this manner is still trivially conservative over the prior theory U just as the virtual theory discussed in the last section was: if we have a proposition in the language of U whose proof makes use of class names, it is a straightforward, mechanical matter to eliminate them and hence convert it into a proof in U of the same proposition. Moreover, the new proof obtained by this mechanical procedure will not be of significantly greater length than the old one. For this reason one might expect the application of the extended theory to be wholly uncontroversial. This is not so, however. What makes this step problematic is that it places a constraint on the quantifiers of our prior theory, which now do not range unrestrictedly over all the objects there are, but only over those that are not classes. But in that case it appears that wholly unrestricted quantification over *everything* is not possible within any formalism, since we can always extend our language as just described, whereupon the paradox will lead us to realize that the quantifiers of our prior theory did not in fact range over everything. We need to pause, therefore, to consider whether *this* should trouble us.

B.3 Classes and quantification

The view we are considering — that our quantifiers are not genuinely unrestricted but range only over the membership of some class $\mathbf{V}$ — seems to have originated with Russell, who persistently identified the ranges of objects which belong to classes with those over which quantification makes sense. 'When I say that a collection has no total, I mean that statements about *all* its members are nonsense.' (Russell 1908, p. 225 n.) But the view has also attracted widespread support since then. Indeed, according to Dummett, 'the one thing we may confidently say that no modern logician believes in' is wholly unrestricted quantification.

The one lesson of the set-theoretic paradoxes which seems quite certain is that we cannot interpret individual variables in Frege's way, as ranging simultaneously over the totality of all objects which could meaningfully be referred to or quantified over. This is why modern explanations of the semantics of first-order predicate calculus always require that a domain be *specified* for the individual variables: we cannot, as Frege supposed, rely on a once-for-all explanation that individual variables are always to be thought of as ranging over the totality of all objects. (1973, p. 567)

It seems unlikely that there was ever, as Dummett claims, a time when *no* logician believed in wholly unrestricted quantification. Moreover, Dummett's own understanding of the limitations on unrestricted quantification is not fully captured by the above quotation, as we shall see later when we turn to a discussion of his views. Nonetheless, there remains, irrespective of what Dummett himself thinks, the question whether it is right to maintain a rigid link between the legitimacy of quantification over some objects and the existence of a class to which they all belong.

In one direction, indeed, the entailment between the two notions seems clear: it is hard to see what objection there could be to quantifying over all the members of some class, and harder still to find an example of a logician who has suggested one. For if a class is, as we have said, a logical entity obtained by abstraction, it is surely incoherent to suppose that it could be so obtainable even though quantification over its members was impossible.

But what the argument in the opposite direction is supposed to be is more problematic. One that crops up persistently in the modern literature is based on the observation that the standard semantics for first-order logic may be couched in class-theoretic terms. A quantified sentence is said to be valid if it is true in every interpretation of the language; and an interpretation is understood as being a structure consisting of a class D (called the *domain* of the interpretation), a member of the domain for each constant of the language, a subclass of D for each unary predicate symbol, a relation on D for each binary predicate symbol, etc. (cf. §4.10). So, the argument goes, if we can interpret a quantified sentence meaningfully, we must be understanding the quantifiers occurring in it as ranging over the members of some domain D of interpretation, i.e. some class.

It should be clear, I hope, how unpersuasive this is. The standard semantics which the argument invokes is available only from the perspective of a metalanguage, and when we adopt this perspective, we are using words such as 'class' in whatever sense our grasp of that language gives to them. In the course of our discussion of classes, by contrast, we are speakers of the object language. That language has a word 'class' too, but it is a different word (because a word in a different language). While we are speaking this object language, we do not invoke a metalanguage semantics in order to explain what we mean: we simply speak the language.

Or, to put it another way, if it is legitimate to conceive of our variables as ranging over absolutely everything, then that is not a conception which *simultaneously* allows for a metalinguistic perspective. What the move to the metalanguage involves is indeed the conception of the world as a limited whole — as, if you wish to speak this way, a class — but that is not yet to say that if we carry on flat-footedly speaking the object language there is anything *inconsistent* about denying that limitation.

But this way of resolving the tension created by the argument from model-theoretic semantics immediately prompts a second, rather subtler account which argues that we cannot quantify over everything because it is in some way indeterminate what 'everything' amounts to. It is important, of course, to be clear what sort of indeterminacy is in question here. It is no doubt true that I have no clear conception of what, according to my representation of the world, there is; and, further, that no such clear conception is to be had because my representation just is insufficiently precise to determine it. But it is not immediately clear what *this* sort of indeterminacy has to do with the set-theoretic paradoxes, for it seems to be an essentially representational indeterminacy, whereas if the set-theoretic paradoxes demonstrate an indeterminacy at all, it is (at least according to the platonist) an indeterminacy in *what there is* and not merely in what we represent there to be. But if it is this last sort of indeterminacy that is in question, we need to say more if we are to explain why the platonist should regard it as a threat to the coherence of unrestricted quantification at all. After all, Russell's paradox does not seem to hint at any inherent vagueness as to which objects are classes.

Perhaps, though, the most convincing sort of argument against unrestricted quantification is one that starts from the natural first thought that universal quantification is analogous to conjunction, and existential quantification to disjunction. Thus if there are only finitely many objects $a_1, \dots, a_n$ in the domain of quantification,

$$(\forall x)\Phi(x) \Leftrightarrow \Phi(a_1) \text{ and } \Phi(a_2) \text{ and} \dots \text{ and } \Phi(a_n),$$
$$(\exists x)\Phi(x) \Leftrightarrow \Phi(a_1) \text{ or } \Phi(a_2) \text{ or} \dots \text{ or } \Phi(a_n).$$

And in the case where the domain of quantification is infinite, perhaps we can imagine there to be corresponding equivalences which only the constraints of our finite language prevent us from expressing. But it would be a mistake to treat these equivalences as expressing the whole *meaning* of the quantifiers, even in the finite case, because they omit to express the vital information that $a_1, \dots, a_n$ really are *all* the objects in the domain. Yet to conceive of them as exhausting the domain is, the thought goes, to conceive of the domain as circumscribed, as being a limited portion of a larger range of objects. And this is just what is required for there to be a class with just $a_1, \dots, a_n$ as its members.

This third sort of argument arises naturally from consideration of Wittgenstein's *Tractatus* (1922), in which it is closely linked with the second sort of argument already considered. For if $a_1, \dots, a_n$ are all the objects there are, then according to the *Tractatus* I would be capable of representing this to be the case only if I could adopt an essentially different (transcendental) perspective from which to see the world (the old world I represented from my former perspective) as a limited whole (and even then, one supposes, only imperfectly); but if I remain resolutely within the bounds of the original perspective, I cannot express (and hence cannot fully comprehend) these limits.

B.4 Classes quantified

If we treat classes as genuine objects and learn to live, however reluctantly, with the consequent realization that our prior quantifiers did not really range over everything, that need not prevent us from introducing a *new* kind of quantifier to range over the new objects. Let us proceed cautiously, however, and investigate first the case where we ban these new variables from occurring in the instances of the two schemes governing the behaviour of classes. (We shall discuss later whether this restriction is well-motivated.)

One interesting feature of this way of extending our theory is that because we are now permitted to quantify over class variables, any axiom scheme in U can be replaced by a single axiom in the new theory $\overline{U}$. The reason this is of interest is that if the original language has only a finite number of non-logical primitives, the extension will be finitely axiomatizable. (This is not quite trivial.) So the theory of classes we are considering provides us with a general method of moving from a first-order theory U axiomatized by finitely many schemes to a finitely axiomatized conservative extension $\overline{U}$.

This undoubtedly represents a significant extension of the theory we had before, for although our new theory $\overline{U}$ with class quantifiers is still conservative over U, the proof that it is is no longer trivial as it was in the case we discussed in §B.2. What the proof of conservativeness involves now is an explicit (finitistic) recipe for converting a proof in $\overline{U}$ of a sentence of the class-free language into a proof of the same sentence in U. But there is some work involved in doing this: we have first to rewrite the proof in cut-free form so that it has an especially simple structure amenable to proof-theoretic analysis. But it is a routine fact of proof theory that such rewritings, although mechanical, may involve a marked (often exponential) increase in length, and so the new proof will not in general be merely a trivial rewriting as in the earlier case. Quantifying over classes, although conservative, is thus a significant step to take because it has the potential to alter the lengths of proofs substantially.

But in a way the very fact that it is a significant step may provide a clue to the reason for taking it. Our new proof may, as we have noted, be markedly different from the old one: in particular, it may be much shorter. If so, our introduction of quantification over classes will have an instrumental justification: we have here, in fact, an example of a successful application of Hilbert's programme. For it was Hilbert (1925) who conceived the general project of instrumentally justifying our use of an extension of an already accepted theory U by proving finitistically that it is conservative. The case which most interested Hilbert was one in which the reason for accepting U is itself finitistic, whereas of course here we are envisaging that U may be a theory of sets such as Z or ZU which goes well beyond what the finitist would accept, but the shape of the justification for the *extension* is the same.

According to Quine's (1948) dictum that to be is to be the value of a variable, by adopting the quantified theory of classes we express a commitment to an ontology of classes. But Hilbert's programme suggests an alternative view, according to which the objects of the new theory may be regarded as no more than convenient fictions whose use is justified by the conservativeness proof: it is useful to argue *as if* such entities exist because doing so sometimes permits shorter proofs, but in every such case we know that if required we could in principle purge the proofs of all mention of the ideal elements so as to be left with only unobjectionable reference to the real objects to which our prior theory already committed us.

B.5 Impredicative classes

So far we have banned class quantifiers from occurring in the formulae in our two class schemes, the abstraction scheme and the epsilon scheme. Suppose now that we relax this restriction. We shall say that the classes which result from this relaxation are *weakly impredicative*. The new theory U_1 is still conservative over U. To see this, suppose that Φ is a sentence in the language of U which is not provable in U. Then by the completeness of first-order logic U has a set-theoretic model in which Φ is false. Now interpret the class variables as ranging over the subsets of the domain D of this model, and interpret any class term $[x : \Psi(x)]$ as referring to the set of those members of D which belong to the interpretation of Ψ in this model. With these definitions the structure becomes a model of U_1 in which Φ is still false, so Φ is not provable in U_1.

Notice that this argument is radically different from the proof of conservativeness of the predicative extension considered in §B.4. That earlier proof was finitistic — it explicitly converted proofs in the enlarged system into class-free proofs in U — whereas now our demonstration is model-theoretic. This has important consequences.

First, the demonstration itself makes use of impredicative set-theoretic methods in the metatheory. It can thus no longer be thought of as an instance of Hilbert's programme as he originally conceived it, since we cannot use it to convince an agnostic of the reliability of such methods.

Second, given a proof of Φ in U_1, our demonstration that there is a proof of Φ in U has been wholly non-constructive and, as is typical of non-constructive demonstrations, gives us no clue how to obtain a proof in U or (even approximately) how long to expect it to be. Indeed, model-theoretic demonstrations of this kind give us only the bare fact of existence: if there is a proof in U_1, there is a proof in U. They do not entitle us to expect *any* relationship between the proof in U_1 and the proof in U: in addition to differing markedly in length, they might also use quite different proof ideas.

B.6 Impredicativity

If we rejected the platonistic attempt to link the legitimacy of quantification with the existence of a collection to act as the domain of quantification, that would not yet be to refute the constructivist's more modest variant of this claim. For the constructivist may claim, as Dummett has famously done, that a domain is required not in order that quantification over the domain should be *comprehensible* but only in order that it should be guaranteed to deliver a determinate truth-value for each such sentence. So sentences quantifying over everything will be possible provided that we are willing to give up the law of excluded middle in application to them.

Now we saw in §2.5 that Dummett bases his argument for this conclusion on the notion of indefinite extensibility. I suggested then that in a wholly general setting it is difficult to see what the argument is. But the current context gives us more resources with which to gain a grip on the idea. The key point is encapsulated in Dummett's remark that

> the criterion for asserting something of all objects falling under a concept is an essential feature of that concept, but is not automatically given with the criterion for a given object's falling under it. (1994*b*, p. 338)

There is, that is to say, something further to our grasp of the universal generalization which goes beyond our grasp of each of its instances.

The clue to understanding this remark lies, I think, in our earlier attempts to find a platonistic argument to connect quantification with collectivization. The constructivist will suggest that those attempts were on the whole unsuccessful because they relied on a distinction (which the platonist of course accepts) between what there is and what I represent there to be. Let us see what happens if we collapse that distinction and identify one with the other.

Our best argument ended, let us recall, by observing that if we treat a universally quantified proposition as a (possibly finite) conjunction, we leave out the information that the objects we refer to in the conjunction really are all the objects there are. But suppose now that *what there is* is itself an indefinitely extensible concept. By hypothesis, then, there is a process which, when applied to all the objects there are, yields another object which is not among all the objects there are. Contradiction. The only way out is therefore to deny that we are capable of a determinate conception of what there is independent of any particular representation. Denying this leads us to see, in other words, that whenever we employ classical quantification, we inevitably quantify only over some limited totality which cannot exhaust everything there is (or indeed everything falling under any indefinitely extensible concept).

What makes this conclusion seem intolerable to the platonist, as we have seen, is that we seem so clearly capable of quantifying over a range that is wider than the membership of any particular limited domain. Indeed the act of seeing that a domain *is* limited seems explicable only by appeal to such a quantification. The Dummettian constructivist, however, has a way of resolving this tension by explaining unrestricted quantification schematically. We may represent this by saying that the unrestricted generalization $(\forall x)\Phi(x)$ expresses our willingness to assert any instance $\Phi(\sigma)$ obtained by instantiating the variable x with a term σ.

Notice, though, that this cannot be intended merely as expressing a formal rule. For if we were to read the scheme as representing only instantiations by the terms of a fixed formal language, we would have failed to capture fully the open-endedness of the conception to which the notion of an indefinitely extensible concept is supposed to lead us. A commitment to the universal generalization is intended to commit us to $\Phi(\sigma)$ for any σ that we may come in the *future* to recognize as a term, even if the formal rules of our language do not yet provide for it.

But it is important, the constructivist maintains, to observe that quantified sentences express something very different when understood in this manner from when they are understood classically, for now they 'can be interpreted only as expressing claims, not as making statements' (Dummett 1994*a*, p. 249). These two ways of interpreting assertoric utterances are, according to Dummett, fundamentally different. 'With a statement is associated a condition for its truth: if the condition is satisfied, the assertion is correct; if not, the statement is false and its assertion incorrect.' A claim, on the other hand, 'is to the effect that a certain intellectual or linguistic feat can be performed'.

> The difference between the two types of assertoric utterance is that the condition for the truth of a statement is independent of the speaker's or anyone else's abilities or epistemic condition, except, of course, where the statement was about him, whereas the condition for a claim to be justified always turns on what the speaker of someone he can call on to act for him is able to do. (Dummett 1994*a*, pp. 246–7)

The contrast Dummett is making here goes back to Ramsey, who referred to the claims expressed by genuinely unrestricted quantification as 'variable hypotheticals'.

> What have they in common with conjunctions, and in what do they differ from them? Roughly we can say that when we look at them subjectively they differ altogether, but

when we look at them objectively, i.e. at the conditions of their truth and falsity, they appear to be the same. $(x).\phi x$ differs from a conjunction because

(a) It cannot be written out as one.

(b) Its constitution as a conjunction is never used; we never use it in class-thinking except in its application to a finite class, i.e. we use only the applicative rule.

(c) (This is the same as (b) in another way.) It always goes beyond what we know or want. ... It expresses an inference we are at any time prepared to make, not a belief of the primary sort.

A belief of the primary sort is a map of neighbouring space by which we steer. It remains such a map however much we complicate it or fill in details. But if we professedly extend it to infinity, it is no longer a map; we cannot take it in or steer by it. Our journey is over before we need its remoter parts. (Ramsey 1931, pp. 237–8)

Dummett claims that although the distinction between statements and claims allows us to resolve our difficulty concerning unrestricted quantification, it does so only at a price: we have to give up the notion that classical logic applies to claims. But the difficulty we face now, as Ramsey recognized, is to explain what our grasp of the variable hypothetical comes to if it is *not* a proposition.

When we ask what would make it true, we inevitably answer that it is true if and only if every x has ϕ; i.e. when we regard it as a proposition capable of the two cases truth and falsity, we are forced to make it a conjunction, and to have a theory of conjunctions which we cannot express for lack of symbolic power.
(But what we can't say we can't say, and we can't whistle it either.)
If then it is not a conjunction, it is not a proposition at all; and then the question arises in what way can it be right or wrong. (Ramsey 1931, p. 238)

This is surely the crux of the matter. Dummett seems to portray the abandonment of the classical law of the excluded middle as both a consequence of adopting the constructivist position and a solution to the problem that led to it. Even if he is right about the former, it is hard to see why he should be right about the latter. Dummett (1994*a*, p. 246) assures us that 'the evaluation of a claim as justifiable is as objective a matter as the evaluation of a statement as true', but why?

B.7 Using classes to enrich the original theory

Taking as our starting point a theory U, we have been considering a series of extensions of U which take ever more seriously the claims to existence of classes obtained by abstracting from the properties formulable in the language of U. What we have not done so far, however, is to let the conception of classes thus obtained infect the original theory.

But consider now the case in which U is axiomatized by schemes. In other words, among the axioms of U are all the sentences of some form $\ldots\Phi\ldots$

for Φ any formula in the language of U. As we remarked earlier, a common motivation for laying down a theory of this form is that we already believe the second-order sentence

$$(\forall X) \ldots X \ldots \tag{1}$$

and adopt the first-order scheme as the best approximation to this that is expressible in our first-order language. But if this is our motivation for assenting to the scheme, then when we extend the language we should automatically extend this assent to cover all the instances in the extended language.

What this amounts to in the current case is that we should allow a second sort of impredicativity into our theory. Not only do we permit class terms $[x : \Phi(x)]$ to be formed in which the formula Φ itself quantifies over classes, but we also assent to all instances of the scheme $\ldots \Phi \ldots$ in which Φ is a formula of the extended language and thus may involve quantification over classes. Let us call the *strongly impredicative* theory thus obtained $\widetilde{U}$. In one sense this is a very natural step to take. We have already at the previous stage, when we adopted the weakly impredicative theory, accepted the coherence of quantification over all classes. So if Φ is any formula of the extended language, we must presumably accept the existence of a property possessed by all and only the objects x such that $\Phi(x)$. And if our acceptance of the scheme in U was based on an acceptance of the second-order principle (1), it follows that we should assent to all the axioms of $\widetilde{U}$ without demur.

Notice, though, where this has got us. The considerations just outlined have led us from a prior acceptance of U to the adoption of $\widetilde{U}$. But — and this is the crucial point — $\widetilde{U}$ will not typically be conservative over U. (More precisely, it will not be conservative unless U is so impoverished that it does not permit the arithmetization of its syntax.) What is remarkable about this is that the process which led us to advance from U to the strictly stronger theory $\widetilde{U}$ involved no more than taking seriously the ontological commitments involved in recognizing the existence of classes, i.e. extensional entities abstracted from properties.

Appendix C

Sets and classes

The versions of set theory to be found in the literature are, despite some family resemblances, enormously varied. But the picture becomes much clearer when one realizes that most of the variations can be classified quite succinctly. One dimension of variation we have already met concerns whether the theory permits there to be any individuals; a second concerns how many levels there are in the hierarchy; and a third concerns how rich our conception is, once a level is given, of the level that follows it. In this appendix we discuss a fourth sort of variation.

There is no set of all sets: that, we know, is a consequence of Russell's paradox (at any rate if we assume separation). But Russell's paradox does not prevent there being something else — a *class* of all sets — which is an extensional entity behaving in some respects as sets do. Theories of sets may thus be categorized according to whether they countenance such entities.

C.1 Adding classes to set theory

In the previous appendix we discussed the notion of a class in isolation from that of a set: everything we said was therefore at a very general level, and we made rather few assumptions about the nature of the prior theory U on which talk of classes was being superimposed. Let us now give up that neutrality and restrict our attention to the case which interests us in this book, namely that in which U is a theory of sets such as ZU or one of its extensions. If we apply any of the procedures we described in appendix B to such a theory, we obtain an extension in which we can talk of both sets and classes. Many authors — most notably von Neumann (1925), Bernays (1937) and Gödel (1938) — have opted for the proposal considered in §B.4 which gives rise to a finitely axiomatizable conservative extension $\overline{\mathsf{ZF}}$ of ZF: various books refer to versions of this theory as VB, or sometimes NBG.

But we may also consider the stronger, impredicative extension mentioned in §B.7. For the axioms of our original set theory, on which the theory of classes is being overlaid, include a scheme, the axiom scheme of separation (and ZF, the most popular version of set theory, includes another, the axiom scheme of replacement). We can consider extending our theory still further by adding to the axioms those instances of the schemes obtained by substituting for Φ formulae of the *extended* language. Few mathematicians have been inclined to adopt this kind of impredicativity. The principal exceptions are Morse (1965) and Kelley (1955), and the theory $\widetilde{\mathsf{ZF}}$ which we get from ZF if we extend it in this manner is known as MK. In MK, that is to say, we assert the axiom schemes of ZF not only, as before, when Φ is a sentence in the language of set theory, but also when it is allowed to contain class variables.

The significance of taking this step is that now, as we noted in §B.7, the extension process is *not* conservative: there are sentences in the language of sets provable in MK but not in ZF. (This is hardly surprising since to get MK we have added new set-existence axioms to our theory.) But if our motivation for including separation among our axioms in the first place was the one suggested in §3.5 — namely that it represents the best approximation expressible in our limited formal language to the second-order separation principle — then it is hard to see what reason we would have to resist extending separation in this manner, since the new version is evidently a closer approximation to the second-order principle than the old.

Thus we have shown how it is possible to extend set theory to allow for proper classes, either conservatively (von Neumann/Bernays/Gödel) or non-conservatively (Morse/Kelley). But should we bother? For most mathematical purposes it matters not at all whether we afforce our theory of sets with the language of classes, virtual or otherwise. The one obvious exception to this is category theory. In principle, the simplest facts about categories can be expressed in set theory without classes, but the methods for doing this are syntactically elaborate and non-intuitive. And we do not have to progress far into the subject to come across statements for which these methods of translation fail.

The authors of Bourbaki wrestled with this problem for some time before eventually abandoning the attempt to add a chapter on category theory to their *Théorie des Ensembles*. Simply adding classes is not the whole answer, however. Even in MK we would have to resort to coding tricks to represent functors between categories. So the easiest way to represent categories in set theory is by means of an intermediate universe of sets capable of belonging to sets higher in the hierarchy. One reason for the formulation of set theory in my 1990 was to facilitate this representation. This suggests that the issues raised by the set-theoretic representation lie in a different plane from the distinction between logical and mathematical concepts of aggregations that we are concerned with here.

C.2 The difference between sets and classes

There is evidently a natural way of associating with every set a the class $\overline{a} = [x : x \,\varepsilon\, a]$, and we then have

$$x \in a \Leftrightarrow x \,\varepsilon\, \overline{a}.$$

We know that we cannot simply identify sets and classes in all cases: the class $\mathbf{V} = [x : x = x]$, for instance, does not correspond to any set. But many mathematicians have been inclined to identify sets with the classes *they* correspond to, leaving the other classes as a special case, called *proper classes*.

If we do this, an obvious simplification of the theory results. Indeed, if we had planned to do this all along, there would have been no need for us to have introduced the first kind of variable, ranging only over sets: we could make do with only the second sort, ranging over classes. And there would be no need to distinguish between the two species of membership $\in$ and ε. Sets would simply be classes belonging to $\mathbf{V}$, and quantification over sets would be represented by restricted quantifiers '$\forall x \in \mathbf{V}$' and '$\exists x \in \mathbf{V}$' ranging over just such classes.

The question that remains to be addressed is whether we are right to make the identification of sets with classes which is required if the non-conservative Morse/Kelley extension is even to be possible. The idea of making some sort of distinction between sets and classes goes back to Cantor, who wrote letters to Dedekind and Hilbert in the late 1890s distinguishing what he called 'consistent' and 'inconsistent' classes. The first published system which adopts a distinction between sets and classes (von Neumann 1925) treats sets as a particular kind of class. But there is nothing in either Cantor or von Neumann to suggest that they conceived of the distinction, as we are doing here, as a categorical distinction between metaphysical and logical objects. And without that distinction the question we are addressing now — whether to identify sets with the classes they correspond to — cannot even arise.

There is, incidentally, a scattering of later writers who draw a distinction between a 'logical' and a 'combinatorial' notion of aggregation, but those who get that far generally do so only in order to make it plain that their remarks are directed only at one notion or the other. Moreover, it is not clear that the distinction as these writers conceive of it is quite the same as the one we want here. Writers who make such a distinction often seem to be interested mainly in the idea of arbitrariness: on the logical conception, a class is derived from a property, and hence all its elements share the property, whereas on the combinatorial conception, any elements whatever form a class, whether or not there is a property characterizing them. But this distinction makes little sense if taken on its own independent of any context of representation. For if there are some objects, however disparate, forming an aggregate in the combinatorial sense, then there *is* a property that characterizes just those things, namely the property of belonging to that aggregate.

The obvious retort is to say that what is meant is a property *expressible in language*. But the difficulty is to say which language. For any fixed formal language we can, by diagonalization, obtain a property not expressible in that language which, once recognized, we are forced to treat as legitimate. Or, to put it in Dummettian terms once more, the notion of property that we are appealing to here is indefinitely extensible (cf. Dummett 1978). For this reason it seems wrong to regard the distinction between sets and classes as simply one of arbitrariness, but that still does not answer our question about whether it is right to identify a set with the associated class.

A short answer is that if we cannot identify *every* class with a set, we should not identify *any*. Slightly less briefly, if we treat sets as a kind of class, we have to give up the explanation of the hierarchy of sets based on the notion of dependency, since that was derived from a conception of the metaphysical nature of sets, whereas classes, being logical, do not — we may suppose — have substantial metaphysical properties independent of the properties of the entities from which they are abstracted. So if we regard the classes $\bar{a}$ and $\bar{b}$ as derived from the sets a and b, we may, if we wish, think of $\bar{a}$ and $\bar{b}$ as inheriting from a and b whatever dependency relations hold between them. But if we simply *identify* a with $\bar{a}$ and b with $\bar{b}$, this collapses, and we are left without a route to the account of dependency which we used in chapter 3 to explain the paradoxes.

We can see the symptoms of this collapse if we reflect on the form of a theory which assimilates sets to classes from the outset. In such a theory we are told that among classes there are some (sets) that belong to other classes and some (proper classes) that do not. But why the difference? Why is there in such a theory no room for such a class as $\{\mathbf{V}\}$, for instance? The identification of sets as a kind of class has deprived us of the resources to respond.

C.3 The metalinguistic perspective

We mentioned in §4.10 the central idea of model theory that a (generalized) structure can be regarded as an interpretation of a formal language: the quantifiers of the language are interpreted as ranging over the members of the domain, a unary predicate symbol is interpreted by means of a subset of the domain, a binary predicate symbol by means of a relation on it, etc.

Now model theory is a branch of mathematics; and, like other branches of mathematics, it can be formalized in set theory. If this is done, the various claims it makes are transformed into statements in the theory of sets. There is a familiar distinction drawn at the beginning of logic between syntax and semantics. If we confine ourselves to *pure* model theory, the part which is concerned solely with semantic concepts, the process of formalization is relatively unproblematic. But we must be careful when we embark on the formalization of syntax as well, for at that point formal languages themselves become the objects of study. The reason this is dangerous is that set theory, the framework within which our study is being conducted, is itself a formal theory, and so the theorems of logic are (apparently) applicable to it. The danger lies in the fact that in order to study our language we de-interpret it — treat it as consisting of strings of symbols without meaning. What we must not do is to confuse these strings of signs with the symbols they become when we read them as part of our own language.

The paradox of the set of all sets provides a ready example of the difficulty. Set-theoretic semantics provides a means of interpreting the sentences of a formal theory in such a way that all the quantifiers in them are relativized to some *set*, the domain of the interpretation. In the *intended* interpretation of set theory itself, the quantifiers range over all sets, so set-theoretic semantics would require the domain of the intended interpretation to be the set of all sets. But there is no set of all sets. Contradiction.

But this is simply the set-theoretic version of an argument we considered for classes in §B.3 and rests on a similarly bad pun. When we use set-theoretic semantics to talk about our language, there is an inevitable shift of perspective and our words change meaning. Our set-theoretic language becomes an object language, while the semantics is now conducted in the metalanguage. In the metalanguage we do indeed interpret all the quantifiers in our object language as ranging over all sets (in the object language sense), but there is nothing contradictory about the idea that there should be a set (in the metalanguage sense) which has them all as members.

So one way of understanding classes is as the 'sets' of the metalanguage. On this way of talking, the models of our object language set theory will be sets only from the metalinguistic perspective: from the object language perspective they are *classes*, not sets.

We may also if we wish talk in the object language about set-theoretic models of 'set theory'. But if we do, we must remember that the 'set theory' of which they are models is not the same as the theory within which our talk is formulated but a formal simulacrum of it which we have synthesized.

Notes

The account of the distinction between sets and classes in this appendix is greatly indebted to Parsons (1974). Other authors who have tried to draw a principled distinction between sets and classes include Mayberry (2000, §3.5), Maddy (1983) and Simmons (2000).

The problems of representing category theory inside set theory, which we gestured towards in §C.1, have been much discussed: see Feferman (1977), McLarty (1992, ch. 12, §1), and Borceux (1994, ch. 1, §1). The solution of assuming the existence of an intermediate universe, which I adopted in my earlier book (1990), is discussed by Mac Lane (1969). For another possible solution see Muller 2001.

Pudlak (1998, §7) summarizes what is known about the relative lengths of proofs in ZF and VB.

References

Abian, A. (1974), 'Nonstandard models for arithmetic and analysis', *Studia Logica,* 33: 11–22

Aczel, P. (1988), *Non-well-founded Sets*, CSLI Lecture Notes, 14, CSLI, Stanford

Aken, J. V. (1986), 'Axioms for the set-theoretic hierarchy', *J. Symb. Logic,* 51: 992–1004

Alexandroff, P. S. (1916), 'Sur la puissance des ensembles mesurables B', *C. R. Acad. Sci. Paris,* 162: 323–5

Aristotle (1971), *Metaphysics: Books* Γ, Δ *and* E, Oxford: Clarendon Press

Artin, E. and Schreier, O. (1927), 'Algebraische Konstruktion reeller Körper', *Abh. math. Sem. Univ. Hamburg,* 5: 83–115

Ax, J. and Kochen, S. (1965), 'Diophantine problems on local fields', *Amer. J. Math.,* 87: 605–30

Bachmann, H. (1955), *Transfinite Zahlen*, Berlin: Springer

Baire, R. (1898), 'Sur les fonctions discontinues qui se rattachent avec fonctions continues', *C. R. Acad. Sci. Paris,* 126: 1621–3

——— (1909), 'Sur la représentation des fonctions discontinues II', *Acta Math.,* 32: 97–176

Balaguer, M. (1998), *Platonism and Anti-platonism in Mathematics*, Oxford University Press

Banach, S. and Tarski, A. (1924), 'Sur la décomposition des ensembles de points en parties respectivement congruentes', *Fund. Math.,* 6: 244–77

Barwise, J., ed. (1977), *Handbook of Mathematical Logic*, Amsterdam: North-Holland

Barwise, J. and Etchemendy, J. (1987), *The Liar*, Oxford University Press

Bell, J. L. (1977), *Boolean-valued Models and Independence Proofs in Set Theory*, Oxford University Press

——— (1994), 'Fregean extensions of first-order theories', *Math. Logic Quart.,* 40: 27–30

Benacerraf, P. (1965), 'What numbers could not be', *Phil. Rev.,* 74: 47–73

Benacerraf, P. and Putnam, H., eds (1964), *Philosophy of Mathematics: Selected Readings*, Oxford: Blackwell

Bendixson, I. (1883), 'Quelques théorèmes de la théorie des ensembles de points', *Acta Math.*, 2: 415–29

Berkeley, G. (1734), *The Analyst, or, A Discourse addressed to an Infidel Mathematician*, London: Tonson (repr. in Ewald 1996, pp. 60–92)

Bernays, P. (1935), 'Sur le platonisme en mathématiques', *Enseignement Math.*, 34: 52–69 (trans. in Benacerraf and Putnam 1964, pp. 258–71)

——— (1937), 'A system of set theory I', *J. Symb. Logic*, 2: 65–77

——— (1942), 'A system of set theory III', *J. Symb. Logic*, 7: 65–89

Bernstein, F. (1905*a*), 'Über die Reihe der transfiniten Ordnungszahlen', *Math. Ann.*, 60: 187–93

——— (1905*b*), 'Untersuchungen aus der Mengenlehre', *Math. Ann.*, 61: 117–55

——— (1908), 'Zur Theorie der trigonimetrischen Reihen', *Berichte Verhandl. Königl. Sächs. Gesell. Wiss. Leipzig, Math.-phys. Kl.*, 60: 325–38

Bettazzi, R. (1896), 'Gruppi finiti ed infiniti di enti', *Atti Accad. Sci. Torino, Cl. Sci. Fis. Mat. Nat.*, 31: 506–12

Birkhoff, G. (1975), 'Introduction', *Hist. Math.*, 2: 535

Blair, C. E. (1977), 'The Baire category theorem implies the principle of dependent choices', *Bull. Acad. Polon. Sci., Sér. Sci. Math., Astron. Phys.*, 25: 933–4

Bochner, S. (1928), 'Fortsetzung Riemannscher Flaschen', *Math. Ann.*, 98: 406–21

Boffa, M. (1969), 'Axiome et schéma de fondement dans le système de Zermelo', *Bull. Acad. Polon. Sci., Sér. Sci. Math.*, 17: 113–15

Bolzano, B. (1851), *Paradoxien des Unendlichen*, Leipzig: Reclam

Boolos, G. (1971), 'The iterative conception of set', *J. Phil.*, 68: 215–31

——— (1975), 'On second-order logic', *J. Phil.*, 72: 509–27

——— (1987), 'A curious inference', *J. Phil. Logic*, 16: 1–12

——— (1989), 'Iteration again', *Phil. Topics*, 17: 5–21

——— (1993), 'Whence the contradiction?', *Proc. Arist. Soc., Supp. Vol.*, 67: 213–33

——— (1994), 'The advantages of honest toil over theft', in George 1994, pp. 27–44

——— (2000), 'Must we believe in set theory?', in Sher and Tieszen 2000, pp. 257–68

Borceux, F. (1994), *Handbook of Categorical Algebra I: Basic Category Theory*, Cambridge University Press

Borel, E. (1898), *Leçons sur la théorie des fonctions*, Paris: Gauthiers-Villars

——— (1905), 'Quelques remarques sur les principes de la théorie des ensembles', *Math. Ann.*, 60: 194–5

Borel, E., Baire, R., Hadamard, J. and Lebesgue, H. (1905), 'Cinq lettres sur la théorie des ensembles', *Bull. Soc. Math. France,* 33: 261–73

Bourbaki, N. (1939), *Théorie des ensembles (Fascicule de résultats)*, Actualités scientifiques et industrielles, 858, Paris: Hermann

——— (1949*a*), 'Foundations of mathematics for the working mathematician', *J. Symb. Logic,* 14: 1–8

——— (1949*b*), 'Sur le théorème de Zorn', *Arch. Math.*, 2: 434–7

——— (1954), *Théorie des ensembles, chs I et II,* Actualités scientifiques et industrielles, 1212, Paris: Hermann

——— (1956), *Théorie des ensembles, ch. III*, Actualités scientifiques et industrielles, 1243, Paris: Hermann

Brandl, J. L. and Sullivan, P., eds (1998), *New Essays on the Philosophy of Michael Dummett*, Amsterdam: Rodopi

Burali-Forti, C. (1896), 'Sopra un teorema del sig. G. Cantor', *Atti Accad. Sci. Torino, Cl. Sci. Fis. Mat. Nat.*, 32: 229–37

——— (1897*a*), 'Sulle classi ben ordinate', *Rend. Circ. Mat. Palermo,* 11: 260

——— (1897*b*), 'Una questione sui numeri transfiniti', *Rend. Circ. Mat. Palermo,* 11: 154–64

Buss, S. R. (1994), 'On Gödel's theorems on lengths of proofs I: Number of lines and speedup for arithmetics', *J. Symb. Logic,* 59: 737–56

Butts, R. E. and Hintikka, J., eds (1977), *Logic, Foundations of Mathematics, and Computability Theory*, University of Western Ontario Series in Philosophy of Science, 9, Dordrecht: Reidel

Campbell, P. J. (1978), 'The origins of Zorn's lemma', *Hist. Math.*, 5: 77–89

Cantor, G. (1872), 'Über die Ausdehnung eines Satzes aus der Theorie der trigonimetrischen Reihen', *Math. Ann.*, 5: 123–32

——— (1874), 'Über eine Eigenschaft des Inbegriffes aller reellen algebraischen Zahlen', *J. reine angew. Math.*, 77: 258–62

——— (1878), 'Ein Beitrag zur Mannigfaltigkeitslehre', *J. reine angew. Math.*, 84: 242–58

——— (1883), *Grundlagen einer allgemeinen Mannigfaltigkeitslehre. Ein mathematisch-philosophischer Versuch in der Lehre des Unendlichen*, Leipzig: Teubner

——— (1886), 'Über die verschiedenen Standpunkte in bezug auf das aktuelle Unendliche', *Zeitschr. Phil. phil. Krit.*, 88: 224–33

——— (1887), 'Mitteilungen zur Lehre vom Transfiniten I', *Zeitschr. Phil. phil. Krit.*, 91: 81–125

——— (1892), 'Über eine elementare Frage der Mannigfaltigkeitslehre', *Jahresber. Deutsch. Math.-Ver.*, 1: 75–8

——— (1895), 'Beiträge zur Begründung der transfiniten Mengenlehre I', *Math. Ann.*, 47: 481–512

——— (1897), 'Beiträge zur Begründung der transfiniten Mengenlehre II', *Math. Ann.*, 49: 207–46

——— (1932), *Gesammelte Abhandlungen*, Berlin: Springer

——— (1991), *Briefe*, Berlin: Springer

Carnap, R. (1931), 'Die logizistische Grundlegung der Mathematik', *Erkenntnis*, 2: 91–105

Cartan, H. (1943), 'Sur le fondement logique des mathématiques', *Rev. Sci. (Rev. Rose)*, 81: 3–11

Cauchy, A. L. (1821), *Cours d'analyse*, Paris: de Bure

——— (1844), *Exercise d'Analyse et de Physique Mathématique*, Vol. 3, Paris: Bachelier

Cegielski, P. (1981), 'La théorie élémentaire de la multiplication est conséquence d'un nombre fini d'axiomes', *C. R. Acad. Sci. Paris*, 290: 351–2

Clark, P. (1993*a*), 'Dummett on indefinite extensibility', *Proc. Arist. Soc., Supp. Vol.*, 67: 235–49

——— (1993*b*), 'Logicism, the continuum and anti-realism', *Analysis*, 53: 129–41

——— (1998), 'Dummett's argument for the indefinite extensibility of set and real number', in Brandl and Sullivan 1998, pp. 51–63

Cohen, P. J. (1963), 'The independence of the continuum hypothesis', *Proc. Natl. Acad. Sci. USA*, 50: 1143–8

——— (1966), *Set Theory and the Continuum Hypothesis*, Reading, MA: Benjamin

——— (1973), 'A large power set axiom', *J. Symb. Logic*, 40: 48–54

Corry, L. (1996), *Modern Algebra and the Rise of Mathematical Structures*, Basel: Birkhäuser

Craig, W. and Vaught, R. (1958), 'Finite axiomatizability using additional predicates', *J. Symb. Logic*, 23: 289–308

Dales, H. G. and Oliveri, G., eds (1998), *Truth in Mathematics*, Oxford University Press

Dauben, J. W. (1990), *Georg Cantor: His Mathematics and Philosophy of the Infinite*, repr. edn, Princeton University Press

Davenport, J. H., Siret, Y. and Tournier, E. (1993), *Computer Algebra: Systems and Algorithms for Algebraic Computation*, 2nd edn, London: Academic Press

Davis, M. (1964), 'Infinite games of perfect information', in M. Dresher, L. S. Shapley and A. W. Tucker, eds, *Advances in Game Theory*, Princeton University Press, pp. 85–101

Dedekind, R. (1872), *Stetigkeit und irrationale Zahlen*, Braunschweig: Vieweg

——— (1888), *Was sind und was sollen die Zahlen?*, Braunschweig: Vieweg (trans. in Ewald 1996, pp. 787–832)

——— (1932), *Gesammelte mathematische Werke*, Braunschweig: Vieweg

Devlin, K. (1984), *Constructibility*, Berlin: Springer

Dieudonné, J. A. (1970), 'The work of Nicholas Bourbaki', *Amer. Math. Monthly*, 77: 134–45

Dirichlet, G. P. L. (1837), 'Beweis des Satzes, dass jede unbegrenzte arithmetische Progression, deren erstes Glied und Differenz ganze Zahlen ohne gemeinschaftlichen Factor sind, unendlich viele Primzahlen enthält', *Abh. Königl. Preuss. Akad. Wiss.*, 34: 45–81

Dixon, A. C. (1905), 'On "well-ordered" aggregates', *Proc. London Math. Soc.*, 4: 18–20

Doets, K. (1999), 'Relatives of the Russell paradox', *Math. Logic Quart.*, 45: 73–83

Drabbe, J. (1969), 'Les axiomes de l'infini dans la théorie des ensembles sans axiome de substitution', *C. R. Acad. Sci. Paris*, 268: 137–8

Drake, F. R. (1974), *Set theory: An Introduction to Large Cardinals*, Amsterdam: North-Holland

——— (1989), 'On the foundations of mathematics in 1987', in H.-D. Ebbinghaus, J. Fernandez-Prian, M. Garrido, D. Lascar and M. R. Artadejo, eds, *Logic Colloquium '87*, Amsterdam: North-Holland, pp. 11–25

Dugac, P. (1976), *Richard Dedekind et les fondements des mathématiques*, Paris: Vrin

Dummett, M. (1973), *Frege: Philosophy of Language*, London: Duckworth

——— (1978), 'The philosophical significance of Gödel's theorem', in *Truth and other enigmas*, London: Duckworth, pp. 186–201

——— (1991), *Frege: Philosophy of Mathematics*, London: Duckworth

——— (1993), 'What is mathematics about?', in *The seas of language*, Oxford: Clarendon Press, pp. 429–45

——— (1994*a*), 'Chairman's address: Basic Law V', *Proc. Arist. Soc.*, 94: 243–51

——— (1994*b*), 'Reply to Wright', in B. McGuinness and G. Oliveri, eds, *The Philosophy of Michael Dummett*, Dordrecht: Kluwer, pp. 329–38

Errera, A. (1952), 'Le problème du continu', *Atti Accad. Ligure Sci. Lett. (Roma)*, 9: 176–83

Ewald, W. B., ed. (1996), *From Kant to Hilbert: A Source Book in the Foundations of Mathematics*, Oxford: Clarendon Press

Feferman, S. (1965), 'Some applications of the notions of forcing and generic sets', *Fund. Math.*, 56: 325–45

——— (1977), 'Categorical foundations and foundations of category theory', in Butts and Hintikka 1977, pp. 149–69

——— (2000), 'Why the programs for new axioms need to be questioned', *Bull. Symb. Logic,* 6: 401–13

Feferman, S. and Levy, A. (1963), 'Independence results in set theory by Cohen's method', *Not. Amer. Math. Soc.,* 10: 593

Field, H. (1998), 'Which undecidable mathematical sentences have determinate truth values', in Dales and Oliveri 1998, pp. 291–310

Fine, K. (1995), 'Ontological dependence', *Proc. Arist. Soc.,* 95: 267–90

Foreman, M., Magidor, M. and Shelah, S. (1988), 'Martin's maximum, saturated ideals and non-regular ultrafilters I', *Ann. Math.,* 127: 1–47

Forster, T. E. (1995), *Set Theory with a Universal Set: Exploring an Untyped Universe,* Oxford logic guides, 31, 2nd edn, Oxford: Clarendon Press

Fraenkel, A. A. (1922*a*), 'Über den Begriff "definit" und die Unabhängigkeit des Auswahlaxioms', *Sitzungsber. Preuss. Akad. Wiss., Phys.-math. Kl.,* pp. 253–7

——— (1922*b*), 'Zu den Grundlagen der Cantor-Zermeloschen Mengenlehre', *Math. Ann.,* 86: 230–7

——— (1967), *Lebenskreise: Aus den Erinnerung eines jüdischen Mathematikers,* Stuttgart: Deutsche Verlags-Anstalt

Fraenkel, A. A., Bar-Hillel, Y. and Levy, A. (1958), *Foundations of Set Theory,* Amsterdam: North-Holland

Frege, G. (1879), *Begriffsschrift, eine der arithmetischen nachgebildete Formelsprache des reinen Denkens,* Halle: Nebert

——— (1884), *Die Grundlagen der Arithmetik,* Breslau: Koebner (trans. as Frege 1953)

——— (1893–1903), *Grundgesetze der Arithmetik,* Jena: Pohle (partially trans. in Frege 1980, pp. 117–224)

——— (1895), 'Kritische Beleuchtung einiger Punkte in E. Schröders *Vorlesungen über die Algebra der Logik*', *Archiv für systematische Philosophie,* 1: 433–56 (trans. in Frege 1980, pp. 86–106)

——— (1906), 'Über die Grundlagen der Geometrie II', *Jahresber. Deutsch. Math.-Ver.,* 15: 293–309, 377–403, 423–30

——— (1953), *The Foundations of Arithmetic,* 2nd edn, Oxford: Blackwell

——— (1980), *Translations from the Philosophical Writings of Gottlob Frege,* 3rd edn, Oxford: Blackwell

Friedman, H. (1971), 'Higher set theory and mathematical practice', *Ann. Math. Logic,* 2: 326–57

——— (1986), 'Necessary uses of abstract set theory in finite mathematics', *Advances in Mathematics,* 60: 92–122

——— (1998), 'Finite functions and the necessary use of large cardinals', *Ann. Math.,* 148: 803–93

——— (2000), 'Normal mathematics will need new axioms', *Bull. Symb. Logic,* 6: 434–46

Gale, D. and Stewart, F. M. (1953), 'Infinite games with perfect information', in *Contributions to the Theory of Games,* Annals of Mathematics Studies, 28, Princeton University Press, pp. 245–66

Gauss, C. F. (1860–65), *Briefwechsel zwischen C. F. Gauss und H. C. Schumacher,* Altona: Esch

Gentzen, G. (1936), 'Die Widerspruchsfreiheit der reinen Zahlentheorie', *Math. Ann.,* 112: 493–565

George, A., ed. (1994), *Mathematics and Mind,* Oxford University Press

Gödel, K. (1931), 'Über formal unentscheidbare Sätze der *Principia Mathematica* und verwandter Systeme I', *Monatsh. Math. Physik,* 38: 173–98

——— (1933), 'Zur intuitionistischen Arithmetik und Zahlentheorie', *Ergebnisse eines mathematischen Kolloquiums,* 4: 34–8

——— (1938), 'The consistency of the axiom of choice and the generalized continuum hypothesis', *Proc. Natl. Acad. Sci. USA,* 24: 556–7

——— (1944), 'Russell's mathematical logic', in Schilpp 1944, pp. 123–53

——— (1947), 'What is Cantor's continuum problem?', *Amer. Math. Monthly,* 54: 515–25

——— (1965), 'Remarks before the Princeton bicentennial conference, 1946', in M. Davis, ed., *The Undecidable,* Hewlett, NY: Raven Press, pp. 71–3

——— (1986–2003), *Collected Works,* Oxford University Press

Goldfarb, W. (1979), 'Logic in the twenties: the nature of the quantifier', *J. Symb. Logic,* 44: 351–68

Good, C. and Tree, I. (1995), 'Continuing horrors of topology without choice', *Topology Appl.,* 63: 79–90

Goodstein, R. L. (1944), 'On the restricted ordinal theorem', *J. Symb. Logic,* 9: 33–41

Grassmann, H. (1861), *Lehrbuch der Arithmetik für höhere Lehranstalten,* Berlin: Enslin

Gray, R. (1994), 'Georg Cantor and transcendental numbers', *Amer. Math. Monthly,* 101: 819–32

Hale, B. and Wright, C. (2001), *The Reason's Proper Study,* Oxford: Clarendon Press

Hallett, M. (1984), *Cantorian Set Theory and Limitation of Size,* Oxford Logic Guides, 10, Oxford: Clarendon Press

Hardy, G. H. (1904), 'A theorem concerning the infinite cardinal numbers', *Quart. J. Math.,* 35: 87–94

——— (1906), 'The continuum and the second number class', *Proc. London Math. Soc.*, 4: 10–17

——— (1910), *Pure Mathematics*, Cambridge University Press

Hardy, G. H. and Wright, E. M. (1938), *An Introduction to the Theory of Numbers*, Oxford: Clarendon Press

Hartogs, F. (1915), 'Über das Problem der Wohlordnung', *Math. Ann.*, 76: 438–43

Hausdorff, F. (1908), 'Grundzüge einer Theorie der geordneten Mengen', *Math. Ann.*, 65: 435–505

——— (1909), 'Die Graduierung nach dem Endverlauf', *Abh. math.-phys. Kl. Königl. Sächs. Gesell. Wiss.*, 31: 295–334

——— (1914), *Grundzüge der Mengenlehre*, Leipzig: Veit

Heck, R. (1993), 'Critical notice of Michael Dummett, *Frege: Philosophy of Mathematics*', *Phil. Quart.*, 43: 223–33

Heine, E. (1872), 'Die Elemente der Functionlehre', *J. reine angew. Math.*, 74: 172–88

Henkin, L., Smith, W. N., Varineau, V. J. and Walsh, M. J. (1962), *Retracing Elementary Mathematics*, New York: Macmillan

Henle, J. M. and Kleinberg, E. M. (1980), *Infinitesimal Calculus*, Cambridge, MA: MIT Press

Henry, D. P. (1991), *Medieval Mereology*, Amsterdam: Grüner

Henson, C. W. and Keisler, H. J. (1986), 'On the strength of nonstandard analysis', *J. Symb. Logic*, 51: 377–86

Hermite, C. (1873), 'Sur la fonction exponentielle', *C. R. Acad. Sci. Paris*, 77: 18–24

Hilbert, D. (1900), 'Mathematische Probleme', *Nachr. Königl. Gesell. Wiss. Göttingen*, pp. 253–97

——— (1925), 'Über das Unendliche', *Math. Ann.*, 95: 161–90

Hilbert, D. and Ackermann, W. (1928), *Grundzüge der theoretischen Logik*, Die Grundlehren der mathematischen Wissenschaften, 27, Berlin: Springer

Hintikka, J. (1998), *Language, Truth and Logic in Mathematics*, Dordrecht: Kluwer

Hobson, E. W. (1905), 'On the general theory of transfinite numbers and order types', *Proc. London Math. Soc.*, 3: 170–88

——— (1921), *The Theory of Functions of a Real Variable, and the Theory of Fourier's Series*, 2nd edn, Cambridge University Press

Hodges, W. (1983), 'Elementary predicate logic', in D. Gabbay and F. Guenthner, eds, *Handbook of Philosophical Logic*, Vol. I, Dordrecht: Kluwer, pp. 1–131

——— (1993), *Model Theory*, Cambridge University Press

——— (1998), 'An editor recalls some hopeless papers', *Bull. Symb. Logic*, 4: 1–16

Hurd, A. E. and Loeb, P. A. (1985), *An Introduction to Nonstandard Real Analysis*, Orlando, FL: Academic Press

Isaacson, D. (1987), 'Arithmetical truth and hidden higher order concepts', in P. L. Group, ed., *Logic Colloquium '85*, Amsterdam: North-Holland, pp. 147–69

——— (1992), 'Some considerations on arithmetical truth and the ω-rule', in M. Detlefsen, ed., *Proof, Logic and Formalization*, London: Routledge, pp. 94–138

Jech, T. J. (1967), 'Non-provability of Souslin's hypothesis', *Comment. Math. Univ. Carolinae*, 8: 291–305

Jech, T. J., ed. (1974), *Axiomatic Set Theory II*, Proceedings of Symposia in Pure Mathematics, 13, Providence, RI: American Mathematical Society

Jensen, R. B. (1966), 'Independence of the axiom of countable dependent choices from the countable axiom of choice', *J. Symb. Logic*, 31: 294

——— (1995), 'Inner models and large cardinals', *Bull. Symb. Logic*, 1: 393–407

Jensen, R. B. and Schröder, M. (1969), 'Mengeninduktion und Fundierungsaxiom', *Archiv Math. Logik Grundlagenforschung*, 12: 119–33

Jourdain, P. E. B. (1905), 'On a proof that every aggregate can be well-ordered', *Math. Ann.*, 60: 465–70

——— (1906), 'On the question of the existence of transfinite numbers', *Proc. London Math. Soc.*, 4: 266–83

Kac, M. and Ullam, S. M. (1968), *Mathematics and Logic*, New York: Praeger

Kanamori, A. (1997), *The Higher Infinite: Large Cardinals in Set Theory from their Beginnings*, Berlin: Springer

Kaufmann, F. (1930), *Das Unendliche in der Mathematik und seine Ausschaltung*, Leipzig: Deuticke (trans. as Kaufmann 1978)

——— (1978), *The Infinite in Mathematics*, Vienna Circle Collection, 9, Dordrecht: Reidel

Kaye, R. (1991), *Models of Peano Arithmetic*, Oxford Logic Guides, 15, Oxford: Clarendon Press

Keisler, H. J. (1976), *Elementary Calculus*, Boston, MA: Prindle, Weber & Schmidt

Kelley, J. L. (1955), *General Topology*, Princeton, NJ: Van Nostrand

Ketland, J. (2002), 'Hume = Small Hume', *Analysis*, 62: 92–3

Kirby, L. and Paris, J. (1982), 'Accessible independence results for Peano arithmetic', *Bull. London Math. Soc.*, 14: 285–93

König, J. (1905), 'Über die Grundlagen der Mengenlehre und das Kontinuumproblem', *Math. Ann.*, 61: 15–160

Korselt, A. (1911), 'Über einen Beweis des Äquivalenzsatzes', *Math. Ann.*, 70: 294–6

Kreisel, G. (1967*a*), 'Informal rigour and completeness proofs', in I. Lakatos, ed., *Problems in the Philosophy of Mathematics*, Amsterdam: North-Holland, pp. 138–86

——— (1967*b*), 'Mathematical logic: What has it done for the philosophy of mathematics?', in R. Schoenman, ed., *Bertrand Russell*, London: Allen & Unwin, pp. 201–72

——— (1971), 'Observations on popular discussions of foundations', in Scott 1971, pp. 189–98

——— (1980), 'Kurt Gödel', *Biog. Mem. Fellows Roy. Soc.,* 26: 149–224

Kunen, K. (1971), 'Elementary embeddings and infinitary combinatorics', *J. Symb. Logic,* 36: 407–13

——— (1980), *Set Theory: An Introduction to Independence Proofs*, Amsterdam: North-Holland

Kuratowski, K. (1921), 'Sur la notion d'ordre dans la théorie des ensembles', *Fund. Math.,* 2: 161–71

——— (1922), 'Une méthode d'élimination des nombres transfinis des raisonnements mathématiques', *Fund. Math.,* 5: 76–108

Kuratowski, K. and Tarski, A. (1931), 'Les opérations logiques et les ensembles projectifs', *Fund. Math.,* 17: 240–8

Lakatos, I. (1976), *Proofs and Refutations: The Logic of Mathematical Discovery*, Cambridge University Press

——— (1978), 'Cauchy and the continuum: The significance of non-standard analysis for the history and philosophy of mathematics', *Math. Intelligencer,* 1: 151–61

Landau, E. (1930), *Grundlagen der Analysis (das Rechnen mit ganzen, rationalen, irrationalen, komplexen Zahlen): Ergänzung zu den Lehrbüchern der Differential- und Integralrechnung*, Leipzig: Akademische Verlagsgesellschaft (trans. as Landau 1960)

——— (1960), *Foundations of Analysis*, 2nd edn, New York: Chelsea

Lear, J. (1977), 'Sets and semantics', *J. Phil.,* 74: 86–102

Lebesgue, H. (1905), 'Sur les fonctions représentables analytiquement', *J. Math. Pures Appl.,* (6)1: 139–216

Leibniz, G. W. (1996), *Schriften zur Logik und zur philosophischen Grundlegung von Mathematik und Naturwissenschaft*, Philosophische Schriften, 4, Frankfurt am Main: Suhrkamp

Lejewski, C. (1964), 'A note on a problem concerning the axiomatic foundations of mereology', *Notre Dame J. Formal Logic,* 4: 135–9

Lennes, N. J. (1922), 'On the foundations of the theory of sets', *Bull. Amer. Math. Soc.,* 28: 300

Leonard, H. S. and Goodman, N. (1940), 'The calculus of individuals and its uses', *J. Symb. Logic,* 5: 45–55

Levi, B. (1902), 'Intorno alla teoria degli aggregati', *Reale Istituto Lombardo Sci. Lett. Rend. (2)*, 35: 863–8

Levy, A. (1969), 'The definability of cardinal numbers', in J. J. Bulloff, T. C. Holyoke and S. W. Hahn, eds, *Foundations of mathematics: Symposium papers commemorating the sixtieth birthday of Kurt Gödel*, Berlin: Springer, pp. 15–38

Levy, A. and Solovay, R. M. (1967), 'Measurable cardinals and the continuum hypothesis', *Israel J. Math.*, 5: 234–48

Levy, P. (1964), 'Remarques sur un théorème de Paul Cohen', *Rev. Métaphys. Morale*, 69: 88–94

Lewis, D. (1991), *Parts of Classes*, Oxford: Blackwell

Lindelöf, E. (1905), 'Remarques sur un théorème fondamental de la théorie des ensembles', *Acta Math.*, 29: 183–90

Lindemann, F. (1882), 'Über die Zahl π', *Math. Ann.*, 20: 213–25

Lindenbaum, A. and Mostowski, A. (1938), 'Über die Unabhändigkeit des Auswahlsaxiom und einiger seiner Folgerungen', *C. R. Varsovie*, 31: 27–32

Lindström, P. (1969), 'On extensions of elementary logic', *Theoria*, 35: 1–11

Liouville, J. (1844), 'Des remarques relative à des classes très-étendues de quantités dont la valeur n'est ni rationelle ni même réductible à des irrationelles algébriques', *C. R. Acad. Sci. Paris*, 18: 883–5

Littlewood, J. E. (1926), *The Elements of the Theory of Real Functions*, 2nd edn, Cambridge: Heffer

Lusin, N. N. (1917), 'Sur la classification de M. Baire', *C. R. Acad. Sci. Paris*, 164: 91–4

——— (1927), 'Sur les ensembles analytiques', *Fund. Math.*, 10: 1–95

——— (1930), *Leçons sur les ensembles analytiques et leurs applications*, Paris: Gauthier-Villars

Mac Lane, S. (1969), 'One universe as a foundation for category theory', in *Reports of the Midwest Category Seminar III*, Lecture Notes in Mathematics, 106, Berlin: Springer, pp. 192–200

——— (1986), *Mathematics: Form and Function*, New York: Springer

Maddy, P. (1983), 'Proper classes', *J. Symb. Logic*, 48: 113–39

——— (1988), 'Believing the axioms', *J. Symb. Logic*, 53: 481–511, 736–64

——— (1990), *Realism in Mathematics*, Oxford: Clarendon Press

——— (1993), 'Does $V = L$?', *J. Symb. Logic*, 58: 15–41

Martin, D. A. (1975), 'Borel determinacy', *Ann. Math.*, 102: 363–71

——— (1976), 'Hilbert's first problem', in F. E. Browder, ed., *Mathematical developments arising from Hilbert's problems*, Proc. Symp. Pure Mathematics, 28, Providence, RI: American Mathematical Society, pp. 81–92

——— (1977), 'Descriptive set theory', in Barwise 1977, pp. 783–815

Martin, D. A. and Steel, J. R. (1988), 'Projective determinacy', *Proc. Natl. Acad. Sci. USA*, 85: 6582–6

Mathias, A. (2001), 'Slim models of Zermelo set theory', *J. Symb. Logic*, 66: 487–96

——— (2002), 'A term of length 4,523,659,424,929', *Synthese*, 133: 75–86

Mayberry, J. P. (1977), 'On the consistency problem for set theory', *British J. Phil. Sci.*, 28: 1–34, 137–70

——— (1994), 'What is required of a foundation for mathematics?', *Phil. Math. (III)*, 2: 16–35

——— (2000), *The Foundations of Mathematics in the Theory of Sets*, Encyclopaedia of Mathematics and its Applications, 82, Cambridge University Press

Mazurkiewicz, S. and Sierpinski, W. (1914), 'Sur un ensemble superposables avec chacune de ses deux parties', *C. R. Acad. Sci. Paris*, 158: 618–19

McIntosh, C. (1979), 'Skolem's criticisms of set theory', *Noûs*, 13: 313–34

McLarty, C. (1992), *Elementary Categories, Elementary Toposes*, Oxford: Clarendon Press

Mendelson, E. (1956), 'Some proofs of independence in axiomatic set theory', *J. Symb. Logic*, 21: 291–303

Méray, C. (1872), *Nouveau Précis d'Analyse infinitésimale*, Paris: Savy

Mirimanoff, D. (1917), 'Les antinomies de Russell et de Burali-Forti et le problème fondamental de la théorie des ensembles', *Enseignement Math.*, 19: 37–52

Montague, R. (1961), 'Semantic closure and non-finite axiomatizability I', in *Infinitistic Methods: Proceedings of the Symposium on Foundations of Mathematics (Warsaw, 1959)*, New York: Pergamon, pp. 45–69

Moore, A. W. (1990), *The Infinite*, London: Routledge

Moore, G. H. (1980), 'Beyond first order logic: the historical interplay between mathematical logic and axiomatic set theory', *Hist. Phil. Logic*, 1: 95–137

——— (1982), *Zermelo's Axiom of Choice: Its Origins, Development and Influence*, New York: Springer

——— (1989), 'Towards a history of Cantor's continuum problem', in D. E. Rowe and J. McCleary, eds, *The History of Modern Mathematics: Ideas and their Reception*, San Diego, CA: Academic Press, pp. 79–121

Morse, A. P. (1965), *A Theory of Sets*, New York: Academic Press

Moschovakis, Y. N. (1980), *Descriptive Set Theory*, Amsterdam: North-Holland

Mostowski, A. (1945), 'Axiom of choice for finite sets', *Fund. Math.*, 33: 137–68

——— (1948), 'On the principle of dependent choices', *Fund. Math.*, 35: 127–30

——— (1952), *Sentences Undecidable in Formalized Arithmetic: An Exposition of the Theory of Kurt Gödel*, Amsterdam: North-Holland

Muller, F. A. (2001), 'Sets, classes, and categories', *British J. Phil. Sci.,* 52: 539–73

Mycielski, J. and Steinhaus, H. (1962), 'A mathematical axiom contradicting the axiom of choice', *Bull. Acad. Polon. Sci.,* 10: 1–3

Mycielski, J. and Swierczkowski, S. (1964), 'On the Lebesgue measurability and the axiom of determinateness', *Fund. Math.,* 54: 67–71

Nathanson, M. (2000), *Elementary Methods in Number Theory,* Berlin: Springer

Newman, M. (1928), 'Mr Russell's "causal theory of perception" ', *Mind,* 37: 137–48

Oliver, A. (1994), 'Are subclasses parts of classes?', *Analysis,* 54: 215–23

——— (1998), 'Hazy totalities and indefinitely extensible concepts: an exercise in the interpretation of Dummett's philosophy of mathematics', in Brandl and Sullivan 1998, pp. 25–50

Parsons, C. (1974), 'Sets and classes', *Noûs,* 8: 1–12

——— (1977), 'What is the iterative concept of set?', in Butts and Hintikka 1977, pp. 335–67

——— (1983), *Mathematics in Philosophy: Selected Essays,* Ithaca, NY: Cornell University Press

——— (1996), 'Hao Wang as philosopher', in P. Hájek, ed., *Gödel '96,* Lecture Notes in Logic, 6, Berlin: Springer, pp. 64–80

Pascal, B. (1665), *Traité du triangle arithmétique,* Paris

Paseau, A. (2003), 'The open-endedness of the set concept and the semantics of set theory', *Synthese,* 135: 381–401

Peano, G. (1889), *Arithmetices principia, nova methodo exposita,* Torino: Bocca

——— (1890), 'Démonstration de l'intégrabilité des équations différentielles ordinaires', *Math. Ann.,* 37: 182–228 (repr. in Peano 1957–9, pp. 119–70)

——— (1906), 'Super theorema de Cantor-Bernstein', *Rend. Circ. Mat. Palermo,* 21: 360–6

——— (1921), 'Le definizione in matematica', *Periodico di matematiche (4),* 1: 175–89

——— (1957–9), *Opera scelte,* Rome: Cremonese

Platek, R. A. (1969), 'Eliminating the continuum hypothesis', *J. Symb. Logic,* 34: 219–25

Poincaré, H. (1906), 'Les mathématiques et la logique', *Rev. Métaphys. Morale,* 14: 17–34, 294–317, 866–8

——— (1913), *Dernières Pensées,* Paris: Flammarion (trans. as Poincaré 1963)

——— (1963), *Mathematics and Science: Last Essays,* New York: Dover

Potter, M. (1990), *Sets: An Introduction,* Oxford: Clarendon Press

——— (1993), 'Critical notice of *Parts of Classes* by David Lewis', *Phil. Quart.*, 43: 362–6

——— (1998), 'Classical arithmetic is part of intuitionistic arithmetic', *Grazer phil. Stud.*, 55: 127–41

——— (2000), *Reason's Nearest Kin: Philosophies of Arithmetic from Kant to Carnap*, Oxford University Press

Presburger, M. (1930), 'Über die Vollständigkeit eines gewissen Systems der Arithmetik ganzer Zahlen', in *Sprawozdanie z I Kongresu Mat. Krajów Slowiańskich*, Warsaw, pp. 92–101

Priest, G. (1995), *Beyond the Limits of Thought*, Cambridge University Press

Pudlak, P. (1998), 'The lengths of proofs', in S. Buss, ed., *Handbook of Proof Theory*, Amsterdam: North-Holland, pp. 547–637

Putnam, H. (1980), 'Models and reality', *J. Symb. Logic*, 45: 464–82

——— (1981), *Reason, Truth and History*, Cambridge University Press

——— (2000), 'Paradox revisited II: Sets', in Sher and Tieszen 2000, pp. 16–26

Quine, W. V. (1937), 'New foundations for mathematical logic', *Amer. Math. Monthly*, 44: 70–80

——— (1940), *Mathematical Logic*, New York: Norton

——— (1948), 'On what there is', *Rev. Metaphys.*, 2: 21–8

——— (1951), *Mathematical Logic*, rev. edn, Cambridge, MA: Harvard University Press

——— (1969), *Set Theory and its Logic*, rev. edn, Cambridge, MA: Harvard University Press

Ramsey, F. P. (1926), 'The foundations of mathematics', *Proc. London Math. Soc.*, 25: 338–84

——— (1931), *The Foundations of Mathematics and Other Logical Essays*, London: Kegan Paul, Trench & Trubner

Rang, B. and Thomas, W. (1981), 'Zermelo's discovery of the "Russell paradox"', *Hist. Math.*, 8: 15–22

Reinhardt, W. (1974), 'Remarks on reflection principles, large cardinals, and elementary embeddings', in Jech 1974, pp. 189–206

Restall, G. (1992), 'A note on naive set theory in LP', *Notre Dame J. Formal Logic*, 33: 422–32

Rieger, A. (2000), 'An argument for Finsler-Aczel set theory', *Mind*, 109: 241–53

Robinson, A. (1961), 'Non-standard analysis', *Nederl. Akad. Wetensch. Proc., Ser. A*, 64: 432–40

——— (1968), 'Some thoughts on the history of mathematics', *Comp. Math.*, 20: 188–93

——— (1974), *Non-Standard Analysis*, rev. edn, Princeton University Press

Robinson, R. M. (1947), 'On the decomposition of spheres', *Fund. Math.*, 34: 246–60

Rosser, B. (1942), 'The Burali-Forti paradox', *J. Symb. Logic*, 7: 1–17

Rubin, H. and Rubin, J. (1985), *Equivalents of the Axiom of Choice II*, Studies in Logic and the Foundations of Mathematics, 116, Amsterdam: North-Holland

Russell, B. (1903), *The Principles of Mathematics*, London: Allen & Unwin

——— (1906*a*), 'Les paradoxes de la logique', *Rev. Métaphys. Morale*, 14: 627–50

——— (1906*b*), 'On some difficulties in the theory of transfinite numbers and order types', *Proc. London Math. Soc.*, 4: 29–53 (repr. in Russell 1973*a*, pp. 135–64)

——— (1908), 'Mathematical logic as based on the theory of types', *Amer. J. Math.*, 30: 222–62

——— (1919), *Introduction to Mathematical Philosophy*, London: Allen & Unwin

——— (1927), *The Analysis of Matter*, London: Kegan Paul, Trench & Trubner

——— (1936), 'The limits of empiricism', *Proc. Arist. Soc.*, 36: 131–50

——— (1973*a*), *Essays in Analysis*, London: Allen & Unwin

——— (1973*b*), 'The regressive method of discovering the premises of mathematics', in *Essays in Analysis* (Russell 1973*a*), pp. 272–83 (written in 1907)

Schilpp, P. A., ed. (1944), *The Philosophy of Bertrand Russell*, Library of Living Philosophers, New York: Tudor Publishing Co.

Schönflies, A. (1905), 'Über wohlgeordnete Mengen', *Math. Ann.*, 60: 181–6

——— (1913), *Entwickelung der Mengenlehre und ihre Anwendungen*, Leipzig: Teubner

Schröder, E. (1890–5), *Vorlesungen über die Algebra der Logik*, Leipzig: Teubner

——— (1898), 'Über zwei Definitionen der Endlichkeit und G. Cantor'sche Sätze', *Abh. Kaiserl. Leopoldin.-Carolin. Deutsch. Akad. Naturfor.*, 71: 303–62

Scott, D. S. (1955), 'Definitions by abstraction in axiomatic set theory', *Bull. Amer. Math. Soc.*, 61: 442

——— (1961), 'Measurable cardinals and constructible sets', *Bull. Acad. Pol. Sci.*, 9: 521–4

——— (1974), 'Axiomatizing set theory', in Jech 1974, pp. 207–14

Scott, D. S., ed. (1971), *Axiomatic Set Theory I*, Proceedings of Symposia in Pure Mathematics, 13, Providence, RI: American Mathematical Society

Selberg, A. (1949), 'An elementary proof of Dirichlet's theorem about primes in an arithmetical progression', *Ann. Math.*, 50: 297–304

Serre, J. (1973), *A Course of Arithmetic*, Berlin: Springer

Shapiro, S. (1985), 'Second-order languages and mathematical practice', *J. Symb. Logic,* 50: 714–42

——— (1991), *Foundations without Foundationalism: A Case for Second-Order Logic*, Oxford University Press

Shapiro, S. and Weir, A. (1999), 'New V, ZF and abstraction', *Phil. Math. (III),* 7: 293–321

Sher, G. and Tieszen, R., eds (2000), *Between Logic and Intuition: Essays in Honor of Charles Parsons*, Cambridge University Press

Shoenfield, J. R. (1977), 'Axioms of set theory', in Barwise 1977, pp. 321–44

Sierpinski, W. (1924), 'Sur l'hypothèse du continu ($2^{\aleph_0} = \aleph_1$)', *Fund. Math.,* 5: 177–87

——— (1934), *L'hypothèse du continu*, Monografie Matematyczne, 4, Warsaw

——— (1965), *Cardinal and Ordinal Numbers*, 2nd edn, Warsaw: Polish Scientific Publishers

Simmons, K. (2000), 'Sets, classes and extensions: a singularity approach to Russell's paradox', *Phil. Studies,* 100: 109–49

Sinaceur, M. A. (1973), 'Appartenance et inclusion: un inédit de Richard Dedekind', *Rev. Hist. Sci.,* 24: 247–54

Skolem, T. (1922), 'Einige Bemerkungen zur axiomatischen Begründung der Mengenlehre', in *Wiss. Vorträge gehalten auf dem 5. Kongress der Skandinav. Mathematiken in Helsingfors* (repr. in Skolem 1970, pp. 137–52)

——— (1931), 'Über einige Satzfunktionen in der Arithmetik', *Skrifter utgitt av Det Norske Videnskaps-Akademi i Oslo, I. Mathematisk-naturvidenskapelig klasse,* 7: 1–28 (repr. in Skolem 1970, pp. 287–306)

——— (1970), *Selected Works in Logic*, Oslo: Skandinavian University Books

Smoryński, C. (1991), *Logical Number Theory*, Berlin: Springer

Solovay, R. M. (1964), '$2^{\aleph_0}$ can be anything it ought to be', in J. W. Addison, L. Henkin and A. Tarski, eds, *The Theory of Models*, Amsterdam: North-Holland, p. 435

——— (1970), 'A model of set theory in which every set of reals is Lebesgue measurable', *Ann. Math.,* 92: 1–56

——— (1974), 'Strongly compact cardinals and the generalized continuum hypothesis', in L. Henkin, ed., *Tarski Symposium*, Proceedings of Symposia in Pure Mathematics, 25, Providence, RI: American Mathematical Society, pp. 365–72

Solovay, R. M. and Tennenbaum, S. (1971), 'Iterated Cohen extensions and Souslin's problem', *Ann. Math.,* 94: 201–45

Souslin, M. Y. (1917), 'Sur une définition des ensembles mesurables B sans nombres transfinis', *C. R. Acad. Sci. Paris,* 164: 88–91

——— (1920), 'Problème 3', *Fund. Math.*, 1: 223

Steel, J. R. (2000), 'Mathematics needs new axioms', *Bull. Symb. Logic,* 6: 422–33

Steinitz, E. (1910), 'Algebraische Theorie der Körpern', *J. reine angew. Math.*, 137: 163–309

Stromberg, K. R. (1981), *An Introduction to Classical Real Analysis*, Belmont, CA: Wadsworth

Sullivan, P. M. (1995), 'Wittgenstein on *The Foundations of Mathematics*, June 1927', *Theoria,* 61: 105–42

Suppes, P. (1960), *Axiomatic Set Theory*, Princeton, NJ: Van Nostrand

Tait, W. W. (1994), 'The law of excluded middle and the axiom of choice', in George 1994, pp. 45–70

——— (1998), 'Foundations of set theory', in Dales and Oliveri 1998, pp. 273–90

——— (2000), 'Cantor's *Grundlagen* and the paradoxes of set theory', in Sher and Tieszen 2000, pp. 269–90

Takeuti, G. (1971), 'Hypotheses on power sets', in Scott 1971, pp. 439–46

Tall, D. (1982), 'Elementary axioms and pictures for infinitesimal calculus', *Bull. IMA,* 18: 43–8

Tarski, A. (1924), 'Sur quelques théorèmes qui équivalent à l'axiome du choix', *Fund. Math.,* 5: 147–54

——— (1948), *A Decision Method for Elementary Algebra and Geometry*, Santa Monica, CA: RAND Corporation

——— (1955), 'The notion of rank in axiomatic set theory and some of its applications', *Bull. Amer. Math. Soc.,* 61: 443

——— (1959), 'What is elementary geometry?', in L. Henkin, P. Suppes and A. Tarski, eds, *The Axiomatic Method*, Amsterdam: North-Holland, pp. 16–29

Teichmüller, O. (1939), 'Braucht der Algebraiker das Auswahlaxiom?', *Deutsche Math.,* 4: 567–77

Tharp, L. (1975), 'Which logic is the right logic?', *Synthese,* 31: 1–21

Tourlakis, G. (2003), *Lectures in Logic and Set Theory*, Cambridge University Press

Truss, J. K. (1997), *Foundations of Mathematical Analysis*, Oxford University Press

Tukey, J. W. (1940), *Convergence and Uniformity in Topology*, Annals of Mathematics Studies, 2, Princeton University Press

Uzquiano, G. (1999), 'Models of second-order Zermelo set theory', *Bull. Symb. Logic,* 5: 289–302

van den Dries, L. (1988), 'Alfred Tarski's elimination theory for real closed fields', *J. Symb. Logic,* 53: 7–19

van der Waerden, B. (1949), *Modern Algebra*, New York: Ungar van Heijenoort, J. (1977), 'Set-theoretic semantics', in R. O. Gandy and J. M. E. Hyland, eds, *Logic Colloquium '76*, Amsterdam: North-Holland, pp. 183–90

van Heijenoort, J., ed. (1967), *From Frege to Gödel: A Source Book in Mathematical Logic, 1879–1931*, Cambridge, MA: Harvard University Press

Vaught, R. (1967), 'Axiomatizability by schemas', *J. Symb. Logic,* 32: 473–9

von Neumann, J. (1925), 'Eine Axiomatisierung der Mengenlehre', *J. reine angew. Math.,* 154: 219–40

Wang, H. (1963), *A Survey of Mathematical Logic*, Studies in Logic and the Foundations of Mathematics, Amsterdam: North-Holland

——— (1974), *From Mathematics to Philosophy*, London: Routledge & Kegan Paul

——— (1977), 'Large sets', in Butts and Hintikka 1977, pp. 309–33

Warner, S. (1968), *Algebra*, Englewood Cliffs, NJ: Prentice-Hall

Waterhouse, W. C. (1979), 'Gauss on infinity', *Hist. Math.,* 6: 430–6

Weir, A. (1998*a*), 'Naive set theory is innocent', *Mind,* 107: 763–98

——— (1998*b*), 'Naive set theory, paraconsistency and indeterminacy, I', *Logique et Analyse,* 41: 219–66

——— (1999), 'Naive set theory, paraconsistency and indeterminacy, II', *Logique et Analyse,* 42: 283–340

Weyl, H. (1949), *Philosophy of Mathematics and Natural Science*, Princeton University Press

Whitehead, A. N. (1902), 'On cardinal numbers', *Amer. J. Math.,* 24: 367–94

Whitehead, A. N. and Russell, B. (1910–13), *Principia Mathematica*, Cambridge University Press

——— (1927), *Principia Mathematica*, 2nd edn, Cambridge University Press

Wiener, N. (1914), 'A simplification of the logic of relations', *Proc. Camb. Phil. Soc.,* 17: 387–90 (repr. in van Heijenoort 1967, pp. 224–7)

——— (1953), *Ex-prodigy: My Childhood and Youth*, New York: Simon & Schuster

Wilkinson, J. H. (1959), 'The evaluation of the zeros of ill-conditioned polynomials', *Num. Math.,* 1: 150–80

Wittgenstein, L. (1922), *Tractatus Logico-Philosophicus*, London: Kegan Paul & Trubner

——— (1976), *Lectures on the Foundations of Mathematics, Cambridge, 1939*, Ithaca, NY: Cornell University Press

Woodin, W. H. (1994), 'Large cardinal axioms and independence: The continuum problem revisited', *Math. Intelligencer,* 16: 31–5

——— (2001*a*), 'The continuum hypothesis, I', *Not. Amer. Math. Soc.,* 48: 567–76

——— (2001*b*), 'The continuum hypothesis, II', *Not. Amer. Math. Soc.,* 48: 681–90

Wright, C. (1983), *Frege's Conception of Numbers as Objects*, Scots Philosophical Monographs, 2, Aberdeen University Press

——— (1985), 'Skolem and the skeptic', *Proc. Arist. Soc., Supp. Vol.,* 59: 117–37

——— (1999), 'Is Hume's principle analytic?', *Notre Dame J. Formal Logic,* 40: 6–30

Zermelo, E. (1904), 'Beweis daß jede Menge wohlgeordnet werden kann', *Math. Ann.,* 59: 514–16

——— (1908*a*), 'Neuer Beweis für die Möglichkeit einer Wohlordnung', *Math. Ann.,* 65: 107–28

——— (1908*b*), 'Untersuchungen über die Grundlagen der Mengenlehre I', *Math. Ann.,* 65: 261–81

——— (1930), 'Über Grenzzahlen und Mengenbereiche', *Fund. Math.,* 16: 29–47

Zorn, M. (1935), 'A remark on method in transfinite algebra', *Bull. Amer. Math. Soc.,* 41: 667–70

List of symbols

Index of definitions

Index of names